Benthic suspension feeders and flow

Benthic suspension feeders and flow

DAVID WILDISH
Biological Station, St. Andrews, Canada
DAVID KRISTMANSON
University of New Brunswick

CAMBRIDGE UNIVERSITY PRESS
Cambridge, New York, Melbourne, Madrid, Cape Town, Singapore, São Paulo

Cambridge University Press
The Edinburgh Building, Cambridge CB2 2RU, UK

Published in the United States of America by Cambridge University Press, New York

www.cambridge.org
Information on this title: www.cambridge.org/9780521445238

First published 1997
This digitally printed first paperback version 2005

A catalogue record for this publication is available from the British Library

Library of Congress Cataloguing in Publication data

Wildish, David, 1939–
Benthic suspension feaders and flow / David Wildish, David Kristmanson.
p. cm.
Includes bibliographical references and index.
ISBN 0-521-44523-X
1. Benthos. 2. Marine biology. 3. Hydrodynamics.
I. Kristmanson, D. D. II. Title
QH91.8.B4W55 1997
577.7′4 – dc21 97-17118
CIP

ISBN-13 978-0-521-44523-8 hardback
ISBN-10 0-521-44523-X hardback

ISBN-13 978-0-521-02347-4 paperback
ISBN-10 0-521-02347-5 paperback

for Ann and Bernice

Contents

Preface

Hydrodynamics and benthic biology represent a very broad field encompassing many possible levels of hierarchical organization and subjects. Thus, it is not possible for any author(s) to have a detailed personal knowledge and experience of all the subjects discussed herein. Nor does the book seek to cover every subject which could be included; for example, there is little on corals or freshwater biology, although we have occasionally mentioned such examples to illuminate the points we are trying to make. The subject matter of each chapter is also limited to matters where hydrodynamics are centrally involved. Hence, in Chapter 3, for example, larval biology is limited to dispersal and settlement where flow is clearly important, and our overview is not intended to replace the more complete reviews of this subject which are available.

It became clear to us early in the genesis of this book that our topic, at most hierarchical levels of benthic biology, was at a very early stage in the learning curve. Indeed, one of the reviewers suggested that we wait for a few years to begin the book in the hope that by then the subject would have reached a firmer theoretical footing. As the evidence in front of you will suggest, we did not take this advice because of our own interests and agenda. It is graduate students who are in the best position to carry on the large amount of work still to be done in the field we have outlined here. Consequently, our book has been shaped specifically towards their needs.

Our collaboration began in 1970 and became interesting soon afterwards during joint studies involving pollution of a marine tidal inlet in the Bay of Fundy. We were both struck by the powerful control that the tidal flow of the Bay appeared to have in shaping the benthic populations which were present there. We realized that we had the means of an interdisciplinary collaboration at our disposal: D. D. Kristmanson's training in chemical engineering and special interest in how flows allow molecules to come together and react, and D. J. Wildish's training in marine biology and special interest in benthic animals, inclusive of suspension feeders. Our modus operandi evolved into one in which new ideas were tested against accepted hydrodynamic theory, before designing specific null and alternate hypotheses for biological experimental testing.

Until recently, most marine biologists were not required to study fluid mechanics during undergraduate training. We believe that such training, as first introduced by Steven Vogel in his book *Life in Moving Fluids: The Physical Biology of Flow*, will become mandatory during the training of undergraduates in marine and freshwater biology.

This book was written in St. Andrews at the Biological Station and in Fredericton at the University of New Brunswick, between 1992 and 1995. So that we could interact better, we have also spent shorter periods together at the Huntsman Marine Science Centre in St. Andrews and in a small office, without a telephone, in Vancouver. At both of the latter locations, we were able to visit a public aquarium and saw some familiar benthic animals. At the Aquarium–Museum of the Huntsman Marine Science Centre in St. Andrews, we saw the solitary lampshell *Terebratulina septentrionalis* and the giant scallop *Placopecten magellanicus*, which looks much more attractive in the living state than as a shell used as an ashtray. At the Vancouver Aquarium in Stanley Park, we saw the beautiful sea pen *Ptilosarcus qurneyi* and the polychaete worm *Eudostylia vancouveri*. We suggest to readers that, unless they are accomplished SCUBA divers, this is the best way to see some of the animals which are the subject of this book.

Acknowledgments

Many people, in various ways, helped in the completion of this book. Without naming all, we thank them for their efforts on our behalf. The following research scientists provided frank and detailed critiques of one or two chapters: Edwin Bourget, Cheryl-Ann Butman, Jim Eckman, Craig Emerson, Marcel Fréchette, Barry Hargrave, C. Barker Jørgensen, Beth Okamura, and J. Evan Ward. All of this group (part of the Northwest Atlantic invisible college of the benthos) have, with the exception of Professor Jørgensen, attended at least one of the Biennial Benthic Workshops held in St. Andrews in October. These meetings began in 1983 to help celebrate the St. Andrews Biological Station's 75th Anniversary. After missing a year in 1987, the Biennial Benthic Workshops are now even-year events, with the next one planned for 1998. The larger invisible college, probably in excess of 200 persons over the years, has been an important source of information exchange, inspiration, criticism, and friendship. We thank all of our mud-grubbing colleagues who have helped to make these meetings so friendly and stimulating to us.

This book could not have been completed without the excellent services available at the St. Andrews Biological Station library and the dedicated help of the library staff, Marilynn Rudi and Joanne Cleghorn. Brenda Best, also of the St. Andrews Biological Station, was able to translate our scribbles or unformatted chapters on disk into readable English efficiently, and Susan Laurie-Borque, of Ottawa, created new or redrafted published art material into the excellent figures included in this book.

The forbearance of the Department of Chemical Engineering of the University of New Brunswick allowed David Kristmanson to pursue the research interests included herein, and the Department of Fisheries and

Oceans, in the persons of Mike Sinclair and Wendy Watson-Wright, encouraged Dave Wildish in this endeavor.

We thank the following for permission:

- Prof. Dr. Klaus Lüning, Biologische Anstalt Helgoland
- Springer-Verlag GmbH & Co. KG for material from Marine Biology, vol. 81, p. 259; vol. 110, pp. 94, 101; vol. 97, p. 119; vol. 119, p. 526; vol. 85, p. 128; vol. 72, p. 128; vol. 105, p. 120; vol. 92, p. 324; from Oecologia (Biol.), vol. 8, p. 26; and from Bivalve Filter Feeders (1993) by R. F. Dame (ed.)
- The Scottish Association for Marine Science
- Dr. Cheryl Ann Butman, Woods Hole Oceanographic Institution
- D. O. Duggins
- Gary L. Taghon
- S. W. Nixon
- Richard Emlet, Oregon Institute of Marine Biology, University of Oregon
- Charles H. Peterson
- Steven Vogel, Department of Zoology, Duke University
- Journal of Phycology
- Walter de Gruyter & Co.
- Annual Reviews Inc. for Figure 5, modified, with permission, from the Annual Review of Fluid Mechanics, Volume 9, © 1977, by Annual Reviews Inc.
- The American Meteorological Society for illustrative material from Jackson, George A., 1984: Internal wave attenuation by coastal kelp stands. J. Phys. Oceanogr., 14, 1300–1306
- *Estuaries* for illustrative material from Hidu, H., and H. H. Haskin. 1978. Swimming speeds of oyster larvae *Crassostrea virginica* in different salinities and temperatures. Estuaries 1:252–255
- NRC Research Press
- Material from Elsevier Science: Reprinted from J. Exp. Mar. Biol. Ecol., vol. 65, p. 121; vol. 83, p. 241; vol. 148, p. 217; vol. 174, p. 68; vol. 181, p. 12; vol. 114, p. 2; vol. 173, pp. 64–5; vol. 155, p. 163; vol. 165, p. 253; vol. 174, p. 76; vol. 113; p. 212; vol. 164, p. 149 with kind permission from Elsevier Science – NL, Sara Burgerhartstraat 25, 1055, KV Amsterdam, The Netherlands
- Academic Press Ltd, London, England, for material from The Biology of Estuaries and Coastal Waters (1974) by E. J. Perkins
- Material from J. Fish. Res. Board. Can., vol. 24, pp. 328, 332; and Can. J. Fish. Aquat. Sci., vol. 40 (Supplement 1), pp. 302, 307, reproduced with permission of the Minister of Supply and Services Canada 1996
- Blackwell Science Ltd., Osney Mead, Oxford OX2 0EL England for material from J. Anion. Ecol., vol. 22, p. 337
- Inter-Research, Nordbünte 23, D-21385 Oldendorf/Luhe, Germany, for material from Mar. Ecol. Prog. Ser. vol. 61, p. 165; vol. 42, p. 164; vol. 28, p. 190; vol. 92, p. 143; vol. 63, p. 213; vol. 34, p. 70
- The Bulletin of Marine Science for material from Bull. Mar. Sci., vol. 39, pp. 290–322
- Material from Bioscience, vol. 28, p. 639, and vol. 29, p. 349, © 1978 American Institute of Biology and Sciences
- Brian Helmuth
- H. G. Verhagen (personal communication), Agricultural University Wageningen, The Netherlands

- Material modified with permission from D. O. Duggins, C. A. Simenstad, & J. A. Estes. Magnification of secondary production by kelp detritus in coastal ecosystems. Science, vol. 245, p. 171. Copyright 1989 American Association for the Advancement of Science.
- Material modified with permission from G. L. Taghon, A.R.M. Nowell, & P. A. Jumars. Introduction of suspension feeding in spionid polychaetes by high particulate fluxes. Science, vol. 210, pp. 562–4. Copyright 1980 American Association for the Advancement of Science.
- Material modified with permission from S. Vogel & W. L. Bretz. Interfacial organisms: passive ventilation in the velocity gradients near surfaces. Science, vol. 175, pp. 210–11. Copyright 1972 American Association for the Advancement of Science.

Let you who enter here
Cast off all cant
Take on the eyes
Of the wondering child.

Let you who enter here
Be blessed with fields
Richly carpeted with
Insight and imagination.

Let you who enter here
Stare in wonder
At this world of beauty
Of which you are a part.

1
Introduction

One objective of benthic ecology is to describe the spatial distribution of living organisms at or near the sediment–water interface, as well as to explain how and why the distribution occurs at that particular location on the seabed. The descriptive part of the job of the benthic ecologist thus becomes a sampling challenge since a depth gradient is involved from the intertidal to the hadal zone – a depth range of 0 to >6000 m (Parsons et al. 1977). Another sampling problem is that of substrate variability, thus hard, e.g. rocks and corals, versus soft substrates, e.g. muds, sands, and gravel, which means that appropriate samplers for each type must be quite different. Because it is considerably more difficult to sample at depth, the bulk of the available sampling results are for the littoral and sublittoral zones down to SCUBA diving depth. Thus, much of our knowledge of the spatial distribution of benthic organisms is heavily biased to these limited depths in the nearshore environment. The benthic animals which we would expect to be present in a grab or core sample of soft sediments in the nearshore region would include micro-, meio-, and macrofauna (Table 1.1). The size ranges shown are arbitrary and based on sampling convenience; thus, if a 0.5- or 0.8-mm mesh was used in sieving, then the lower size limit for macrofauna becomes 500 or 800 μm.

Because of a need to limit the size of this book, we have had to limit its subject matter to marine or estuarine benthic macrofauna and exclude micro- and meiofauna from detailed consideration. This is simply for presentation convenience and it is obvious to us that flow will have many important direct influences on micro- and meiofauna also. Thus, if barnacle cyprids can distinguish between microbial films developed on hard substrates in either low or high shear stress flows (see the section, Hard substrates, Chap. 3), then flow itself must be an important factor in shaping the microbial association which develops there.

Table 1.1. *Arbitrary size divisions of benthic fauna.*

Name	Size range (μm)	Examples
Microfauna	1–100	Some smaller Protozoa
Meiofauna	100–1000	Some Protozoa, e.g. Foraminifera; nematodes; harpacticoid copepods; gastotrichs; isopods; turbullarians; small macrofauna
Macrofauna	>1000	Amphipods, mysids, echinoderms, sponges, brachiopods, corals, hemichordates, echinoderms, molluscs, polychaetes
Megafauna	?	Cannot be adequately sampled with grab or corer, e.g. the mud star *Ctenodiscus crispatus* or benthic fish

Source: Based on Parsons et al. (1977).

The reader should be aware that there is a parallel interest among freshwater biologists (e.g. Hildrew and Giller 1994) in the effect of flow on river or stream benthos. For the same reasons of space limitation, we have not considered this work, except incidentally where research described presents a particularly good example of a pertinent interdisciplinary study.

This book is concerned only with those macrofauna characterized as suspension feeders that live on either soft or hard substrates of the seabed and feed in a particular way, that is by filtering microscopic particulates, or seston, transported to them by ambient flows. Our focus is the interactions which occur between individuals, populations of suspension feeders, or communities and ecosystems containing suspension feeders and the various types of water movement characterized as flow in the last paragraph of this section.

Functionally, benthic macrofauna can be classified as *suspension feeders*, *deposit feeders*, carnivores, omnivores, and algal scrapers (Chap. 7, the section, Benthic limitation by flow). The two major trophic types of macrofauna can be distinguished by their mode of feeding – deposit feeders are limited to soft sediments and ingest sediment which has already been deposited after seawater transport, while suspension feeders feed by capturing *seston*, a mix of microscopic particles which may contain detritus, bacteria, microalgae, small animals, and sediments, suspended in and transported by seawater flows. If the seston is carried by flow directly to the capture surface, the suspension feeder is described as *passive*. If, on the other hand, seawater containing seston is pumped by a ciliary or muscular pump to the capture surface, the suspension feeder is described as *active*. Further classification of passive and active suspension feeders is considered in Chapter 4 (see the section, Background).

Suspension feeders as a group comprise a diverse assortment of taxa drawn from most of the major invertebrate phyla inclusive of corals, hydrozoans, bryozoans, brachiopods, some polychaetes and bivalve molluscs, and a few echinoderms and crustaceans. Taxa which are found on hard substrates, e.g. barnacles, are generally *epifaunal*, living firmly attached to the rock or coral reef surface, and are said to be sessile. A few specialized species are able to burrow into some hard substrates, e.g. the rock-boring bivalve *Petricola pholadiformis*, which is also a suspension feeder. Soft sediment macrofaunal taxa may be either epifaunal and sessile, e.g. tube-living polychaete worms such as many spionids, or *infaunal*, that is to say, able to live and burrow within the sediment matrix, e.g. free-living polychaete burrowers such as *Chaetopterus variopedatus*. Again, both taxa are suspension feeders.

Flow is the general term that we have used for all the types of water movement which can occur in the marine or estuarine environment. Seawater movements include those caused by the action of winds directly on the sea surface (*wind–wave effects*), by the variable gravitational attraction of both moon and sun on seawater (*tidal currents*), or by density differences between different water masses due either to inequalities of heat exchange across the air–seawater interface or to differences in salinity (Tait 1968).

Asking the right question

Two general types of question apply in biology: *proximate* and *ultimate* (Alcock 1989). The proximate type equates to the "how does it work" questions of Table 1.2 and requires an answer based on a reductionist experimental program of inquiry. By contrast, the ultimate type is a "why" question involving knowledge of evolutionary mechanisms and history of evolutionary trends, often over significant geological periods.

Typical ultimate questions listed in Table 1.2 clearly demonstrate that most cannot be answered by reductionist experimental methods. Thus,

Table 1.2. *Some general questions related to the disciplines listed and referable to suspension feeders or the communities in which they live.*

Discipline	No.	Proximate	Ultimate
Physiology	1	How do suspension feeders filter feed?	Why did active suspension feeders evolve from passive ones (or vice versa)?
	2	How can energy costs of ciliary pumping be estimated?	Why did a ciliary pump evolve?
Ethology	3	How do competent larvae choose substrates suitable for initial colonization?	Why have some sessile suspension feeders evolved a short, and others a long, period of planktonic life?
	4	How do epifaunal suspension feeders become aggregated?	Why do hydrodynamic factors cause epifaunal suspension feeders to aggregate?
	5	How do some bivalve molluscs swim?	What adaptive imperatives resulted in swimming bivalves?
Ecology	6	How does larval retention in an estuary occur?	How do larvae with specific responses to environmental variables, e.g. light, gravity, and flow, evolve?
	7	How do mixed benthic assemblages dominated by suspension feeders occur?	What causes a mixed benthic assemblage dominated by suspension feeders to evolve?
	8	How are materials cycled between pelagic and benthic parts of an ecosystem?	Why do ecosystems evolve over geological time?

for the first physiological ultimate question of Table 1.2, we have no direct knowledge of the ancient environments which may have resulted in the development of an active suspension feeder, nor any idea of the evolution of its ciliary pump since they are soft structures which are not preserved very well in the geological record. Although circumstantial evidence from the geological record may help, e.g. general climatic conditions deduced from stratigraphic deposition or from the morphology and physiology of extant ciliary pumps from "primitive" to "advanced" forms, such deductive methods often do not provide an adequate answer for the ultimate questions. For this reason, we have limited the questions addressed in Chapters 3 to 8 to those that are clearly proximate in nature. In Chapter 9, where possible future benthic biological questions are posed, we include consideration of ultimate ones.

As pointed out by Fenchel (1987), biology as a scientific discipline is organized hierarchically. At the most complex organizational level is the *ecosystem*, by which is meant a community of living organisms, together with the physical and chemical environmental factors present interacting together to form a recognizable ecological unit, e.g. an estuary. Descending the list of hierarchical order, we thus arrive at successively lower levels – ecosystem, community, population, individual, organ, cells, organelles, and molecules (Fenchel 1987). As pointed out by Fenchel, these levels of organization correspond to the specialized subdisciplines of biology: *ecology*, inclusive of ecosystems, communities, or populations; *ethology*, inclusive of social groups and individuals; *physiology*, inclusive of individuals and organs; *cell biology*, inclusive of cells and organelles; and *biochemistry*, inclusive of organic molecules. From this list we have selected the three higher hierarchical levels and thus subdisciplines of biology – ecology, ethology, and physiology – to investigate benthic interactions with flow.

Where the focus of concern is the individual animal as in ethology and physiology, it is easy to appreciate how a reductionist research program could be developed into an experimental study of any suspension feeder. In ecology, where the unit of study ranges from a population to an ecosystem, that is to say, a mussel reef to the whole estuary in which the mussels live, it is much more difficult to see which questions are appropriate. Thus, it becomes necessary to determine exactly what are valid ecological questions. Fenchel (1987) defines ecology as "the study of the principles which govern temporal and spatial patterns of assemblages of organisms." This succinct wording is a good encapsulation but does not quite convey the depth of the subject matter, and a more complete snapshot of ecology, at least in the mid-1980s, can be gleaned from an opinion survey of 645 members of the British Ecological Society (Cherrett 1989). This group was composed of university trained ecologists with mixed backgrounds and employment, who suggested 236 different concepts as fundamentally important in ecology. Considerable overlap in the concepts existed and, using the data provided by Cherrett (1989), we suggest that they can be reduced to the seven shown in Table 1.3.

Table 1.3. *Important concepts in ecology derived from a survey conducted by the British Ecological Society.*

No.	Description	Examples
1	Spatial	Studies involving the ecosystem, community, and niche
2	Temporal	Studies involving succession, climax, and diversity
3	Energy flow process study	Energy flow, materials cycling, and food webs
4	Biological interactions	Competition, density dependence, and predator–prey studies
5	Limiting factors	Physical limitations in ecology at various hierarchical levels
6	Evolution	Life history strategies, adaptations, and co-evolution
7	Ecosystem management	Conservation, ecosystem fragility, and maximum sustainable yield

Source: Cherrett (1989).

In this book we have excluded the long-term, temporal perspective in line with our resolve to exclude ultimate questions (number 6, Table 1.3), and because succession, climax, and diversity concepts (number 2, Table 1.3) have not been studied in relation to flow. This means that our attention is given entirely to numbers 1, 3, 4, and 5, but omitting 7, in Table 1.3. This is not as ruthless as it may sound since there is only a small amount of work published on evolutionary aspects of benthic biology. In any case, we return briefly to pertinent questions concerning numbers 6 and 7 (Table 1.3) in Chapter 9.

Benthic biology is a relatively young scientific discipline. Its quantitative study dates from the work of the Danish investigator Petersen (1911), who studied those factors which regulate benthic macrofaunal density and biomass in order to predict the productivity of benthic feeding and commercially important groundfish. It is only since the early 1970s that the focus of benthic biological research has moved from observation to inferential experimentation, with a more rigorous effort at field or laboratory experimentation. This may help explain why benthic biology has so few widely applicable predictive models to offer the newcomer to this field.

Table 1.4. *Major types of scientific method.*

Type	Description
Observation	*Simple*: involves direct experience, e.g. looking down a microscope. Includes descriptive taxonomy, natural history observations. *Complex*: sampling or measurement which may involve bias, e.g. benthic grab sampling involving different sieve mesh sizes in sorting the animals.
Theory/model	*Reductionist*: a way of looking at a subfield of benthic biology which can be represented verbally or by a simple conceptual model, e.g. the benthic limitation by flow theory (see Chap. 8). *Holistic*: as above, except that because of complexity or the nature of the model, formal mathematical representation is required, e.g. an ecosystem simulation of a bivalve reef based on energy flow.
Inferential experimentation	Laboratory or field experimental tests of formulated null, H_0, and alternate, H_1, hypotheses.
Criticism	Logical refutation of any conclusions or constructs derived from any of the types of scientific method described here.

Scientific methods

Science is concerned with the creation and communication of knowledge. Concerning the first of these, scientific knowledge is created by one of the four major methods shown in Table 1.4.

Observation, in both the field and laboratory, is fundamentally important to the scientific enterprise. During this process, realistic questions and, hence, theories, models, or hypotheses are formulated. Such formulations are possible constructs determined by logical reasoning concerning the relations of benthic animals to the observed external world.

As pointed out by Peters (1991), there is no consensus on how words like theory, model, and hypothesis are used, and he recommends using them synonymously. In our account, we define *hypothesis* as a working explanation of observations proposed in such a way that an experimental test can choose between two contrasting constructs – null and alternate. The hypothesis may apply fully, or only partially, to the theory or model constructs which are conceived to be at a higher hierarchical level. The

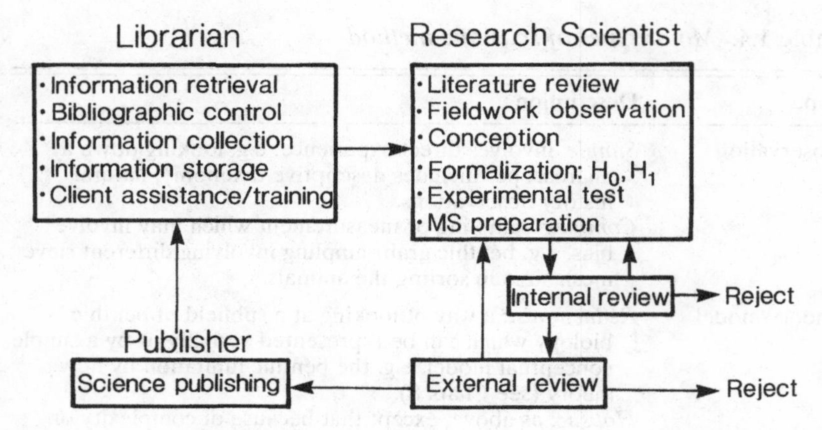

Figure 1.1 Outline of the process used in science to conceive, experimentally test, disseminate, and store information.

alternative and null hypotheses are tested by inferential experimentation with the desired outcome that one or other of the constructs can be discarded as a result of the test. As pointed out by Sprintall (1990), care must be taken to distinguish between post facto and experimental research. The former concerns field studies where independent and dependent variables are not assigned before the work begins and, hence, causation or rejection cannot be established. In experimental research, the independent and dependent variables are established before the work begins and, if the results warrant, can lead logically to the rejection or establishment of cause. If a theory or model can be represented by a simple conceptual model, we refer to it as a reductionist model, in contrast to a complex model, which requires mathematical formalism to express it and which is referred to as an holistic model.

Criticism, either from one's self, using accumulated experience, or from others and formally expressed as in a written review, informally as during a workshop, and as part of the peer review process before primary publication (Fig. 1.1), is important to the well-being of the scientific knowledge database. It is during the critical process that judgements of hypotheses are made.

An outline of the process used to conceive, test, and store scientific information is shown in Fig. 1.1. Peters (1991) has characterized hypothetico-deductive science as an alternation between creation and criticism. For the research scientist, it is in the private or synthetic phase of the process shown in Fig. 1.1 (first three items in the box) where the individual creates a new theory and/or new hypothesis referable to an already established theory. Exactly what is involved in the synthetic phase of science has proved difficult to define (Peters 1991). The creative process involves keen observation and intuition, an ability to juxtapose two apparently unrelated ideas, or use of accepted theories from related disciplines. It is the latter process which is the main approach used in this book with benthic biology as the recipient of theory from the well established discipline of hydrodynamics.

It is frequently necessary to use multiple hypotheses to solve a scientific question, e.g. in determining the precise mechanism of filtration in suspension feeders (Chap. 5), and this approach is championed by Platt (1964) and Chamberlin (1965). Quinn and Dunham (1983) drew attention to the difficulty of applying Baconian experimental tests to pairs of mutually exclusive hypotheses in many ecological questions where multiple alternative hypotheses represent the usual situation. As an example, we site the growth or production of a suspension feeder population. Here, the null hypothesis might be that flow does not affect suspension feeder growth, but is swamped by multiple alternative environmental factors, e.g. seston concentration, seston quality, temperature, or growth-limiting flow at high velocity. According to Simberloff (1983), multiple alternative hypotheses should be met by increased ingenuity from researchers in framing unambiguous hypotheses capable of dealing with multiple alternatives and the potential interactions between them.

Scientific communication

Concerning the communication of science results mentioned earlier, the active participants in science tend to form into natural groups which are referred to as an "invisible college" because of shared research interests and problems. Communication is the most important function of the invisible college and various ways have been tried to optimize it. These include personal communication (often electronic), workshops, symposia, published periodicals, and books.

In regard to periodicals, there are four distinct types: *science magazines*, which provide interpretive articles for a multidisciplinary audience, but not original research articles (e.g. *New Scientist, Scientific American*); *technology transfer periodicals*, whose purpose is to transfer science to a commercially active field such as aquaculture (e.g. *World Aquaculture*) but which does not usually carry original research; a few *multidisciplinary journals*, notably *Science* and *Nature*, which contain original abbreviated research articles thought to be of wide general interest and from all fields of science, including physics, chemistry, biology, geology and medicine; and single discipline or *primary journals*, which contain original research articles from a specific area of research or subdiscipline and therefore are targeted to a specialized audience. As a result of recent advances in electronic publishing and the widespread use by research scientists of personal computers, the future possibility of online journals is real (Maddox 1992). Nevertheless, in 1997, the printed word remains the pre-eminent means of communication for the aquatic sciences.

The creation and communication of scientific knowledge in printed multidisciplinary or primary journals involve three different types of professional (Fig. 1.1). The research scientist creates scientific knowledge and describes the work in research articles or evaluates the work of other research scientists in review articles. The publisher prepares copy for the journal consisting of a series of research articles and is responsible for preparing a sufficient number of copies for wide dissemination among the extended invisible college. Wide dissemination of printed copy is less important today because of the availability of online electronic abstracting services with which, and by using key words, relevant articles can be obtained. The librarian assists the research scientist by facilitating access to published sources by collecting, storing, and retrieving articles involving bibliographic control inclusive of indexing, abstracting, and classification.

Newcomers to the interdisciplinary field of hydrodynamics/benthic biology may need assistance in finding reliable literature sources which explain the physics of flow. Some of the sources which we have found helpful may be of interest. There are two excellent introductions to hydrodynamics for biologists: Vogel (1981), *Life in Moving Fluids: The Physical Biology of Flow*, recently revised and expanded in a second edition (Vogel 1994); and Denny (1993), *Air and Water: The Biology and Physics of Life's Media*. Both provide an introduction to the fundamental principles of hydrodynamics and many applications of interest in benthic biology. Denny (1988) has also provided a similar introduction and also an account of flow phenomena on the wave-swept shore. These books interpret the solid body of accepted hydrodynamic theory developed by mathematicians, physicists, and engineers over a period of more than 150 years. For the more mathematically adept, there are many introductory texts in fluid dynamics which might be consulted. We have found Daugherty et al. (1985), *Fluid Mechanics with Engineering Appli-*

cations, to contain clear descriptions of most of the ideas of interest to benthic biologists. When fluid flows past a solid boundary, it is arrested at the boundary and a sheared zone develops. This flow, which has been studied in great detail by aerodynamicists, is called a "boundary layer." In benthic biology, boundary layers influence forces on submerged objects and the flow around them. They also influence the processes by which nutrients and other dissolved molecules are transferred between biological surfaces and the ambient flow. The standard reference for such flows is Schlichting (1979), *Boundary-Layer Theory*. Boundary layers over smooth or uneven marine sediments are described by an extension of engineering boundary layer theory. A lucid introduction to this subject is given in Mann and Lazier (1991). Since the terminology will be unfamiliar to many biologists, we have included a Glossary beginning after Chapter 9. It contains brief definitions of those words which we consider to be the most important in the benthic interdisciplinary field considered in this book. The Glossary includes terms used in hydrodynamics, benthic biology, and surficial geology.

With regard to the biological aspects of the interdisciplinary field of hydrodynamics/benthic biology, Jørgensen (1966), *Biology of Suspension Feeding*, is concerned with the wide taxonomic range of invertebrates which are suspension feeders. This book pioneered the physiological and ethological study of feeding by suspension feeders. A later book by Jørgensen (1990) presents selected updates on suspension feeding physiology and ecology. This includes the Danish school's elegant description of the bivalve gill as a ciliary pump and a novel hypothesis regarding the mechanism of seston capture on the bivalve gill surface. In addition, we have found the biological dictionary of Lincoln, Boxshall, and Clark (1982) to be useful and it can be used profitably when reading this book.

Among the plethora of aquatic, multidisciplinary, and primary journals, we have selected those which are most frequently used to carry original research articles in the English language on benthic suspension feeders and flow. The results of an analysis of the references used in Chapters 2 through 8 are shown in Table 1.5. It is of interest that the first four journals in rank order carry nearly 40% of all 799 cited articles. Some journals, e.g. *Limnology and Oceanography* and the *Canadian Journal of Fisheries and Aquatic Sciences*, clearly stick to ecological rather than physiological research articles. Table 1.5 also demonstrates another point: that the relevant research articles are spread over a wide range of journals, as is evidenced by the "Other journals" entry. The

Table 1.5. *Most cited journals in Chapters 2–8.*

Journal	Chapters							Total citations	Rank order
	2	3	4	5	6	7	8		
J. Exp. Mar. Biol. Ecol.	16	17	21	4	16	13	15	102	1
Mar. Ecol. Prog. Ser.	11	14	13	8	7	13	14	80	2
Mar. Biol.	15	10	14	11	13	7	8	78	3
Limnol. Oceanogr.	24	15	1	2	6	8	7	63	4
Biol. Bull.	7	1	7	6	5	0	1	27	5
Can. J. Fish. Aquat. Sci.	7	3	1	0	5	7	4	27	6
J. Mar. Res.	5	6	1	0	3	5	4	24	7
J. Mar. Biol. Assoc. U.K.	6	4	5	1	2	2	0	20	8
Science	0	4	3	3	4	1	5	20	9
Can. J. Zool.	2	3	0	2	6	0	0	13	10
Helgol. Meersunters.	4	1	1	0	1	3	1	11	11
Ophelia	3	2	2	1	0	1	1	10	12
Oceanogr. Mar. Biol. Ann. Rev.	1	4	0	1	0	3	0	9	13
Nature	3	3	0	0	2	0	1	9	14
Bull. Mar. Sci.	1	7	0	0	0	0	0	8	15
Neth. J. Sea Res.	2	2	0	0	0	2	1	7	16
Other journals	36	31	11	16	36	25	41	196	—
Books, Proceedings, etc.	18	12	9	11	19	13	13	95	—
Totals	161	139	89	66	125	103	116	799	

total of 196 citations here represents 25% of the overall number of cited articles.

Aims and structure

Our central aim is to present interdisciplinary studies which use hydrodynamic theory as an explanation for selected aspects of the benthic biology of suspension feeders. We have deliberately considered as wide a range of hierarchical levels of biology as possible, from organs and physiology to ecosystems and ecology. We wished to consider the widest range possible because we feel that it is at the interfaces of these levels that interesting new insights may be discovered. We also hope to demonstrate that hydrodynamics is an important factor at all hierarchical levels of biology.

Some of the earliest writers in modern biology had realized the importance of water movement in enhancing growth. Thus, Darwin (1842) observed how coral reefs grew fastest where wave energy was most intense. The question as to how growth and productivity were enhanced by water movement, and similar basic questions in other subdisciplines

Table 1.6. *Pioneering modern interdisciplinary studies involving hydrodynamics and benthic biology.*[a]

Subdiscipline	Chapter	Authors and date
General	All	Vogel (1981)
Larval biology	4	Crisp (1955) Eckman (1979) Hannan (1984)
Feeding physiology	5	Kirby-Smith (1972) Jørgensen et al. (1986)
	6	Rubenstein and Koehl (1977) Jørgensen (1983)
Ethology	7	Leversee (1976) Gruffyd (1976) LaBarbera (1977)
Ecology	8	Wolff et al. (1976) Wildish and Kristmanson (1979)
	9	Odum (1971) Odum et al. (1979) Rowe (1971)

[a] *Note*: Full citations are given at the end of the appropriate chapter.

(Table 1.6), did not begin to be adequately addressed until the 1970s, except by Crisp (1953). Thus, most of the research that will be discussed is less than 25 years old.

The decision to limit the questions considered to proximate ones we believe to be justifiable and liberating. It is liberating because it is then possible to use the same experimental approach of strong inference introduced by Francis Bacon in 1620, rounded out by the method of multiple hypotheses advocated by Platt (1964), for most of the studies reviewed herein. Purely descriptive or theoretical/model studies, unless they have been tested by inferential experimentation, have been omitted or de-emphasized. The exclusion of evolutionary questions is also liberating because it absolves us from the responsibility to devise satisfactory methods of study. That is not to say that benthic biologists should not be interested in questions involving neo-Darwinian evolution (see Chap. 9).

Another difficulty we faced during the preparation of the book was, Whom are we writing it for, and what should be its style? We view our audience as primarily active research scientists, either already professionals in the interdisciplinary field of hydrodynamics/benthic biology or recently graduated students whom we would like to attract to it. As regards the style, we did not set out to produce a definitive review of each chapter topic but rather an eclectic overview of the literature up to the middle of 1995. We are painfully aware that a different author or combination of authors might produce a quite different account. Specific reasons for taking the overview – more general – approach are that because of the wide range of subjects we wished to introduce we could hardly claim to be comprehensive in each of them and because it became clear to us during the book's preparation that many important questions in this field remain to be formulated and many formulated questions have not yet been satisfactorily answered.

We believe that an explanation of how the nine chapters are organized will help the reader to use this book. The first chapter is a general introduction to the subject matter considered. In Chapter 2, we present an overview of currently used methods to study flow and benthic biological questions, in both the field and the flume laboratory, with selected examples of the special methods involving flow in larval biology, feeding physiology, ethology, and ecology. The final chapter, Chapter 9, is a discussion of the possible future research directions that could be taken in the field of flow-related aspects of the benthic biology of macrofaunal suspension feeders.

Each chapter has its own bibliography, which we hope will make locating references to a particular subject convenient for the reader. One unfortunate effect of this is that there is some repetition of references between chapters.

Chapters 3 through 8 bear two similarities in their organization that distinguish them from the others. The first is that each chapter (except Chap. 5) is organized around a group of hypotheses which are considered to be the major ones in the subfields concerned. Much of each chapter involves a discussion of the experimental tests which have been reported in connection with each hypothesis. The second similarity is that each of the six chapters also contains a summary, including a judgement on each null and alternative hypothesis. Each alternative hypothesis is designated as accepted, rejected, and status unclear or further work needed.

The subject matter of Chapters 3 though 8 is as follows: In Chapter 3 we deal with those aspects of larval biology most affected by flow – larval dispersal and larval settlement/recruitment processes. Two chapters concern those parts of the physiology of feeding by suspension feeders which are most affected by flow: seawater transport to the filtration surface

(Chap. 4) and capture mechanisms at the filtration surface (Chap. 5). In Chapter 5 it proved difficult to erect a suitable null hypothesis for the multiple possibilities of the aerosol theory, and an attempt to organize this chapter around pertinent hypotheses had to be abandoned. In Chapter 6 we present an introduction to studies involving behavioral responses to flow, and although a number of different questions are raised, the dearth of published work means that not enough suitable hypotheses have yet been proposed. The next two chapters deal with ecology and flow at different hierarchical levels. The first refers to populations of suspension feeders and flow (Chap. 7) and the second to selected ecosystems and flow (Chap. 8).

References

Alcock, J. 1989. Animal behaviour: an evolutionary approach. 4th ed. Sinnauer Associates, Sunderland, Mass. p. 596.

Chamberlin, T. C. 1965. The method of multiple working hypotheses. Science 148: 754–759.

Cherrett, J. M. 1989. Key concepts: the results of a survey of our members' opinions, p. 1–16. *In* J. M. Cherrett (ed.) Ecological concepts: the contribution of ecology to an understanding of the natural world. Blackwell Scientific, Oxford.

Crisp, D. J. 1953. Changes in the orientation of barnacles of certain species in relation to water currents. J. Anim. Ecol. 22: 331–343.

Darwin, C. 1842. The structure and distribution of coral reefs. Smith, Elder, London.

Daugherty, R. L., J. B. Franzini, and E. J. Finnemore. 1985. Fluid mechanics with engineering applications. 8th ed. McGraw-Hill, New York.

Denny, M. W. 1988. Biology and the mechanics of the wave swept environment. Princeton University Press, Princeton, N. J.

1993. Air and water: the biology and physics of life's media. Princeton University Press, Princeton, N. J.

Fenchel, T. 1987. Ecology – potentials and limitations. Ecology Institute, Oldendorf/Luhe.

Hildrew, A. G., and P. S. Giller. 1994. Patchiness, species interactions and disturbance in the stream benthos, p. 21–62. *In* P. S. Giller, A. G. Hildrew, and D. F. Raffaelli (eds.) Aquatic ecology, scale, pattern and process. Blackwell, Oxford.

Jørgensen, C. B. 1966. Biology of suspension feeding. Pergamon Press, London.

1990. Bivalve filter feeding: hydrodynamics, bioenergetics, physiology and ecology. Olsen and Olsen, Fredensborg, Denmark.

Lincoln, R. J., G. A. Boxshall, and P. F. Clark. 1982. A dictionary of ecology, evolution and systematics. Cambridge University Press, Cambridge.

Maddox, J. 1992. Electronic journals have a future. Nature 356: 559.

Mann, K. H., and J. R. N. Lazier. 1991. Dynamics of marine ecosystems: biological–physical interactions in the oceans. Blackwell Scientific, Boston.

Parsons, T. R., M. Takahashi, and B. Hargrave. 1977. Biological oceanographic processes. 2nd ed. Pergamon, Oxford.
Peters, R. H. 1991. A critique for ecology. Cambridge University Press, Cambridge.
Petersen, C. G. J. 1911. Valuation of the sea. 1. Animal life of the sea-bottom, its food and quantity. Rep. Dan. Biol. Sta. 20: 1–79.
Platt, J. R. 1964. Strong inference. Science 146: 347–353.
Quinn, J. F., and A. E. Dunham. 1983. On hypothesis testing in ecology and evolution. Am. Nat. 122: 602–617.
Schlichting, H. 1979. Boundary-layer theory. 7th ed. McGraw-Hill, New York.
Simberloff, D. 1983. Competition theory, hypothesis testing, and other community ecological buzzwords. Am. Nat. 122: 626–635.
Sprintall, R. 1990. Basic statistical analysis. 3rd ed. Prentice Hall, Englewood Cliffs, N. J.
Tait, R. V. 1968. Elements of marine ecology: an introductory course. Butterworths, London.
Vogel, S. 1981. Life in moving fluids: the physical biology of flow. Willard Grant Press, Boston.
1994. Life in moving fluids: the physical biology of flow. 2nd rev. ed. Princeton University Press, Princeton, N. J.

2

Methods of study

The purpose of this chapter is to present an overview of the methods available to research scientists involved in field observations and inferential field and laboratory experiments concerning benthic suspension feeders. As well as hydrodynamics, the disciplines we consider are larval biology, physiology, ethology, and ecology. Because of space limitations, the methods described for each discipline could not be dealt with comprehensively. Instead, we have selected a few themes for each which we believe will be most useful for the novice research scientist trying to set up an interdisciplinary hydrodynamic/benthic biology study.

Previous reviews concerned with some aspects of hydrodynamics/ benthic biological methodology involving suspension feeders have been presented in Vogel (1981, 1994), Denny (1988), and Jorgensen (1990). The content of each was briefly considered in Chapter 1, and we suggest that the appropriate book should be consulted for suitable methods if they are not included here.

A common difficulty for a research scientist in this field is to determine whether a laboratory or field experiment, or both, is the best approach to use in a particular research project. Thus, laboratory flume experiments are reductionist and yield simplified environments which are questionably realistic. On the other hand, field experiments provide realistic environmental conditions, but many of the variables cannot be controlled or replicated, rendering the results unique. Resolution of this difficulty, of generality versus special conditions, depends on the nature of the question. Hence, it is more likely to be field experimentation the higher in the biological hierarchy that the question is framed. Generality versus uniqueness is considered by Sulkin (1990) from the point of view of larval orientation mechanisms.

The subjects dealt with here include how to simulate flow in both field and laboratory flumes. We also describe various ways of measuring velocity, as well as methods of sampling seston quantitatively. The final section is concerned with specialized methods associated with the interdisciplinary fields of flow/benthic biology studies related to larvae, feeding physiology, behavior, and ecology of suspension feeders.

Flow simulation

It is required to simulate flow so as to perform experiments in which flow speed (unidirectional flow) or flow speed and periodicity of flow reversal (oscillating flows) are controllable variables. The flume requirements for physiological and behavioral experiments are often less rigorous than the full dynamic similarity required for realistically simulating sediment–water interface conditions within flumes.

Although for convenience we have separated methods of simulating flow into those in the field and those in the laboratory, the distinction is not always easy to make. This is because there is a gradual change from one to the other, with the difference in the degree of control of the important variables: velocity, salinity, and temperature.

Field flows

Flow simulators suitable for ecological studies in the field have been designed to measure growth rates of small populations of bivalves, or to determine ecosystem level fluxes of materials, e.g. seston, plant nutrients, carbon, or dissolved oxygen. The former are required to reproduce natural flows, so that growth rate is a function of independent variables such as bulk velocity or bottom shear stress (U^*), unhindered by wall effects or an undeveloped boundary layer. For ecosystem fluxes, the channels require well-defined flows so that the sampling at inlet and outlet gives a good estimate of the fluxes of the materials of interest. All of these are flowthrough devices and use a natural seawater supply. Flow velocities used are either natural tidal currents or simulated ones using a submersible pump or head tank and gravity flow.

The Kirby-Smith growth apparatus consists of eight 1.5-m-long acrylic tubes which receive seawater from a constant level head tank (Fig. 2.1). In our experiments (Wildish and Kristmanson 1985), bivalves were supported in the mid-centre line by plastic mesh inserts which minimally impeded flow. Flow rates through the tubes could be regulated by changing the diameter of the outlet tube fitting. The bulk flow velocities are

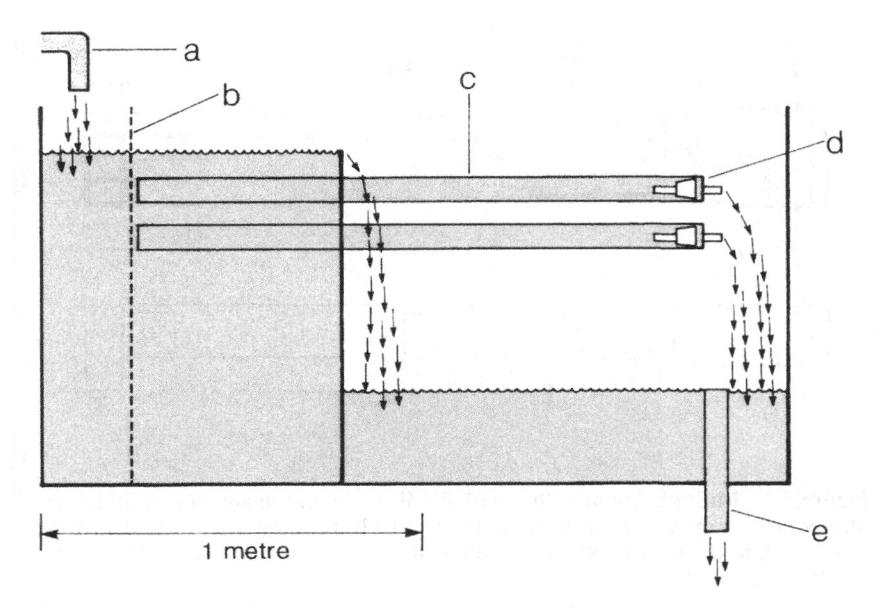

Figure 2.1 Kirby-Smith growth tube apparatus as used by Wildish and Kristmanson (1985): a, inlet seawater; b, screen; c, growth tube; d, outlet tube; e, main outlet.

calculated by timing collections of the volumetric flow and then dividing by the tube cross sectional area:

$$U = \frac{Q}{A} \tag{2.1}$$

where

U is the bulk velocity in $cm \cdot s^{-1}$
Q is the volumetric flow in $cm^3 \cdot s^{-1}$
A is the cross sectional tube area, cm^2

Pipe flows are well known hydrodynamically (Schlichting 1979). Mean velocity profiles and turbulence characteristics in empty pipes can be related to the mean pipe velocity and diameter. However, the pipes used in the Kirby-Smith apparatus are too short for a fully developed profile and the presence of inserts and bivalves disturbs the flow. Thus, the bulk velocity measures recorded may not estimate the velocity experienced by an experimental subject very well and may be apparatus-specific in determining their responses.

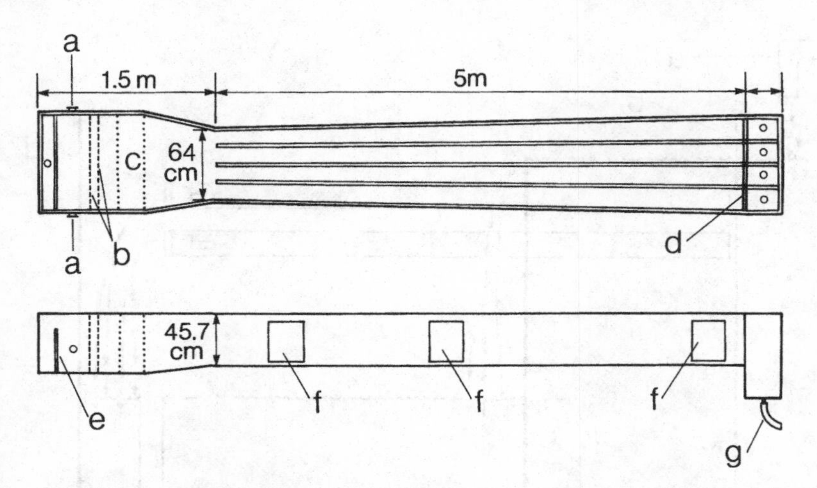

Figure 2.2 Multiple-channel flume of Wildish and Kristmanson (1988): upper, plan view; lower, elevation view; a, inlet port; b, perforated plywood; c, brass screens; d, perforated bulkhead; e, adjustable weir; f, acrylic plastic (Plexiglas) windows; g, drain.

The hydrodynamics of flow in a tube also differs from that in a natural benthic boundary layer. Test animals which are suspended at the centre of the tube would not experience the same flow field as those placed on the tube wall in the boundary layer flow. Flumes offer a better simulation for measuring the growth of suspension feeders than tubes since within each channel a benthic boundary layer develops on the bottom and benthic animals experience shear flows of similar magnitude to that in ambient flows. In most flumes discussed here, however, the benthic boundary layer never becomes as fully developed as it does in the sea.

A multiple-channel flume was designed by Wildish and Kristmanson (1988) to measure the population growth rate of bivalves while varying the unidirectional velocity. The multiple-channel flume (Fig. 2.2) allowed for four different velocities to be run at the same time, each with the same initial seston concentration at the inlet end of each channel. Seawater is pumped to the intake box by a submersible pump where a constant head is maintained by an adjustable weir. Velocity in each channel is controlled by rubber stoppers which can be removed from the perforated bulkhead. Hydrodynamic observations in this flume suggested that the flow had a developing boundary layer of ca. 20 cm thickness at the downstream end, and, therefore, this simulation provides flows which are more like the turbulent shear of the benthic boundary layer than can be achieved in tube flow.

Multiple-channel flumes have also been used in field conditions so the natural substrate is present. The flume used by Grizzle, Langan, and Howell (1992) consisted of four channels, 3.5 m long by 0.2 × 0.2 m, operated as three units side by side so that 12 different flow combinations were possible. The aim of this work was to determine the effect of velocity on representative siphonate and non-siphonate bivalve growth rates. Each unit was supplied with natural seawater from a constant head tank and the flow to each channel controlled by a ball valve. A plastic mesh flow straightener was included at the inlet and a sediment box near the outlet to permit experiments with burrowing bivalve infauna. Judge, Coen, and Heck (1992) describe a multiple-channel flume for use in the field with natural water movement which was similar to the single-channel flumes used in ecosystem studies described later. Its purpose was also to measure bivalve growth rates as a function of velocity. It consisted of nine adjacent channels with walls 7 m long by 1.2 m high, made of plywood sheets which were dug 0.15 m into the sediment during deployment. The flume was aligned channel end to the major flow in a seagrass meadow so that flows were regulated by size of the openings, which were equal at either end. Channel width was varied from 0.6 to 1.2 m and was not constant along its length. A 0.25-m^2 experimental plot was located mid-way down the channel so that it was beyond any wall boundary layer flow effects.

The other major purpose of field flumes was to measure fluxes of materials in subsystems, such as bivalve reefs, which are part of larger estuarine ecosystems. The simple tunnel designed by Dame, Zingmark, and Haskin (1984) utilized natural tidal currents, but it required precise orientation and did not include wind–wave forces unless the wind direction was of the same orientation. The benthic ecosystem tunnel of Dame et al. (1984) was constructed of 1/4-in-thick acrylic sheet, supported by a fibreglass strip and neoprene skirt to prevent leaking (Fig. 2.3).

A much larger, single-channel field flume of 40 m × 2 m width × 0.66 m was also used in the North Carolina salt marshes to measure material fluxes from the high to low marsh (Wolaver et al. 1985). The flume walls of the device could be removed after use and consisted of corrugated fibreglass panels. A similar apparatus – the Sylt flume (Fig. 2.4) – of 70 m length was constructed by Asmus and Asmus (1991) with 2-m-high walls and a channel width of 2 m. Plastic sheet formed the walls and could be removed from the wooden frame when not in use. To determine the flux of materials, e.g. oxygen, which occurs as a result of the canalized water within the flume at inlet (C_a, g $O_2 \cdot m^{-3}$) and outlet (C_b, g $O_2 \cdot m^{-3}$), the

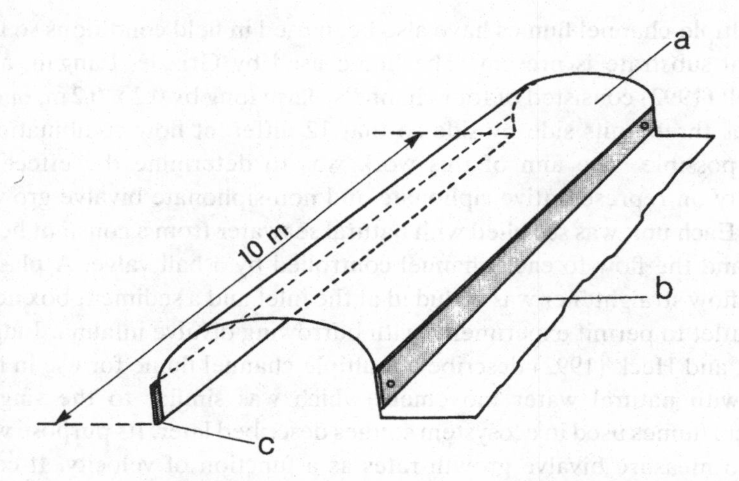

Figure 2.3 The benthic ecosystem tunnel of Dame et al. (1984): a, acrylic sheet; b, neoprene skirt; c, reinforcement rod.

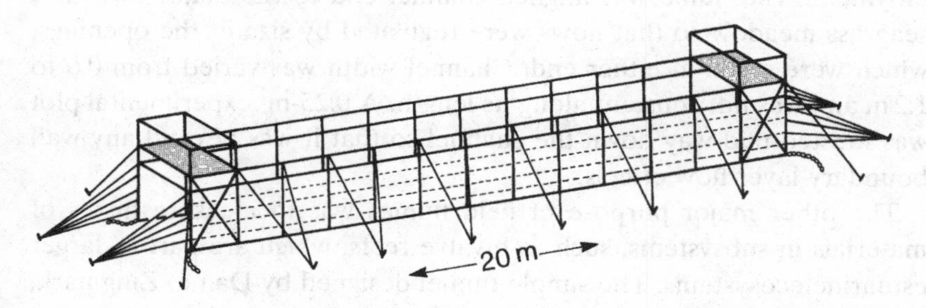

Figure 2.4 Diagram of the Sylt flume. The plastic sheeting can be removed when the flume is not in use (Asmus and Asmus 1991).

instantaneous rate of oxygen uptake, I, in units of g $O_2 \cdot m^{-2} \cdot h^{-1}$, is calculated:

$$I = \frac{Q(C_a - C_b)}{LW} \tag{2.2}$$

where

Q is the volumetric flow over the bed in $m^3 \cdot h^{-1}$
L is the distance, in m, between a and b
W is the path length of sediment, in m, over which the canalized water flows

This equation can only be used if the flow is well mixed with respect to dissolved oxygen concentrations at the inlet and outlet. Otherwise, the inputs and outputs, R, must be determined by integration of the form

$$R = \frac{1}{A}\int_A vcda \tag{2.3}$$

where v and c are the velocity and concentration of dissolved oxygen, respectively, at the infinitesimal element of area normal to the flow, dA, and A is the cross sectional area at inlet or outlet. Under favorable circumstances, similar studies can be conducted without the use of a tunnel, e.g. when the flow is shallow and vertically well mixed. In this case, the benthic boundary layer is depth limited and an estimate of the outlet concentration can be made as in equation (2.2) without the need for vertical positioning in obtaining inlet and outlet samples. This method was followed by some authors (Nixon et al. 1971; Wright et al. 1982) in determining dissolved oxygen uptake by mussel reefs. In deeper conditions when the benthic boundary layer does not reach the water surface, the bottom water may be depleted of dissolved oxygen. In these conditions, it will be necessary to use equation (2.3), requiring multiple vertical and coincident velocity sampling.

Surficial geologists (e.g. Scoffin 1968) have used an enclosed channel firmly pegged to the sediment (3 m × 10 cm × 13 cm) through which water is drawn by a propeller to determine the initiation of sediment motion by flow. The propeller was driven by a sealed dynamo motor, and it was necessary to use SCUBA divers to observe the threshold velocity for resuspension. The Seaflume consists of a channel 4 m × 15 cm high × 61 cm wide (Young 1977) and a propeller pump to create the flows required. A camera and strobe light allow close-up photography to observe the erosional behavior of the sediments.

Detailed hydrodynamic studies have not been made for any of the field flow simulators described. Narrow growth tubes do not provide a flow environment like the benthic boundary layer in the ocean and flumes do a better job of simulating natural flows. However, laboratory and field flumes, as well as tunnels, suffer from the same limitation – that crossflows which are not parallel to the flume walls and are created by wind effects, for example, are excluded. Additionally, the tunnel suffers from a restriction in vertical mixing which could result in lower fluxes. A study by Asmus et al. (1992), comparing fluxes measured in a benthic ecosystem tunnel and the Sylt flume at the same intertidal site, supports this view, although not conclusively, because no replication and hence statis-

Table 2.1. *Recirculating flumes designed for measuring physiological or behavioral responses of suspension feeders to unidirectional flows.*

Flume type	Volumetric capacity, L	Free stream U_{max}, $cm \cdot s^{-1}$	Taxa	Reference
Water tunnel, low speed	200	30	Marine plants	Charters and Anderson (1980)
Vogel–LaBarbera, low speed	20–800	40–50	"General purpose"	Vogel and LaBarbera (1978) Vogel (1981)
Raceway	60	50	Gorgonian coral	Leversee (1976)
Vogel–LaBarbera	115	50	Gorgonian coral	Sponaugle and LaBarbera (1991)
Vogel–LaBarbera	98	50	Octocoral	Patterson (1991)
Vogel–LaBarbera	?	50	Sea pen	Best (1988)
Holland	11	4	Crinoid echinoderm	Leonard (1989)
Vogel–LaBarbera	276	50	Bryozoa	Okamura (1984)
Raceway	?	?	Phoronids	Emig and Bechernii (1970)
Belle W. Baruch Institute Field Lab	?	?	Polychaete	Luckenbach (1986)
Vogel–LaBarbera	15	?	Brachiopods	LaBarbera (1977)
Saunders–Hubbard	200	100	Bivalves	Wildish and Saulnier (1993)
Denny, high speed	1500	500	"Wave-exposed" taxa	Denny (1988)

tical analysis was included. A further restriction for field flumes is that they can only operate intertidally where the tidal prism is less than a few metres, whereas tunnels could operate subtidally at least to SCUBA diving depths.

Unidirectional laboratory flows

Attempts have been made with laboratory flow devices to create a continuous and adjustable unidirectional flow, an oscillating flow to simulate wind–wave effects, and realistic simulation of sediment–water interface conditions.

Simple flows in which physiological or behavioral responses of suspension feeders can be determined do not require the large flumes found in civil engineering or naval laboratories. The design considerations of suitable smaller flumes are discussed by Vogel and LaBarbera (1978), and this work has resulted in many flumes which have housed a variety

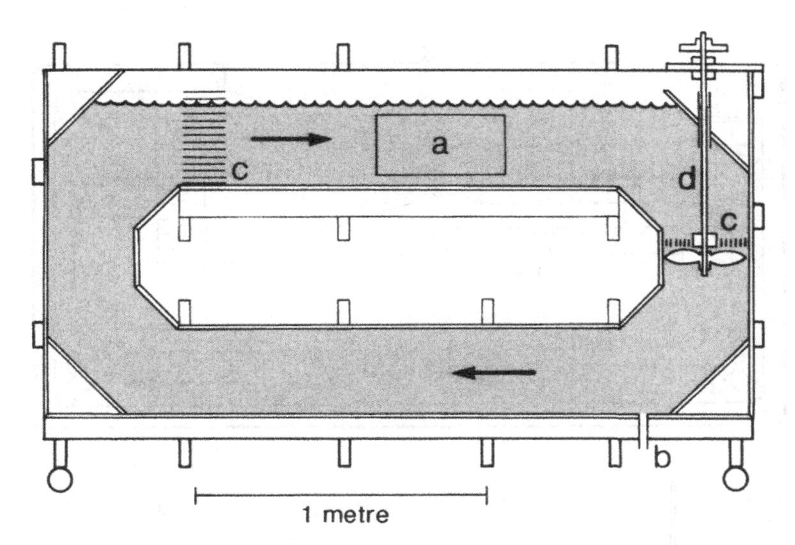

Figure 2.5 Vogel–LaBarbera recirculating flume in side view is constructed of plywood and wooden beam supports: a, viewing window; b, outlet; c, collimators; d, propeller shaft (Vogel and LaBarbera 1978).

of taxa (Table 2.1). Vogel (1981) pointed out that round pipe, rather than the original square type of the Vogel and LaBarbera flume, is preferable, and it has been used by subsequent authors listed in Table 2.1. All of the flumes shown in Table 2.1 are recirculating, meaning that a fixed volume of seawater is propelled through a closed, or partially open, circular arrangement of conduits. If filtration rates are being measured on the basis of the temporal change in seston concentration, the necessity is to limit the volume appropriately to the filtration power of the suspension feeder subjects. A corollary is that this limits the exposure time available for experimental purposes with suspension feeders because of a buildup of excretory substances.

Vogel–LaBarbera flumes are limited to flows $\leq 50\,\mathrm{cm \cdot s^{-1}}$, because at greater velocities, air bubbles begin to appear as a result of cavitation effects at the propeller. Because the flow entrance and exit conditions of the Vogel–LaBarbera type flume (Fig. 2.5) do not reduce the instabilities introduced by forcing the flow around four corners, it causes wave formation, which limits its usefulness at higher flows. If higher flows are required, then two flumes, the Saunders and Hubbard and Denny high speed, are available – the former useful to 100 and the latter to $500\,\mathrm{cm \cdot s^{-1}}$. The large Denny high speed flume is expensive to build, while

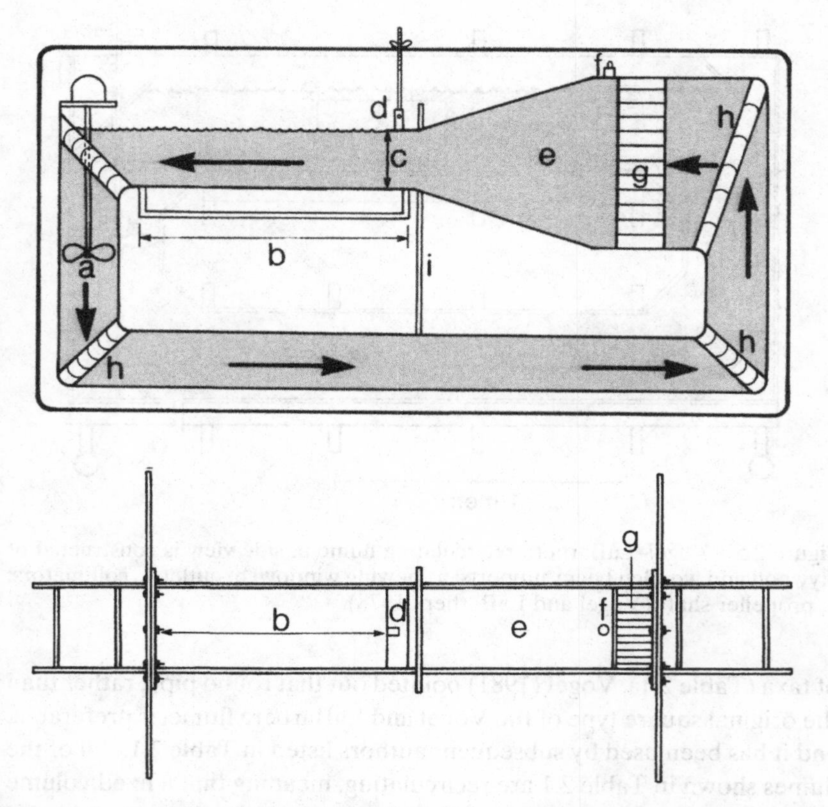

Figure 2.6 Saunders–Hubbard recirculating flume: upper, side view; lower, plan view; a, propeller; b, flume working section with false floor; c, fixed flow depth; d, flap for controlling surface flow; e, throat section; f, air removal; g, collimator; h, turning vanes (Wildish and Saulnier 1993).

the Saunders and Hubbard (Fig. 2.6) is mid-way in expense between the Denny and Vogel–LaBarbera flumes. The design improvements of the Saunders and Hubbard and Denny flumes which allow higher flows free of waves and aeration were borrowed from naval architects (Saunders and Hubbard 1944). For both flumes, they include a throat section choke and turning vanes at each of the corners.

An annular, open channel flume or raceway was used by Emig and Becherini (1970) and Leversee (1976), and this design appears to have similar maximum velocity characteristics to the Vogel and LaBarbera flumes.

The constant head flow tube or water tunnel of Charters and

Table 2.2. *Oscillating flow flumes for measuring physiological and behavioral responses of suspension feeders to bidirectional flows.*

Flume type	Volumetric capacity (L)	Wave character: Oscillation period (s)	Wave character: U_{max} ($cm \cdot s^{-1}$)	Reference
Lofquist	>1000	3–25	30–150	Lofquist (1977)
Lofquist	>1000	3–30	30–150	Hunter (1989)
Vogel–LaBarbera	15	1–>3600	?	Trager et al. (1990)
Lofquist	>1000	3–30	30–150	Turner and Miller (1991), Miller et al. (1992)

Anderson (1980) is a specialized apparatus for measuring the effect of flow on macroalgal physiology.

Oscillating laboratory flow

Svoboda (1970) considered various possibilities of simulating the oscillatory flows present near the sediment–water interface which are the result of wind action at the seawater surface. A consideration of laboratory methods of creating surface waves, primarily of interest to engineers, and problems such as beach erosion control are presented by Sorensen (1993). Oscillating flow tables or eccentrically attached, swinging aquaria were thought to be unsatisfactory, and a water tunnel in which a piston at one end drove seawater through a horizontal tube which acted as the working section was used (Svoboda 1970). Lofquist (1977) also employed this principle to simulate wave action for surficial sediment studies (Table 2.2) characteristic of all but the most exposed natural conditions. Although it is expensive to construct, the Lofquist oscillating flow flume (Fig. 2.7) has been used by biologists (Table 2.2) to determine oscillating flow effects on a wide taxonomic range of suspension feeders. The flume consisted of a water tunnel of U tube design with the vertical part made of paired cylinders at each end, one of which contained tight-fitting pistons, the other acting as a reservoir. A 3/4 horse power motor moved the pistons, either in unison or separately, so that two flow simulations could be run together. The device produces a sinusoidal flow

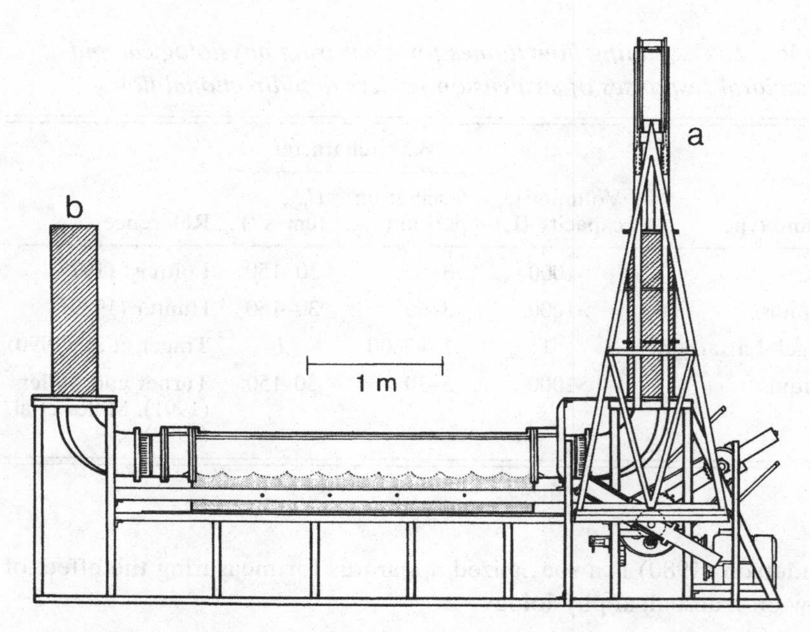

Figure 2.7 Lofquist oscillating flow flume (from Lofquist 1977): a, paired, horizontal cylinders with drive pistons; b, paired horizontal reservoir cylinders.

with periods adjustable between 3 and 25 s and peak velocities of $30\,cm\cdot s^{-1}$. The working section was divided by a removable partition which allowed either single- or double-channel use.

A flume of much smaller volumetric capacity (Table 2.2) was used by Trager et al. (1990) in studies with barnacles. The design of this system required a specialized electronic circuit to control the motor-driven propeller in either direction. With this system, it was possible to produce linear- and sine-wave accelerating and decelerating seawater flows. The maximum velocities achievable with the system described by Trager, Hwang, and Strickler (1990) are probably $<50\,cm\cdot s^{-1}$ because of the Vogel and LaBarbera flume used.

Sediment–water interface simulation

Theoretical considerations needed in designing a flume to simulate sediment–water interfaces have been considered in reviews by Muschenheim, Grant, and Mills (1986) and Nowell and Jumars (1987). They indicate that the flume conditions required to simulate sediment–

Table 2.3. *Sediment–water interfaces and unidirectional flow flumes.*

Flume type	Dimensions	Reference
Unidirectional, flowthrough	3-m-Long working section × 0.35 m wide	Muschenheim et al. (1986)
Annular – lab	2.0 m Outside diameter × 0.1 m wide	Taghon et al. (1984)
Annular – lab	2.0 m Outside diameter × 0.1 m wide	Wainwright (1990)
Annular – field	2.0 m Outside diameter × 0.15 m wide	Amos et al. (1992)
Recirculating flume	3-m-Long working section × 0.6 m wide	O'Riordan et al. (1993)

water interactions realistically are the most demanding of the flume designer's art. Wall effects can be minimized by placing the walls as wide apart as possible and preferably with a width/depth ratio ≥5:1. Nowell and Jumars (1987) point out that the flume must be capable of simulating the following flow and flow-sediment characteristics as they occur in the field:

- Reynolds number
- Froude number
- Strouhal number
- Roughness Reynolds number
- Shields parameter
- Rouse parameter

assuming a uniform bedform and a steady, unidirectional seawater flow. Further information on some of the listed variables is available in the reviews cited at the beginning of this section.

A selection of the types of flume which have been used in determining sediment–water interface interactions is shown in Table 2.3. The flume described by Muschenheim et al. (1986) was designed to investigate suspension–deposit feeding of spionid polychaetes in realistic initial sediment motion conditions as affected by turbulent boundary layer flows. Annular flumes are more suitable for determining continuous sedimentary transport, such as transport distance, but because of secondary circulations inherent in annular flow are not ideal for determining sedimentary transport rates. Flow is driven in the annular flumes by a floating ring, with or without paddles, propelled from a small motor operating in air. A field version of the annular flume – the Sea Carousel – from Amos et al. (1992) was designed to measure in situ sediment erodibility of shallow inter- or subtidal locations. For this purpose, it is equipped with optical backscatter sensors, a video camera, and an electromagnetic flow meter.

The interaction of siphonal exhalant currents generated by bivalves with the development of the concentration boundary layer was studied by introducing rhodamine dye into the exhalant flow by O'Riordan et al. (1993). The effect of different boundary layer velocities on the concentration field was determined by using a non-intrusive snapshot of the dye concentration field. The dye was illuminated with a thin sheet of laser light, and a charged coupled device camera was used to capture images of the dye concentration. As the fluorescent light intensity is proportional to the dye concentration, the method indicates rhodamine concentration.

Velocity measurement

In order to choose a suitable velocity measurement system, answers to some key questions must be obtained. The questions include, Is the measurement to be of ambient velocity in the field or in the laboratory flume?, At what spatial and temporal scales is velocity to be determined? and At what hierarchical level of biological organization are the velocity measurements to apply?

If the hierarchical level of biological organization of interest is ecological, then the standard techniques of physical oceanography, such as electromagnetic current meters, are applicable. Such current meters produce a continuous temporal record of velocity, and often current direction, stored electronically and retrievable with a laptop computer without moving the sensor (Table 2.4). The interested biologist, by either luck or design, may also be able to use a predictive physical oceanographic model to input mean and maximum tidal velocities (e.g. Wildish, Peer, and Greenberg 1986).

A limitation of electromagnetic current meters, and the models they give rise to, is that they usually do not measure orbital motions due to wind–wave effects. Studies by Denny (1982) on wave-exposed rocky shores were based on transducers to measure the wave force received by models of suspension-feeding organisms, e.g. barnacles. The force transducer output was converted to a frequency modulated (FM) radio signal

Table 2.4. *Partial list of commercial suppliers of equipment required for flow measurement.*

Equipment type	Supplier
Electromagnetic current meters	Marsh-McBirney, Inc., 8595 Grovemont Circle, Gaithersburg, Maryland, 20760 USA InterOcean Systems Inc., 3540 Aero Court, San Diego, California, 92123 USA
Acoustic current meters	EG&G Neill Brown Instruments Systems, 1140 Route 28A, P.O. Box 498, Cataumet, Massachusetts, 02534 USA
Miniature flow meters	Nixon Instrumentation, Cirencester Road, Cheltenham, Gloucestershire, UK GL53 8DZ Omega Engineering Inc., One Omega Drive, P.O. Box 4047, Stamford, Connecticut, 06907 USA
Thermistors	Victory Engineering Corp., 1-T Victory Rd., Springfield, New Jersey, 07081 USA
Laser Doppler flow meters	DISA Electronics Inc., 779 Susquehanna Ave., Franklin Lakes, New Jersey, 07417 USA TSI Inc., Fluid Mechanics Instrument Div., P.O. Box 64201, St. Paul, Minnesota, 55164 USA
Flow meters	Blue White Industries, 14931-T Chestnut St., Westminster, California, 92683 USA

received and stored at a shore-based laboratory (Denny 1982). The transducers gave information on the magnitude and maximum force exerted by waves, but further interpretation (Denny 1985) provided information on drag on the model, flow acceleration near the model, lift forces to which the model was subject, velocity and acceleration of waves, as well as the timing of each wave. This system is obviously expensive to construct, with data difficult to interpret and the instrumentation not very portable. A simpler, cheaper method was designed by Bell and Denny (1994) to determine maximum velocity during a passing wave. The equipment (Fig. 2.8) consisted of a spring and a hollow ball attached by monofilament line which acts as a drag element. As the wave passes the hinge-anchored instrument, the ball and spring are stretched, thus moving a rubber tube stretched over the line, indicating its maximum extension. From the linear relationship between the force and spring extension distance, the drag force on the ball or maximum wave

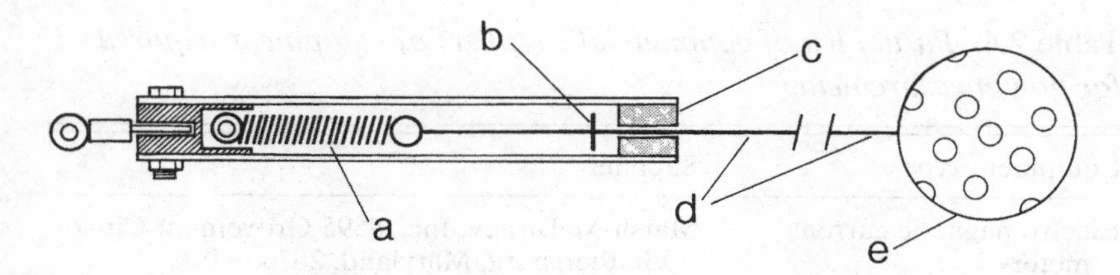

Figure 2.8 Maximum force recorder of Bell and Denny (1994): a, spring; b, rubber sleeve indicator of maximum force; c, upper plug; d, monofilament line; e, practice golf ball.

velocity can be determined. The spring extension response is non-linear with respect to wave velocity and maximum velocities measurable, limited to ≥2.5 $m \cdot s^{-1}$ as a result of inertia of the spring used.

The gypsum dissolution method can be used as a crude means of indicating water movement. The method incorporates any water motion due to tidal as well as wind–wave effects in the same integrated measure, although only recently have attempts (Thompson and Glenn 1994) been made to provide an absolute measure in seawater. Further advantages of the gypsum dissolution method are that it is inexpensive, the results are simple to interpret, and it is widely applicable for use in nearshore continental shelf environments at all levels of biological organization. The rate limiting step for dissolution of gypsum $CaSO_4 \cdot 2H_2O$ is transport through the diffusive sublayer (Lui and Nancollas 1971). Thus, increases of water movement near the diffusive sublayer enhance diffusion because of a decrease in the thickness of the boundary layer and/or an increase in the concentration gradient across it due to rapid transport of Ca^{2+} and SO_4^{2-} away from their site of dissolution.

The dissolution rate of gypsum follows a first order kinetics equation (Lui and Nancollas 1971). Thus, for any shape of gypsum cast, the rate of loss of mass is given by

$$\frac{dm}{dt} = KA(C_s - C) \tag{2.4}$$

where

m = mass of the cast in kg at time t
K = $m \cdot s^{-1}$, rate constant for a specified shape
A = m^2, area exposed to flow
C_S = $kg \cdot m^3$, saturation constant of gypsum
C = $kg \cdot m^3$, concentration of dissolved gypsum

Following Petticrew and Kalff (1991), this equation can be reformulated as a flux, F, as follows:

$$F = \frac{D}{\delta}\left(C_s - C\right) \tag{2.5}$$

where

F = transfer rate of gypsum per unit area, $kg \cdot m^2 \cdot s^{-1}$
D = diffusion coefficient for $CaSO_4$ in water, $m^2 \cdot s^{-1}$
δ = concentration boundary layer thickness, m

The dissolution rate of gypsum is sensitive to temperature for two reasons: The diffusivity increases with temperature and decreases in viscosity are caused by increases in temperature, which, in turn, reduce boundary layer thickness, δ.

A number of shapes and materials have been used to measure dissolution rates – small rings, actually candy (Lifesavers) – sewn to a kelp surface (Koehl and Alberte 1988) to "clod cards" of gypsum cast in plastic ice cube trays (Doty 1971; Jokiel and Morrissey 1993; Thompson and Glenn 1994) to cylinders (Wildish et al. 1997). The ideal shape for such measurements is a sphere, since flow from any direction should equally influence the dissolution process. A capped right cylinder will also be insensitive to the direction of flow as long as its main axis is oriented perpendicular to the flow. The weight initially (M_0) and after a measured exposure period, drying, and reweighing (M_1) may be used in a comparative or absolute measure of water movement. The dimensionless index proposed by Doty (1971), the diffusion factor, DF, is calculated as

$$DF = \frac{\left[M_0 - M_1\right] \text{ flow}}{\left[M_0 - M_1\right] \text{ still}} \tag{2.6}$$

where the weight loss, determined in the field for a standard time, is divided by the weight loss in still seawater of the same salinity and temperature.

In order to relate gypsum cylinder weight losses to flow in an absolute measure of velocity, it is necessary to calibrate flume flows and relate weight loss to salinity, temperature, and velocity. In the field, replicate cylinders may be deployed (Fig. 2.9) on a frame which includes a swivel and vane so that the gypsum cylinders face the current.

Five general techniques have been used to measure velocity in labora-

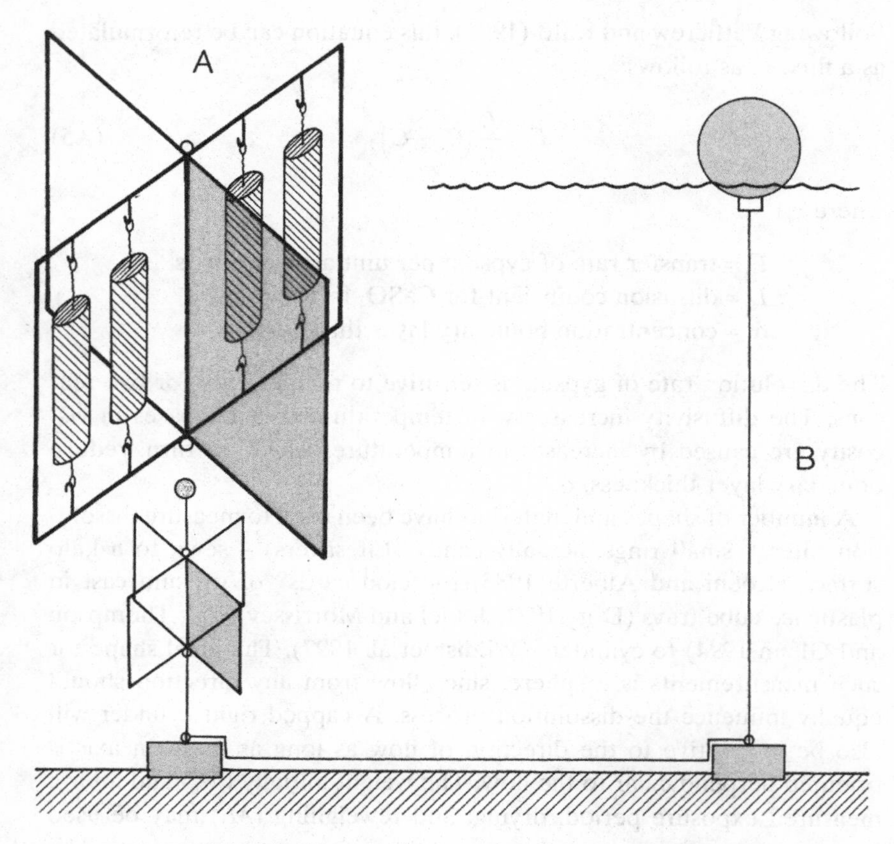

Figure 2.9 Deployment system for gypsum cylinders: A, frame constructed of steel wire for four gypsum cylinders held in place by rubber bands; B, buoy and anchor system for deployment and retrieval from a small boat.

tory flumes: electrochemical, mechanical, thermistor anemometry, laser Doppler velocimetry, and flow visualization. The method used will depend on the purpose of the velocity measurement, the velocity range required, as well as the size of the animal and flume size used in the experiment.

Electrochemical methods are usually not suitable for studies involving living animals because the chemicals involved may be toxic (Muschenheim et al. 1986). The hydrogen bubble method developed for use in freshwater (Schraub et al. 1965) is one exception. It was successfully adapted for measuring flume boundary layer velocities above a mussel reef in seawater by Wildish and Kristmanson (1984), who constructed an electronic square wave simulator in which the electrical frequency and pulse width to the electrode could be adjusted. Velocity was determined by finding the pulse width of known time which just filled a gap of known distance with a continuous sheet of hydrogen bubbles, followed by calculating unit distance/time. Use of a very fine platinum wire as the cathode is required to encourage the formation of sufficiently small and hence slowly rising bubbles. However, because the electrical current densities required to overcome the polarization voltage in seawater are high, the bubbles tend to be large and, consequently, rise too rapidly towards the surface.

The standard mechanical method of hydrodynamics is the Pitot tube (Bradshaw 1964), but applicable only where a steady uniform flow is available. The method measures the dynamic pressure where the flow is sampled and velocity is calculated from the Bernoulli equation (Vogel 1994). At $20\,cm \cdot s^{-1}$-flows, the Pitot tube pressure differential is ca. 0.2 cm freshwater. The precision of the velocity measurement will depend on the precision of the manometer used, but will not likely give satisfactory results below $20\,cm \cdot s^{-1}$. Electronic pressure transducers may yield better results at lower velocities.

The other common mechanical method of measuring flume velocity is by miniaturized current meters with cup or rotors in special bearings which minimize frictional torque. We have used a Streamflo model 403 probe, which is 47 cm long and has a rotor diameter of 11 mm, protected by a metal ring (Table 2.4). The rotor measures the integrated velocity in an area of $95\,mm^2$ and is thus suitable for measuring free-stream flows or determining velocity related to physiological or behavioral experiments of the same size, e.g. of bivalves or larger-sized organisms. The square wave signal generated has a pulse rate proportional to the rotor speed and hence ambient velocity. The indicator used with the rotor probe has a three-digit light-emitting diode (LED) display in hertz with continuous or time-integrated sampling of 1 and 10 s possible. The output can also be displayed on a recorder or imported to a personal computer (PC) spreadsheet. The rated velocity range of the model 403 probe is 2.5–$150\,cm \cdot s^{-1}$ with an accuracy of ±1% for most of the range, except at $<15\,cm \cdot s^{-1}$, the accuracy declines to ±5% at $2.5\,cm \cdot s^{-1}$.

Suitable thermistors for constructing a homemade thermistor bead velocimeter according to the temperature compensated resistor bridge circuit diagram given in Vogel (1981) are available from the Victory

Engineering Corporation (Table 2.4). Complete thermistor anemometer systems are available from DISA Electronics Incorporated (Table 2.4). Because of the small size of the sensing heads or beads, typically 0.1–2.0 mm in diameter, the thermistors are suitable for resolving compressed flume boundary layer flows in steps of 1 mm. Charriaud (1982) has used a thermistor probe of 0.1-mm diameter mounted on a micromanipulator to measure velocity profiles across the actively pumping cloacal siphon of an ascidian in steps of 1 mm. Each thermistor probe used requires separate determination of a probe constant and is tedious to calibrate. The amplifier output is a curvilinear response of velocity, so the useful range may be limited. Further details of thermistor anemometry can be found in Gust (1982).

The Doppler effect can be used to profile flow velocities within the boundary layer with a laser Doppler velocimeter (Table 2.4). Further details of these systems can be found in Penner and Jerskey (1973). Optical receivers measure the scattering of laser light by natural particles or fluorescent dyes (Koochesfahani and Dimolokis 1985) within the ambient flows; thus this is a specialized version of the flow visualization method.

The flow visualization method has the distinct advantage of being non-invasive; therefore, the sensor itself does not influence local flows. An early example of the flow visualization method was reported by Vogel and Feder (1966), in which low speed flows seeded with plastic particles illuminated by pulses of light (12 flashes$\cdot$s^{-1}) from a strobe source were recorded photographically. Carey (1983) used denatured egg white particles in the size range 0.25–1.00 mm in a seawater flume. Miron, Pelletier, and Bourget (1995) used polyvinyl chloride particles of 125–300 µm in diameter in flume studies of scallop spat collectors. Time-lapse photography through the acrylic plastic wall of the flume allowed calculation of particle trajectories and velocity by determining the particle streak length for the known time-lapse period.

The use of natural particles such as seston as the flow streak line monitor is a considerable improvement on the previous methods. Trager et al. (1990) used such a system with the small oscillating flow flume described in the section, Oscillating laboratory flow. The optical part of the apparatus consisted of a fiber-optic white lamp and expanded collimated laser beam of 740 nm, both focused onto the barnacle subject in the 10-cm-wide flume channel. A video camera fitted with a dark field lens system, capable of magnifications up to 50 times, was used to record the images. The latter, inclusive of the sestonic particles, appear white on

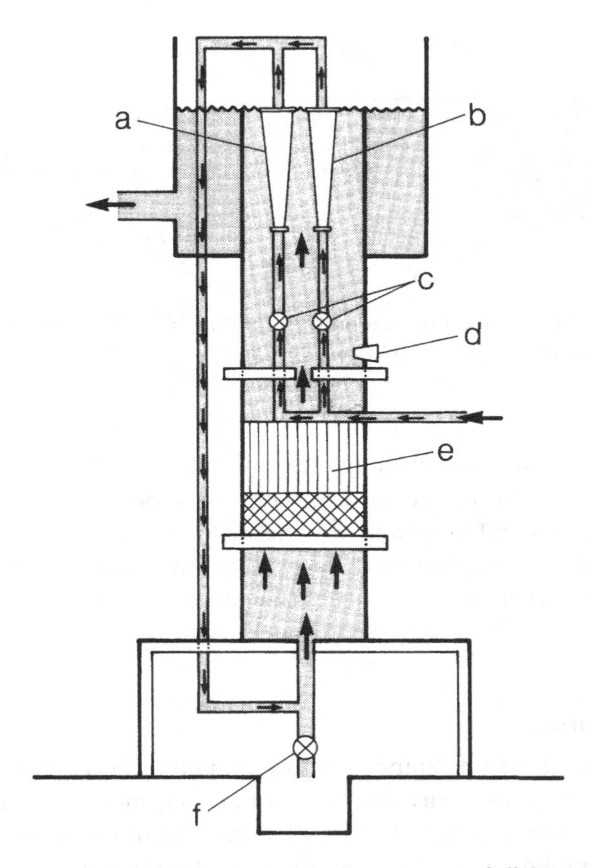

Figure 2.10 Velocity calibrator: a and b, flow meters of different capacity; c, velocity controls; d, position for insertion of probe (Streamflo 403); e, flow formers; f, seawater drain valve.

a black background and, after frame-by-frame analysis, were used to provide data on particle velocity as well as feeding movements of the barnacles. This system was originally described by Strickler (1985) and Leonard (1989).

A general purpose velocity calibrator design is shown in Fig. 2.10. It consists of a 20-cm-diameter acrylic tube and an overflow tank. Seawater from a piped supply is passed through one or both adjustable volumetric flow meters (Table 2.4) and so into the bottom of the acrylic tube. The collimated flow passes through a 5-cm-diameter orifice, just above which test probes such as the model 403 Streamflo can be introduced so that the rotor is normal to the flow. Factory-calibrated Streamflo probes were

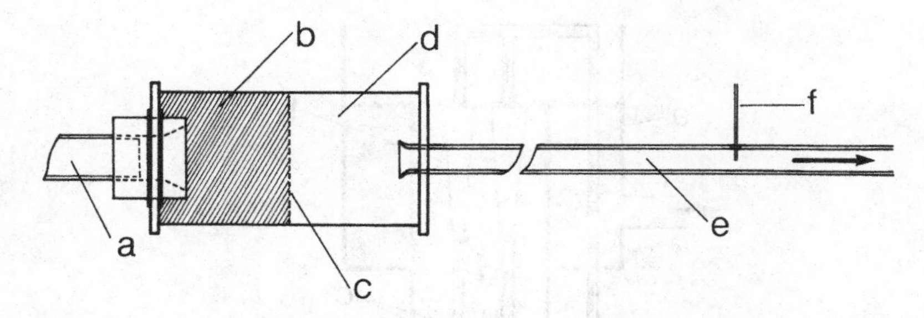

Figure 2.11 Thermistor flow calibrator: a, inlet; b, baffle chamber; c, Nytex screen; d, entrance chamber; e, flared acrylic tube; f, thermistor (Muschenheim et al. 1986).

used to confirm that velocity at the test point was equal to the volumetric flow divided by the orifice area. Calibration plots are made of probe output in hertz as a function of volumetric flow rate.

A similar but miniaturized version of the design was used by Muschenheim et al. (1986) to calibrate small thermistor bead probes (Fig. 2.11).

Seston sampling

The purpose of seston sampling is to determine the concentration and composition of seston particles in seawater. During feeding by suspension feeders, it is necessary to take grab samples of seawater with small diameter sampling tubes, because of the localized nature of the concentration gradient of seston within the benthic boundary layer.

Three general types of seston sampling may be used to determine seston concentration and composition: isokinetic, non-isokinetic, and active suspension feeder–biased. Non-isokinetic is the default option, where sampling is arbitrary, not considering flow and concentration boundary layer conditions. Isokinetic is the appropriate means of sampling for passive suspension feeders, and it, or species-biased sampling to simulate the live animal, is appropriate for suspension feeders with an active ciliary or muscular pump.

Isokinetic sampling

A theory of sampling solid particles from flowing fluids has been developed for aerosols in industrial applications (Hinds 1982). Sampling errors may arise in sampling flowing fluids to measure the solids content as a result of the following:

- A mismatch between ambient and sampling tube velocities
- Variations in settling velocities of the sestonic particles
- The orientation of the sampling tube with respect to the flow

With regard to the latter, Hinds (1982) showed that as the orientation of the sampling tube was moved from opposing the flow (0°) to an increasing angle until it was normal to the flow (90°), so sampling errors due to mismatch of flows also increased. Estimates of this error are given in graphical form in Hinds (1982).

The key hydrodynamic parameter for sampling is the dimensionless Stokes number, Stk

$$\mathrm{Stk} = \frac{W_s U_0}{gD} \tag{2.7}$$

where

W_s = passive terminal settling velocity of seston, $\mathrm{cm \cdot s^{-1}}$
U_0 = ambient velocity, $\mathrm{cm \cdot s^{-1}}$
g = gravitational constant, $981\,\mathrm{cm \cdot s^{-2}}$
D = diameter of sampling tube, cm

The Stokes number indicates the likelihood that the particle will follow the fluid streamlines, with higher values of Stk giving greater sampling error. Thus, if $\mathrm{Stk} < 0.01$ and the velocity mismatch between ambient and sampling flow is within ±4 times the isokinetic rate, then the sampling error will be <5% (Hinds 1982). The maximum terminal settling velocity of a sestonic particle at a fixed $\mathrm{Stk} = 0.01$ can be calculated from

$$W_s < \frac{9.8D}{U_0} \tag{2.8}$$

Assuming $D = 0.64\,\mathrm{cm}$, we can calculate at $U_0 = 5\,\mathrm{cm \cdot s^{-1}}$ that seston with a settling velocity of $\sim 1.25\,\mathrm{cm \cdot s^{-1}}$ will be collected without significant sampling error. Similarly, at $10\,\mathrm{cm \cdot s^{-1}}$, but otherwise the same conditions, the limit for collection bias is particles of settling velocity of $\sim 0.63\,\mathrm{cm \cdot s^{-1}}$. These considerations suggest that sestonic particles such as typical phytoplankton cells with $W_s < 0.01\,\mathrm{cm \cdot s^{-1}}$ would be collected without sampling error (see also Chap. 5). However, if the seston load contained denser sedimentary particles >100 μm, then sampling errors, if strict isokinetic sampling were not practised, might occur.

Table 2.5. *Matching flume flows,* U_0*, and sample tube flow at the isokinetic point.*[a]

U_0 ($cm \cdot s^{-1}$)	Calculated isokinetic point ($cm^3 \cdot s^{-1}$)	Static head (cm)	Time to empty 180-cm-long tube (s)
3	0.05	18	65
5	0.08	24	40
8	0.12	35	24
10	0.15	42	20
12	0.19	48	16
15	0.23	60	13
18	0.28	65	11
20	0.31	77	10
25	0.39	95	8

[a] Tubing is $D = 0.140\,cm$, $A = 0.0154\,cm^2$. PE200 Clay-Adams Intramedic tubing, Becton and Dickson, Parsipanny, NJ 07054, USA. Static head required to deliver matched flows is shown with the time required to empty the total tube volume completely.

The position of the passive or active suspension feeder's sampling surface or inhalant opening with respect to boundary layer flow is critical in determining where sampling should take place. A simple method of sampling from a flume is to deploy a tube near the sampling surface or inhalant of the experimental animal and then allow the other end to siphon. The distance from the free flume surface to the outlet end of the tube is the static head and determines the sampling velocity and volumetric flow rate. Different tube diameters and lengths require experimental calibration (e.g. Table 2.5), which is conveniently achieved with a volumetric cylinder and stopwatch. The static head can then be varied to obtain a range of different exhalant flow velocities. Isokinetic sampling requires sampling tubes opposing the flow and matching the sampling tube flow to the ambient flow at the appropriate boundary layer height. Alternatively, the isokinetically matched flow can be achieved with an adjustable clamp and pump.

Non-isokinetic sampling

From the preceding discussion, we conclude that non-isokinetic sampling will rarely lead to serious sampling errors in collecting seston. Only where sedimentary particles are resuspended during feeding observations or experiments, as was the case in a study of a facultative

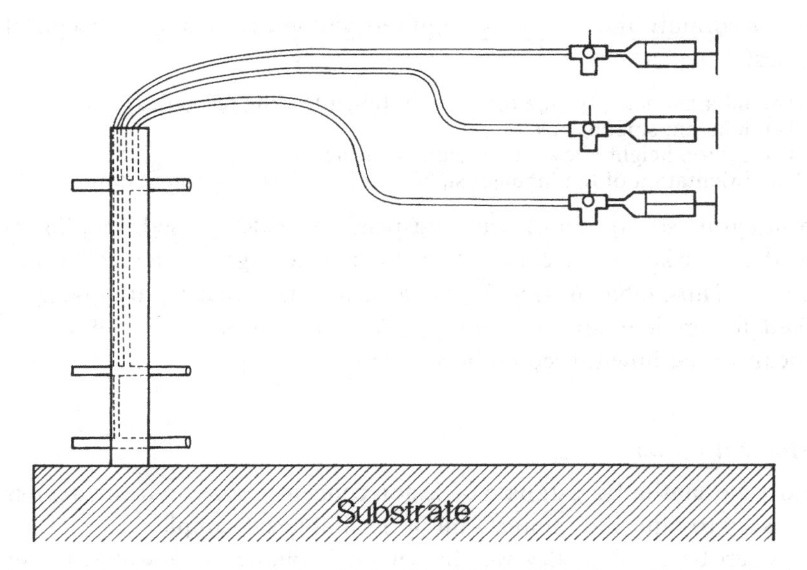

Figure 2.12 Active suspension feeder biased sampling. Sampling ports at 1, 5, 15 cm above the substrate and with internal diameter of 6.4 mm to simulate a 6-cm *Mercenaria mercenaria* (Judge et al. 1993).

suspension–deposit feeding polychaete by Muschenheim (1987), for example, could non-isokinetic sampling lead to serious seston sampling errors.

Active suspension feeder–biased sampling

In active suspension feeder–biased sampling an attempt is made to simulate the localized hydrodynamics of a specific suspension feeder in its initial feeding responses. Judge et al. (1993) described an original attempt to do this with the hard clam *Mercenaria mercenaria*. The sampling device used to simulate hard clam filtration is shown in Fig. 2.12. Simultaneous samples were collected at three depths to simulate clam position on eelgrass stems (1, 5, and 15 cm above the sediment) by manual operation of syringes so that the intake velocity matched that previously reported in hard clams (Walne 1972). Sampling tube diameter, $D = 0.64$ cm, was chosen to follow closely that of the inhalant siphon diameter of a 6-cm-valve-length hard clam.

Whether this form of sampling is more widely applicable to other active suspension feeders will depend on whether it is possible to determine accurately the following required variables for a wide range of species:

- The inhalant velocity (see the section, Initial feeding responses)
- The inhalant siphon area
- The siphon height above the sediment surface
- The orientation of the inhalant siphon with respect to ambient flow

An integral assumption of active suspension feeder–biased sampling is that these variables are constants, yet evidence suggests that this is not the case. Thus, inhalant siphon area and filtration capacity are inversely linked to ambient flow in scallops (Wildish and Saulnier 1993); it is difficult to see how this could be simulated.

Seston collection

Before a quantitative estimate of seston concentration can be made, it is necessary to filter the sample – all or a suitable subsample – so that the result can be given as dry weight per unit volume of seawater. Other measures of concentration commonly used include chlorophyll a, determined either fluorometrically or spectrophotometrically (Strickland and Parsons 1968); adenosine triphosphate (ATP), as an indication of all living cells or tissue (Holm-Hansen and Booth 1966); and the acridine orange epifluorescence direct counting technique of Hobbie, Daley, and Jaspar (1977) to indicate numbers of live bacteria (Wildish and Kristmanson 1984). Electronically sizing and counting sestonic particles (Strickland and Parsons 1968) are other frequently used options in determining seston concentration. More recently, the wider availability of the flow cytometer has meant that this technique has been used (e.g. Shumway et al. 1985; Boucher, Vaulot, and Partensky 1991) to measure particle numbers and fluorescence simultaneously. Shumway et al. (1985) showed that even in mixed phytoplankton cultures containing three species, this technique could indicate the proportional presence of each species. The flow cytometric technique as it applies to seston in seawater is reviewed by Yentsch et al. (1983); also see the review of particle measurement in seawater edited by Demers (1991).

All of the measures of seston concentration described – except the simplest, dry weight determination – are selective, and none of them is capable of indicating the quality of the seston as a food source to a named suspension feeder.

Seston quality

In a study of blue mussel initial feeding responses with a unialgal culture, Wildish and Miyares (1990) found that two factors influenced filtration rate – velocity and chlorophyll a content per cell. In these experiments, the mussels were all fed the same initial cell density of the cultured microalga, but, as a result of growth/time related differences in different cultures, the chlorophyll a content varied, with evidence that the mussels preferred those cells with higher levels of chlorophyll a. This same method cannot be used with a mixed culture or with wild-caught phytoplankton because of differences in amounts of chlorophyll a typical of a species as well as the intraspecific variances already noted. Use of flow cytometric techniques may allow more species of cultured microalgae in a mixed culture to be resolved (e.g., see Olson and Zettler 1995), even where preferential clearance of one species over another is present. However, it too may be of limited use in field samples, where many phytoplankton species are usually present and only a few of them are being ingested.

The nutritional quality of seston for suspension feeders could not be adequately expressed by gross biochemical measures of the seston (Navarro and Thompson 1995). Navarro and Thompson proposed a dimensionless food index based on the proportion of protein, carbohydrate, and lipid content of the total seston available. They demonstrated seasonal changes of the food index, but did not relate them to the growth of any named suspension feeders.

The nutritional value of seston to suspension feeders can only be determined in standardized feeding growth bioassays. The availability of methods to create microparticulate diets (Jones, Mumford, and Gabbott 1974; Langdon, Levine, and Jones 1985; Southgate, Lee, and Nell 1992) with known types and quantities of nutrients suggests that for larvae, juveniles, and adults it is possible to create a standard diet for genetically standardized suspension feeders. Such a diet would be of great value in comparing the quality of the available natural seston as a food for suspension feeders.

Specialized methods

Many different methods have been used in studying the larval biology, physiology, ethology, and ecology of benthic suspension feeders. Because there are so many, we have selected a few methods for each of the four disciplines concerned and present them here.

Larval swimming, feeding, settlement, and recruitment

One advantage of the small size of the ciliated larvae of suspension feeders is that live specimens can be studied individually in relatively small apparatuses. This represents a considerable saving, for example, in the amount of live culture necessary to maintain the experimental stock and run feeding experiments. In reviewing the literature on swimming velocity of larvae, Chia, Buckland-Nicks, and Young (1984) described two general methods in which swimming was observed directly with a microscope, or indirectly with tethered larvae followed by high speed photography and frame-by-frame analysis of the exposed film. From the latter, the larval swimming speed can be calculated as one-half that of the ciliary tip speed. An important consideration with this method is that at such low Reynolds numbers, the drag experienced by a swimming larva is influenced by walls as far away as several hundred body diameters (Chia, Buckland-Nicks, and Young, 1984).

The use of tethered larvae was introduced in feeding studies by Strathman, Jahn, and Fontsia (1972); also see Strathman and Leise (1979) and Gallagher (1988). Tethering involves attaching the larva to a hair or thin glass rod by cyanoacrylate adhesive or with a fine suction pipet to immobilize it. A small recirculating flow tank of low volumetric capacity, inclusive of the head tank (Fig. 2.13), was designed by Emlet (1990) for examining the effect of tethering in flows up to $4\,mm \cdot s^{-1}$. Pediveliger larvae of *Crassostrea gigas* were one species used in these experiments with film exposed at 200 frames$\cdot s^{-1}$ (Emlet 1990).

Fluorescent microparticles have been used to label food particles in experiments on settling with living bivalve larvae (Lindegarth, Jonsson, and André 1991). In this method, the larvae are fed polyvinylchloride particles with densities similar to those of the larvae so that they do not influence sinking rate after ingestion. Before the microparticles are fed to the larvae, they are coated with a fluorescent pigment so that if illuminated with a suitable light source, the particles fluoresce, and a camera fitted with an appropriate light filter can detect the fluorescent label. The label remains visible in the larval gut for a short time – a half-life ~10 h – so that experiments must be short term if a single pulse feeding is employed.

Laboratory experiments involving larval settlement over soft

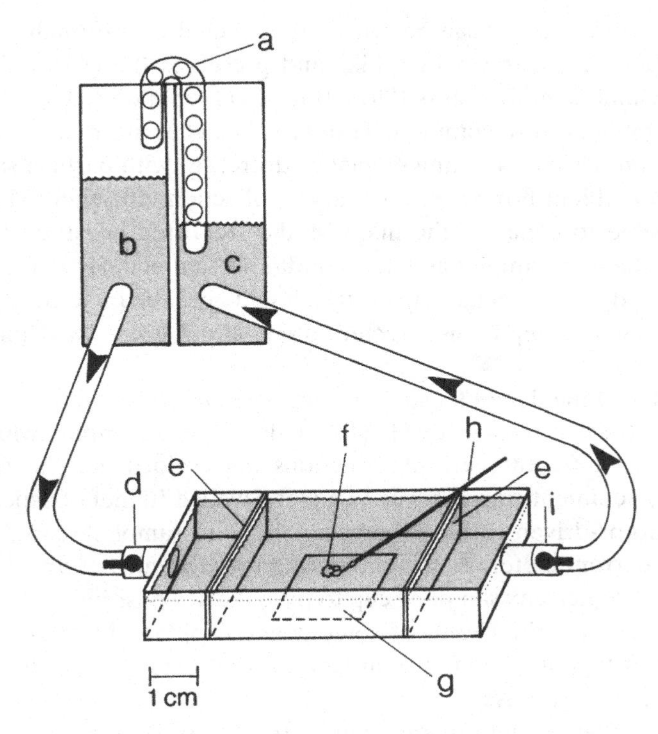

Figure 2.13 Miniature flow tank for observing tethered larvae (after Emlet 1990): a, air lift; b, upstream reservoir; c, downstream reservoir; d, inlet with control valve; e, collimator; f, larva; g, coverslip; h, tether; i, exit port with control valve.

sediments require full dynamic similarity of the sediment–water interface in the flume (see the section, Sediment–water interface simulation) with that in the field. Examples of laboratory experiments reaching this degree of control in flume setups are those of Eckman (1979), Hannan (1984), Butman, Grassle, and Webb (1988) and Grassle, Snelgrove, and Butman (1992).

Field sampling of larvae has been achieved with plankton nets or sediment traps. It was shown by Hargrave and Burns (1979) that the trap shape and sampler height/mouth diameter, or aspect ratio, affected the amount of sediment collected for a given sampling area. For high aspect ratio sediment traps, the amount of particulate matter collected varied asymptotically with the amount of turbulent mixing. Similar high aspect ratio traps are also suitable for sampling larvae if a means of preventing active larval swimming and advective removal from the trap is provided. The latter can be achieved by including a formalin layer in the bottom of the trap which kills and preserves the captured larvae (Yund, Gaines, and Bertness 1991). Butman (1986) pointed out that trap sampling biases were common: Trap larval catches are relative and dependent on design, sampling efficiency decreases with particle size, and different ambient flows influence larval collection efficiency. Thus, it is not possible to compare the many studies reported because of differences in shape of sampler and flow conditions sampled. However, within a given study, a particular trap design should give adequate comparative data as long as sampling is directed to one larval size class (Yund et al. 1991).

Because of the dependence of particulate catch rates on the velocities near the trap, Hargrave et al. (1994) designed a sampler which can respond to different current directions and velocities. The current-activated sediment trap consists of a steep-walled funnel of aspect ratio 1:5, a motor-driven carousel carrying up to 12 sampling cups, and an acoustic current meter. A microcomputer uses the current direction and velocity to switch any sampling cup to the collector position, according to pre-programmed thresholds of velocity and compass direction. A sampler of this type may be useful in larval trap design in order to remove velocity/direction biases.

Studies of larval recruitment on hard, rocky substrates can be achieved by allowing the larvae to settle on a cleared portion of the natural substrate or on a specially designed flat settlement plate. The type of plate – thin, faired; bluff; split or angled – was used by Mullineaux and Butman (1990) and Mullineaux and Garland (1993) to manipulate flows over the plate as a means of investigating larval settlement caused by differences in small-scale boundary layer flows.

In the deep sea where it is required to conduct larval settlement studies, various types of larval mud boxes, lowered by submersible or wire and retrieved by transponder and a flotation device, have been designed (Snelgrove, Butman, and Grassle 1995). These devices were tested in flume experiments to show the hydrodynamic artifacts which can be associated with them. According to Snelgrove et al. (1995), careful design consideration can limit these flow artifacts.

Larval settlement rates have been shown to be affected by the density of previously settled larvae. Thus, in field studies where natural rock surfaces were used, Minchinton and Scheibling (1991) showed how the timing of removal of larvae in the low intertidal zone influenced the densities estimated for recruitment. Both recruitment and post-settlement mortality rates decreased exponentially as the sampling frequency was decreased. This finding suggests that measured recruitment densities will not be comparable unless both studies incorporate the same sampling frequency. It is of obvious importance in studying the supply side ecology of aggregating suspension feeders.

Initial feeding responses

The physiological responses of feeding, growth, and elimination in suspension feeders are defined in Chapter 4. Because we are interested only in those feeding responses influenced by flow, we limit discussion here to initial feeding responses. The initial feeding of suspension feeders involves rate processes defined as follows:

- Seston uptake rate, S, the mass of seston removed from the water column in unit time by a standard weight of a suspension feeder
- Clearance (= filtration) rate, R, the calculated volume of seawater filtered, assuming complete removal of sestonic particles, in unit time and for a standard weight suspension feeder

S is not necessarily equal to the seston consumed, e.g. bivalves may reject some seston as pseudofaeces. Hence,

- Feeding rate, F, the mass of seston ingested in unit time by a standard weight of a suspension feeder

R cannot always be calculated from the volumetric flow rate through the trophic fluid transport system because the retention efficiency of the filtration surface will rarely be 100% or known. It may also temporarily change, e.g. by gill bypassing in bivalves (Famme and Kofoed 1983). The efficiency of the filtration surface in retaining sestonic particles is also of interest for studying mechanisms of particle capture. Thus:

- Retention efficiency, RE, is the proportion of seston particles captured at the filtration surface in relation to the total of those which approach it

The amount of energy-requiring work done in order to pump seston-containing seawater through the trophic fluid transport system of active suspension feeders is presumably related to the rate of this process:

- Pumping rate, P, which is the volume of seawater that actually passes through the trophic fluid transport system of a standard weight animal, irrespective of retention efficiency, in unit time

The clearance rate should not be confused with the pumping rate, and these terms are related by the retention efficiency thus:

$$R = P(\text{RE}) \tag{2.9}$$

Seston particles which differ in size or chemical composition will vary in retention efficiency in bivalves (Famme and Kofoed 1983), so that the clearance rate can be said to be particle-type-specific.

Historically, *indirect methods* for measuring S or R have been most commonly used. Such measurements have been made in closed tanks of fixed volume ("static") or in variously shaped flowthrough vessels. The difference between the initial, C_1, and final, C_2, concentrations of seston, or the inlet and outlet concentrations is used to estimate S or R. The volume of the static system, V, and the mass of animals, M, must be known as well as the elapsed time of the experiment, T, to calculate S:

$$S = (C_1 - C_2) \cdot \frac{V}{MT} \tag{2.10}$$

Typical units for this equation might be $\text{g} \cdot \text{seston} \cdot \text{g}^{-1}$ dry weight $\cdot\, \text{h}^{-1}$.

The clearance rate for a unit weight of animal can be estimated from

$$R = SC_1^{-1} \tag{2.11}$$

with reasonable accuracy if C_1 and C_2 are not greatly different (say, $C_2 > 0.9\ C_1$). Unacceptable inaccuracies may arise because the equation is based on the assumption that the seston concentration of the filtered seawater remains at C_1 during the experiment. Hildreth and Crisp (1976) recognized this difficulty and suggested that C_1 in equation (2.11) be replaced by the actual seston concentration around the suspension feeder throughout the experimental test.

If C_1 divided by C_2 is relatively large, the formula of Coughlan (1969) is preferable:

$$R = \frac{V}{MT} \ln\left(\frac{C_1}{C_2}\right) \tag{2.12}$$

If a flowthrough system is employed, the V/T term in equations (2.10) and (2.12) can be replaced by the volumetric flow rate. Then the terms C_1 and C_2 will denote concentrations in inlet and outlet flows, respectively. The same difficulties in specifying the concentration term in equation (2.11) – that is, the actual concentration of seston present in inhaled seawater – will also be present. This concentration will be between C_1 and C_2, its exact value depending on the mixing characteristics of the flow-through vessel.

The standard indirect methods of determining *S* or *R* currently in use suffer a number of drawbacks:

- Static containers or experimental boxes cannot simulate flow near the suspension feeder, which, in some cases, may directly influence *S* or *R*.
- Seston concentration depletes during the experimental test and this may change the suspension feeders' response to it. Winter (1973) used a constant volume apparatus for measuring *R* at a constant concentration. Again, the mixing characteristics are critically important in determining whether the results are simply apparatus-specific or more universally applicable.
- The indirect method is based upon the assumption, which cannot be fully satisfied, of instantaneous back-mixing of exhalant flow with the rest of the seawater within the experimental chamber (see Hildreth and Crisp 1976; Riisgård 1977). The actual extent of back-mixing is dependent on the geometry of the experimental chamber and the flow rates.
- Errors arising from incomplete mixing can be reduced by arranging experiments with relatively small changes in C_1, but this is a useful strategy only if C_1 and C_2 can be measured with high precision. The standard error of the estimate of *R* is mostly determined by that associated with $C_1 - C_2$. If C_1 and C_2 are close, large standard errors result in calculating *R* even with relatively small errors in *C*. For example, if $C_1 = 10.0$ and $C_2 = 9.0$ with standard errors for each of 0.1, the greatest fractional error in $C_1 - C_2$ is 0.14 (Topping 1957).

A number of different *direct methods* of determining filtering or pumping characteristics have been tried. The constant level apparatus for bivalve molluscs originated with Galstoff (1928) and has been improved by a number of subsequent authors (e.g. Drinnan 1964). The method involves separating the inhalant and exhalant flows when the pumping rate can be directly determined from timed collections of the exhaled water. Filtration retention efficiency (RE) can be determined by comparing the seston concentrations at the inhalant and exhalant. It is necessary to ensure that hydrostatic pressure differences between the inhalant and exhalant sides do not influence the pumping rate (see Chap. 4). The method is difficult to set up and bivalves may respond to the flow divider by reducing pumping, although considerable use has been made of the constant level apparatus introduced by Famme, Riisgård, and Jørgensen (1986). The water level in the exhalant chamber is monitored so that any pressure difference between the two chambers can be maintained with the shunt closed (Fig. 2.14).

Other direct ways of measuring pumping rates include micromethods to determine velocities or pressure changes in the exhalant flow. Because of the small size of the exhalant siphon, it is important to use probes of small dimensions to prevent interference with the flow. Amouroux, Revault d'Allonnes, and Rouant (1975) described a hot film probe with a 0.2- × 0.4-mm tip for measuring local velocity; Foster-Smith (1976) and

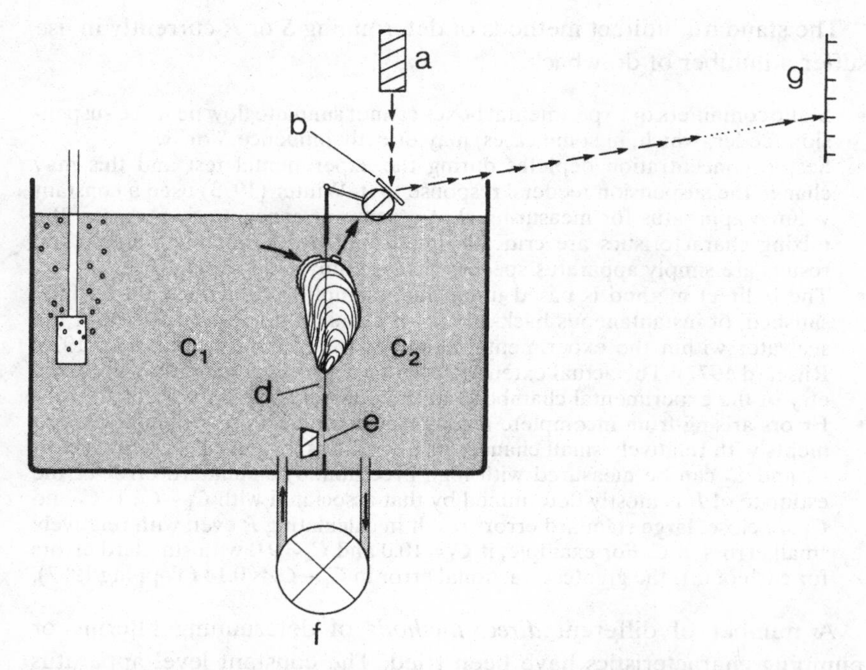

Figure 2.14 Direct method of determining the pumping rate of a mussel (after Famme et al. 1986): a, laser source; b, mirror; c_1 and c_2, inhalant and exhalant sides of a divided chamber; d, membrane; e, shunt; f, pump; g, scale.

Jones and Allen (1986) describe a pressure sensor with a <1-mm tip for sampling inhalant/exhalant pressures. Because of the complex flow and pressure fields within the exhalant flow, it is critical to position the sensor precisely. Because of these exacting requirements, such methods have not often been successfully used.

Mohlenberg and Riisgård (1978) describe an apparatus which can be used in non-flowing conditions to sample the exhalant flow of active suspension feeders isokinetically. The sampling flow is adjusted so that the seston concentration is minimal, at which point the flows are considered to be isokinetic. Mohlenberg and Riisgård's (1978) method could not be used satisfactorily in flume experiments because of the complicating ambient flow around the bivalve (Wildish and Saulnier 1993); direct exhalant flow sampling (Wildish and Saulnier 1993) was substituted. The bivalve gill retention efficiency (RE) can be directly determined from the concentration of inhalant (C_1) compared to exhalant seston (C_2):

$$RE = 1 - \frac{C_2}{C_1} \tag{2.13}$$

The value of R can be calculated as in equation (2.11) and the pumping rate, P, can be calculated by dividing R by RE, provided the estimates are for the same periods.

Suspension feeding bioassays ideally require that seston be maintained at a constant concentration during the experiment. Winter (1973) originally used a gently stirred tank of 10-L capacity containing blue mussels. Improvements to this technique have been suggested by Riisgård and Mohlenberg (1979) and a less expensive version proposed by Haupt (1979). All of these methods use a photometer to monitor the particle concentration by differences in light transmission, without distinguishing the preferred microalga from other sestonic particles offered to the mussel. As the mussels consume the seston, the increasing light, through a relay, switches on a dosing valve connected to a microalgal culture for a predetermined period. This allows the culture to be added to the tank at a rate to maintain a pre-set seston concentration. The rate of addition of the microalgae is also a direct measure of mussel uptake or clearance, after a suitable allowance for microalgal settling or growth and cell division has been made in a control run with mussels absent but for the same period as the experiment. Photometric determination is not specific for microalgal particles and any dense particle, such as faeces or pseudofaeces, reduces the light reaching the photoelectric sensor.

A more specific method for microalgal cells involves fluorescence measurements with two suitable filters, providing exciting light at 430 nm and emitted light at 650 nm, which is suitable for continuous measurement of in vivo chlorophyll concentration (Lorenzen 1966) in a continuous flow fluorometer (Table 2.6; Fig. 2.15). Some limitations of the in vivo fluorescence method should be kept in mind (Loftus and Seliger 1975), e.g. high concentration quenching or direct absorption of light by the microalgae, when using this method. Bayne, Widdows, and Newell (1977) were successful in using flow fluorometry to measure the feeding rate of blue mussels fed unialgal cultures. A suitable setup for maintaining a constant seston concentration with feeding bivalves present in a flume flow is shown in Fig. 2.15. One unsatisfactory feature of this method is that, because of the different fluorescent properties of each microalgal species, it is not possible to maintain the strict relationship

Table 2.6. *Partial list of commercial suppliers of equipment and supplies required for flow-related biological observation or experimentation.*

Type	Supplier
Temperature measurement	Hewlett Packard, 6877 Goreway Drive, Ontario, Canada L4V 1M8
Photoperiod control	Suntracker, Paragon Electric Canada Ltd., 221 Evans Ave., Etobicoke, Ontario, Canada M8Z 1J5
Three-dimensional positioner made from Unislide assembly parts	Velmex Inc., East Bloomfield, NY, 14443 USA
Spaghetti tags for identifying experimental subjects, gypsum cylinders, etc.	Floy Tag & Manufacturing Inc., 4616 Union Bay Place, N.E., Seattle, Washington, 98105 USA
Flow fluorometer, AU-005	Turner Designs, 845 W. Maude Ave., Sunnyvale, California, 94086 USA
Melamime formaldehyde particles	Radiant Color, Richmond, California, 94809 USA
Endoscope	Olympus Corp., Medical Instrument Div., 4-T Nevada Drive, Lake Success, New York, 11042 USA
Sony video system with miniature camera	PEMTEK Technologies, 201 Brownlow 33B, Dartmouth, Nova Scotia, Canada B3B 1W2
Schlieren systems	Aerolab, 9580 Washington Blvd., Laurel, MD, 20723 USA
Dissolved oxygen probes	Yellow Springs Instrument Co. Inc., Box 279, Yellow Springs, Ohio, 45387 USA

observed between fluorescence and cell density if two or more species are present (see also Bayne et al. 1977), particularly if the bivalve is preferentially feeding on a particular microalga, as is likely to be the case. A review of the in vivo fluorescence technique to measure the clearance rates of bivalves was presented by McLatchie (1992).

Other means of controlling seston particle concentrations in this same general way have been tried, e.g. electronic particle sizing and counting (Klein, Breteler, and Laan 1993), and presumably flow cytometry could also be used for this purpose. The former method is not readily adaptable to continuous measurement as are photometry, flow fluorometry,

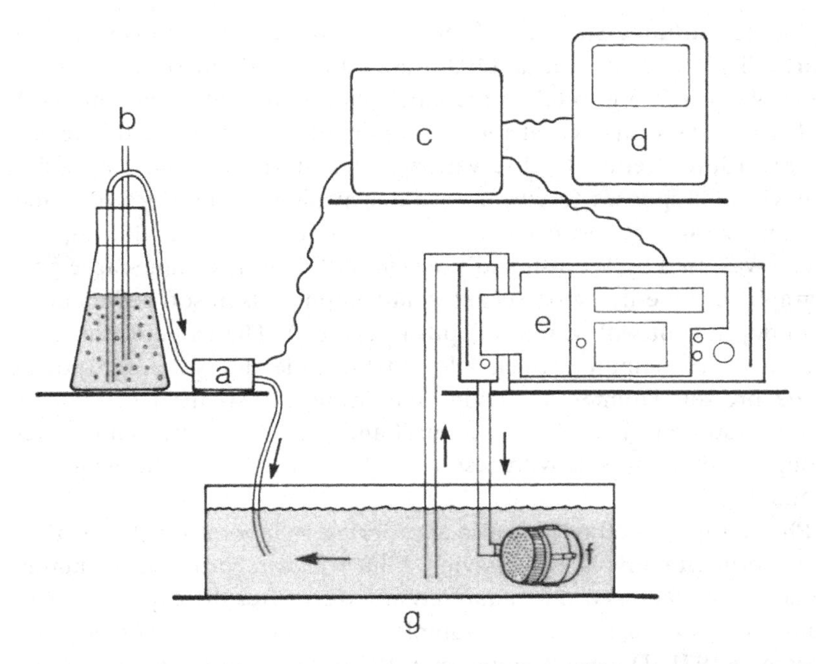

Figure 2.15 Constant seston concentration in flume flow: a, metering pump controlled by c; b, aerated algal supply; c, personal computer controlled data logger with switch relays; d, personal computer; e, Model AU-005 field fluorometer (Turner Designs); f, Little Giant submersible pump; g, flume.

and cytometry (Kolber and Falkowski 1993). Flow cytometry is capable of resolving the identity of microalgal species on the basis of particle sizing and resolution of the fluorescence spectra of in vivo cells.

Carey (1989) employed melamine formaldehyde particles of 3–6 μm and specific gravity = $1.4\,g \cdot cm^{-3}$ to study benthic animal–sediment dynamics (Table 2.6). The particles are treated with dyes so that they can be determined fluorometrically. Perhaps such a technique could be adapted by combining microparticulate, formulated diet particles treated with dyes and determined by flow fluorometry to estimate seston uptake or clearance rates by suspension feeding animals.

Behavioral responses and models

A central method of observing behavior in both swimming and sessile suspension feeders uses various forms of photography. In the field, the use of SCUBA with hand-held underwater still or video cameras has furthered our understanding of the swimming behavior of bivalves (e.g. Hartnoll 1967). Meyer et al. (1984) used a moored video camera in an underwater housing to take time-lapse photographs at ~1-min intervals of feather stars on a coral reef at depths of 3–13 m. With this means of recording frequent observations, short-term crawling episodes, behavioral responses to currents, and spawning events were detected. An underwater stereo camera system mounted on a benthic support frame was used by Wildish and Lobsiger (1987) to take time-lapse photographs of tube-living polychaetes and amphipods in soft sediments at 80 m depth (but with a rated depth to 3000 m). The camera was fitted with a microcomputer which allowed the time interval for exposure to be pre-set. Dolmer, Karlsson, and Svane (1994) used underwater stereophotography to observe natural and experimentally manipulated groups of blue mussels with respect to flow in a 9-m-depth man-made channel.

Photographic methods are also appropriate in laboratory observations or experiments involving behavior. Film which records the swimming behavior of scallops can be analyzed at leisure after the experiments by frame analysis to estimate, for example, the swimming speed (Moore and Trueman 1971; Dadswell and Weihs 1990). Underwater video footage can be analyzed with a video cassette recorder able to analyze frame by frame. Specialized equipment for computerized analysis of multiple-channel events recorded on video tapes is available (Krauss, Morrel-Samuels, and Hochberg 1988) with suitable software (Noldus 1991) to analyze the data obtained.

Video viewing through the acrylic plastic walls of a flume during feeding experiments of scallops showed that the exhalant opening area was controlled by the ambient velocities (Wildish and Saulnier 1993). A similar technique was used by Dolmer et al. (1994) to measure the valve gape of blue mussels exposed to flow/no-flow conditions.

Flow patterns over individual scallops were investigated by using models consisting of cleaned valves filled with plaster of paris (Thorburn and Gruffyd 1979). Flow lines were visualized by video analysis of fine hydrogen gas bubbles or condensed milk diluted with seawater. To improve contrast, white light illumination was provided against a black background, including painting the valves black. This dark field illumination technique was also used by Leonard (1989) to observe tube feet feeding by crinoids.

A trend in the study of suspension feeding has been to observe at higher magnification in the hope of seeing the detailed mechanisms of this process. Strickler (1985) suggested the use of a Schlieren optical pathway and a collimated light beam focused by a condensing lens which makes seston particles appear white on a black background. Trager et al. (1990) used a similar system with magnification of 10–50 times to obtain video images of particles and barnacle feeding appendages. The endoscope, developed for medical purposes, has been used to observe some of the details of the suspension-feeding mechanisms, e.g. by Hunt and Alexander (1991) in barnacles and by Ward, MacDonald, and Thompson (1993) in bivalves.

Physical models of suspension feeders can be used in flume experiments where all environmental conditions can be closely controlled and the experiments repeated without danger of tiring the subjects. One disadvantage of using models is that any subtle behavioral responses to flow may be missed, so the method must be used judiciously and in conjunction with observations of natural behavior. A good example of an appropriate use of this method is the right cylindrical models of tube-living epifauna studied by Eckman and Nowell (1984). These authors were interested in testing tube height and diameter as they affected flow patterns and erosion–sedimentation phenomena close to the tubes.

Simple models of brachiopods based upon propped-open empty shells and with a catheter tube containing rhodamine dye were used by LaBarbera (1977) to visualize flow patterns in and around the shells. A similar technique was used by Johnson (1988) to assess the local flow around simple models of phoronids at different ambient velocities. Bivalve siphon models with the siphon apertures at a fixed opening diameter and adjustable height above the sediment–water interface were prepared by Monismith et al. (1990). Siphon jetting and mixing near the bivalve were visualized by video recording of a fluorescent dye pumped continuously into the exhalant siphon. A limitation of this technique, as a behavioral test, is that the siphons could not be adjusted to open and close partially or fully as occurs in real life in many bivalves.

In studying the factors involved in the dislodgement of epifaunal suspension feeders by flowing or wind-induced seawater movement, life-sized models can be used in suitable flume experiments (e.g. Gruffyd 1976; Millward and Whyte 1992; Wildish and Saulnier 1993). For bivalves, the dried valves can be used to prepare suitable models or complete new models cast with plaster of paris. Millward and Whyte (1992) mounted scallop models on an adjustable arm, while Gruffyd (1976) suspended scallop models directly in flow by means of three threads. Wildish and Saulnier (1993) first determined the buoyant weight of live scallops by weighing a suspended scallop in motionless seawater from the arm of a balance (see Lowndes 1942). Model scallops were then prepared by the addition of plasticine to cleaned valves so that the model had the same buoyant weight as the live scallop.

As pointed out by Vogel (1981), it is possible to study hydrodynamic phenomena in air, rather than in seawater, since at the same temperature, air is 15 times more kinematically viscous. This means that for an object of the same dimensions, air flows must be 15 times greater than seawater velocity to achieve the same Reynolds number. Vogel (1985) employed wind tunnel experiments to show that flow-induced forces assisted the elasticity of the hinge ligament of the scallop *Argopecten irradians* in reopening the left valve during coordinated swimming movements.

Reef metabolism and production

Because ecology is such a vast subject area, we have concentrated only on methods to measure oxygen demand by reefs and to calculate production potentials for benthic populations or communities (see Chaps. 7 and 8).

The mechanisms involved in oxygen exchanges across the sediment–water interface, specifically that of primarily suspension-feeding populations such as a bivalve reef, include diffusion from seawater into sediments caused by the physical gradient, increased exchange of seawater with sediments due to the bioturbation activities of infaunal burrowers, as well as photosynthetic production of molecular oxygen by surface-living diatoms. From within the sediment biological oxygen demands from aerobic microorganisms, meiofaunal and macrofaunal respiration, and chemical oxygen demands due to reduced chemical compounds produced by anaerobic microorganisms all result in utilization of dissolved oxygen in sediment pore water and overlying seawater. Thus, in measuring dissolved oxygen fluxes in the seawater above a reef, we are integrating the resultant of all these processes.

The effect of water movement and bottom surface roughness on the flux of dissolved oxygen across the sediment–water interface was appreciated by the benthic ecologists whose work is discussed in Chapter 8 in the section, Reef metabolism. Their results should be considered as crude measures of reef oxygen demands. A notable problem is that the sampling points chosen are arbitrary and not based on the particular hydrodynamic conditions prevailing at a specific site. In the case of flow canalization caused by the benthic ecosystem tunnel, advective water motion from above is prevented from mixing with canalized seawater. Recently, Dade (1993) has introduced suitable theory for estimating the distribution and flux of dissolved substances such as oxygen within the benthic boundary layer.

Many of the commonly used methods of measuring sediment oxygen demand (Bowman and Delfino 1980) are not adaptable for obtaining accurate field estimates of reef oxygen demands influenced by ambient water movements. The most commonly used enclosed chamber technique – a cylinder pushed into the sediment to enclose a known volume of seawater, <1 – 38 L (Boynton et al. 1981) – cannot readily be adapted to simulate ambient water movement. It has been suggested that the passage of waves in nearshore waters creates differential hydrostatic pressures in the sediment, which, in turn, cause pumping of seawater in and out of the sediment. This phenomenon has been experimentally examined by Malan and McLachlan (1991). They provided a small, unstirred chamber with a flexible membrane (rendered impermeable by bonding sheets of polyethylene and nylon) sealed on the open top of the device. This allowed wave–induced interstitial seawater to enter the chamber through the sediment as a result of compensatory movements of the membrane. A comparison of unstirred flexible-topped and solid-topped chambers deployed in the field showed a 73% increase in oxygen uptake in the former, supporting the view that the passage of waves increased dissolved oxygen fluxes.

We consider that two sensing devices – commercially available dissolved oxygen probes (Table 2.6) and oxygen microelectrodes (Revsbech et al. 1980; Revsbech and Jørgensen 1983) – could be used to provide accurate estimates of oxygen fluxes from suspension-feeding reefs. Advantages of the microelectrode are that the sensing tips are small, 5- to 10-μm diameter, and that 50- to 100-μm sections of the sediment profile can be resolved. From dissolved oxygen profiles within the sediment and application of Fick's law of diffusion, the flux of oxygen at the sediment–water interface can be determined (Revsbech and Jørgensen 1983). These results, coupled with benthic boundary layer theory and detailed contemporary observations by a physical oceanographer, could help to determine the whole reef flux. This method is limited by the necessary assumption that only Fickian diffusion is involved. This assumption is reasonable in fine sediments of low permeability, but in sandy sediments of higher permeability it is not. For the latter, the assumption is invalid because of the passage of waves (Malan and

Table 2.7. *Conversions and predictions of potential benthic production.*

Conversions

Wet weight, g, to kcal, multiply by:
- 1.0 (Steele 1974)
- 0.6 × fresh weight (Mills and Fournier 1979)
- 0.5 (Crisp 1975)

kcal to kJ, multiply by:
- 4.186

kJ to g carbon, divide by:
- 50

Predictions

Production/biomass, $P:B$, ratios may be determined by multiplying W, wet body mass at maturity, g, or lifespan in years, based on the following equations:

Applicability	*Equation*	*Reference*
Macroafauna + meiofauna	$P:B = 0.971W^{-0.167}$	Schwinghamer (1981)
Macrofauna	$P:B = 0.563W^{-0.302}$	Schwinghamer (1981)
Invertebrates, 5°–20°C	$P:B = 0.640W^{-0.370}$	Banse and Mosher (1980)
Invertebrates, less insects	$P:B = 0.617W^{-0.390}$	Banse and Mosher (1980)
Macrofauna	$P:B = 4.571$ lifespan $(\mathrm{yr})^{-0.726}$	Robertson (1979)

Production of suspension feeders, $P(\mathrm{g\cdot m^{-2}\cdot yr^{-1}})$, from mean tidal velocity, $U(\mathrm{m\cdot s^{-1}})$, in the Bay of Fundy:

$\log_{10} P = 5.862U - 1.155$ (Wildish and Kristmanson 1979)

McLachlan 1991) in sediments and/or presence of roughness elements of biological or mechanical origin (Huettel and Gust 1992) at the sediment surface, which destroys the well defined laminar sublayer.

Production estimates of benthic communities are required in ecosystem analysis, e.g. in pelagic–benthic coupling studies. The units in which such estimates are made are convertible (Table 2.7). The crudest estimate of benthic production is made from quantitative grab or core sampling data as biomass ($\mathrm{g\cdot m^{-2}}$). Empirically derived relationships based on the annual turnover ratio, $P:B$, from a series of field studies in temperate climates (Table 2.7) are based on either lifespan (Robertson 1979) or mature biomass (Banse and Mosher 1980; Schwinghamer 1981). These relationships can be used to compute crude annual production estimates by multiplying $P:B$ and the observed mean annual biomass for a species.

It is possible to produce a more accurate measure of benthic production by obtaining detailed information for each cohort of each species in the population. This method requires periodic estimates of population density and estimation of mean growth rates throughout the life of the cohort. Cohort production is determined by summing the time period estimates by one of two computation methods – cohort summation of losses or the Allen curve method (Crisp 1984) – or, if it is impossible to distinguish the cohorts, a size–frequency method (Wildish and Peer 1981). Because of extensive work required to complete the measurements and analysis even for one species population, it is usually only possible to complete it for a few of the most dominant ones. Probably because of the time costs involved, only a few studies using this methodology have been completed (e.g. Buchanan and Warwick 1974). Because of the often near-monoculture conditions of a bivalve reef, such a method is well suited to it.

Ecosystem-level estimates of benthic production require a detailed and accurate map of the predominant benthic community. The development and use of digital multibeam bathymetric systems by surficial geologists to provide detailed and rapidly obtained maps of the seabed (see Courtney and Fader 1994) may prove to be important in indicating benthic communities and, hence, lead to more accurate charts of benthic production than hitherto available. Commercially available systems which adapt ships' echo sounders to produce sedimentary roughness and hardness estimates and plot it on a chart have also been used to identify epibenthic communities, e.g. Magorrian, Service, and Clarke (1995).

Some attempts have been made to determine the limiting factors involved in controlling suspension feeder–dominated communities in natural conditions (e.g. Wildish and Kristmanson 1979) or in bivalve suspension culture (e.g. Emerson et al. 1994). Thus, in the lower Bay of Fundy, a simple linear relationship (Table 2.7) was found between mean tidal velocity and production calculated on the basis of the Robertson lifespan method. Further field studies in the upper Bay of Fundy (Wildish et al. 1986), where tidal velocities and wind–wave effects were more energetic, showed that tidal velocities did not provide a good predictor of production, suggesting that the relationship was not universal. Emerson et al. (1994) measured monthly growth rates of the local scallop *Placopecten magellanicus* and found that up to 68% of growth variation could be explained by temperature and seston quality measured at biweekly intervals. This suggests that a crude estimate of productivity at a specified site might be predicted with just these two environmental variables, although uncertainties remain as to how to measure seston quality.

References

Amos, C. L., J. Grand, G. R. Daborn, and K. Black. 1992. Sea carousel: a benthic annular flume. Estuar. Coast. Shelf Sci. 34: 557–578.

Amouroux, J. M., M. Revault d'Allonnes, and C. Rouant. 1975. Sur la mesure directe du débit de filtration chez les mollusques lamellibranches. Vie Milieu (Ser. 2B) 25: 339–346.

Asmus, H., R. M. Asmus, T. C. Prins, N. Dankers, G. Francés, B. Maas, and K. Reise. 1992. Benthic-pelagic flux rates on mussel beds: tunnel and tidal flume methodology compared. Helgolander Meeresunters. 46: 341–361.

Asmus, R. M., and H. Asmus. 1991. Mussel beds: limiting or promoting phytoplankton? J. Exp. Mar. Biol. Ecol. 148: 215–232.

Banse, K., and S. Mosher. 1980. Adult body mass and annual production/biomass relationships of field populations. Ecol. Monogr. 50: 355–379.

Bayne, B. L., J. Widdows, and R. I. E. Newell. 1977. Physiological measurements on estuarine bivalve molluscs in the field, p. 57–68. *In* B. F. Keegan, P. Ó. Céidigh, and P. J. S. Boaden (ed.) Biology of benthic organisms. Pergamon, Oxford.

Bell, E. C., and M. W. Denny. 1994. Quantifying "wave exposure": a simple device for recording maximum velocity and results of its use at several field sites. J. Exp. Mar. Biol. Ecol. 181: 9–29.

Best, B. A. 1988. Passive suspension feeding in a sea pen: effects of ambient flow on volume flow rate and filtering efficiency. Biol. Bull. 175: 332–342.

Boucher, N., D. Vaulot, and F. Partensky. 1991. Flow cytometric determination of phytoplankton DNA in cultures and oceanic populations. Mar. Ecol. Prog. Ser. 71: 75–84.

Bowman, G. T., and J. J. Delfino. 1980. Sediment oxygen demand techniques: a review and comparison of laboratory and *in situ* systems. Water Res. 14: 491–499.

Boynton, W. R., W. M. Kemp, C. G. Osborne, K. R. Kaumeyer, and M. C. Jenkins. 1981. Influence of water circulation rate on *in situ* measurements of benthic community respiration. Mar. Biol. 65: 185–190.

Bradshaw, P. 1964. Experimental fluid mechanics. Pergamon Press, Oxford.

Buchanan, J. B., and R. M. Warwick. 1974. An estimate of benthic macrofaunal production in the offshore mud of the Northumberland coast. J. Mar. Biol. Assoc. U.K. 54: 197–222.

Butman, C. A. 1986. Sediment trap biases in turbulent flows: results from a laboratory flume study. J. Mar. Res. 44: 645–693.

Butman, C. A., J. P. Grassle, and C. M. Webb. 1988. Substrate choices made by marine larvae settling in still water and in a flume flow. Nature 333: 771–773.

Carey, D. A. 1983. Particle resuspension in the benthic boundary layer by a tube-building polychaete. Can. J. Fish. Aquat. Sci. 40: 301–308.

1989. Fluorometric detection of tracer particles used to study animal–particle dynamics. Limnol. Oceanogr. 34: 630–634.

Charriaud, E. 1982. Direct measurements of velocity profiles and fluxes at the cloacal siphon of the ascidian *Ascidiella aspersa*. Mar. Biol. 70: 35–40.

Charters, A. C., and S. M. Anderson. 1980. A low velocity water tunnel for biological research. J. Hydronaut. 14: 3–4.
Chia, F., J. Buckland-Nicks, and C. M. Young. 1984. Locomotion of invertebrate larvae: a review. Can. J. Zool. 62: 1205–1222.
Coughlan, J. 1969. The estimation of filtering rate from the clearance of suspensions. Mar. Biol. 2: 356–358.
Courtney, R. C., and G. B. J. Fader. 1994. A new understanding of the ocean floor through multibeam mapping, p. 9–14. *In* A. Fiander (ed.) Science review 1992 and 1993. Department of Fisheries and Oceans, Bedford Institute of Oceanography, Dartmouth, Nova Scotia, Canada.
Crisp, D. J. 1975. Secondary productivity in the sea, p. 71–89. Productivity of world ecosystems. National Academy of Sciences, Washington, D.C.
1984. Energy flow measurements, p. 284–372. *In* N. A. Holmes, and A. D. McIntyre (ed.) Methods for the study of marine benthos. IBP Handbook 16. 2nd ed. Blackwell, Oxford.
Dade, W. B. 1993. Near-bed turbulence and hydrodynamic control of diffusional mass transfer at the sea floor. Limnol. Oceanogr. 38: 52–69.
Dadswell, M. J., and D. Weihs. 1990. Size-related hydrodynamic characteristics of the giant scallop, *Placopecten magellanicus* (Bivalvia: Pectinidae). Can. J. Zool. 68: 778–785.
Dame, R. F., R. G. Zingmark, and E. Haskin. 1984. Oyster reefs as processors of estuarine materials. J. Exp. Mar. Biol. Ecol. 83: 239–267.
Demers, S. (Editor). 1991. Particle Analysis in Oceanography. NATO ASI Series G. Vol. 27. Springer-Verlag, Berlin.
Denny, M. W. 1982. Forces on intertidal organisms due to breaking ocean waves: design and application of a telemetry system. Limnol. Oceanogr. 27: 178–183.
1985. Wave forces on intertidal organisms: a case study. Limnol. Oceanogr. 30: 1171–1187.
1988. Biology and the mechanisms of the wave-swept environment. Princeton University Press, Princeton, N. J.
Dolmer, P., M. Karlsson, and I. Svane. 1994. A test of the rheotactic behaviour of the blue mussel *Mytilus edulis* L. Phuket. Mar. Biol. Cent. Spec. Publ. 13: 177–184.
Doty, M. S. 1971. Measurements of water movement in reference to benthic algal growth. Bot. Mar. 14: 32–35.
Drinnan, R. E. 1964. An apparatus for recording the water-pumping behaviour of lamellibranchs. Neth. J. Sea Res. 2: 223–232.
Eckman, J. E. 1979. Small-scale patterns and processes in a soft-substratum, intertidal community. J. Mar. Res. 37: 437–457.
Eckman, J. E., and A. R. M. Nowell. 1984. Boundary skin friction and sediment transport about an animal-tube mimic. Sedimentology 31: 851–862.
Emerson, C. W., J. Grant, A. Mallet, and C. Carver. 1994. Growth and survival of sea scallops *Placopecten magellanicus*: effect of culture depth. Mar. Ecol. Prog. Ser. 108: 119–132.
Emig, C. C., and F. Becherini. 1970. Influence des courants sur l'ethologie alimentaire des phoronidiens. Etude par séries de photographies cycliques. Mar. Biol. 5: 239–244.
Emlet, R. B. 1990. Flow fields around ciliated larvae: effects of natural and artificial tethers. Mar. Ecol. Prog. Ser. 63: 211–225.
Famme, P., and L. H. Kofoed. 1983. Shunt water flow through the mantle

cavity in *Mytilus edulis* (L.) and its influence on particle retention. Mar. Biol. Lett. 4: 207–218.

Famme, P., R. U. Riisgård, and C. B. Jørgensen. 1986. On direct measurements of pumping rates in the mussel, *Mytilus edulis*. Mar. Biol. 92: 323–327.

Foster-Smith, R. L. 1976. Pressures generated by the pumping mechanism of some ciliary filter feeders. J. Exp. Mar. Biol. Ecol. 25: 199–206.

Gallagher, S. M. 1988. Visual observations of particle manipulation during feeding in larvae of a bivalve mollusc. Bull. Mar. Sci. 43: 344–365.

Galstoff, P. S. 1928. Experimental study of the function of the oyster gills and its bearing on the problems of oyster culture and sanitary control of the oyster industry. Bull. U.S. Bur. Fish. 44, Doc. 1035: 1–39.

Grassle, J. P., P. V. R. Snelgrove, and C. A. Butman. 1992. Larval habitat choice in still water and flume flows by the opportunistic bivalve *Mulinia lateralis*. Neth. J. Sea Res. 30: 33–44.

Grizzle, R. E., R. Langan, and W. H. Howell. 1992. Growth responses of suspension-feeding bivalve molluscs to changes in water flow: differences between siphonate and nonsiphonate taxa. J. Exp. Mar. Biol. Ecol. 162: 213–228.

Gruffyd, L. D. 1976. Swimming in *Chlamys islandica* in relation to current speed and an investigation of hydrodynamic lift in this and other scallops. Nor. J. Zool. 24: 365–378.

Gust, G. 1982. Tools for oceanic small-scale high-frequency flows: metal clad hot wires. J. Geophys. Res. 87: 445–447.

Hannan, C. A. 1984. Planktonic larvae may act like passive particles in turbulent near-bottom flows. Limnol. Oceanogr. 29: 1108–1115.

Hargrave, B. T., and W. M. Burns. 1979. Assessment of sediment trap collection efficiency. Limnol. Oceanogr. 24: 1124–1136.

Hargrave, B., G. Siddal, G. Steeves, and G. Awalt. 1994. A current-activated sediment trap. Limnol. Oceanogr. 39: 383–390.

Hartnoll, R. G. 1967. An investigation of the movement of the scallop. Helg. Wiss. Meeresunters. 15: 523–533.

Haupt, K. 1979. A simple device to control concentrations of food algae in feeding experiments with filter feeding organisms. Veroff. Inst. Meeresforsch, Bremerhaven 17: 241–244.

Hildreth, D. I., and D. J. Crisp. 1976. A corrected formula for calculation of filtration rate of bivalve molluscs in an experimental flowing system. J. Mar. Biol. Assoc. U.K. 56: 111–120.

Hinds, W. C. 1982. Aerosol technology. John Wiley & Sons, New York.

Hobbie, J. E., R. Daley, and S. Jasper. 1977. Use of nucleopore filters for counting bacteria by fluorescence microscopy. Appl. Environ. Microbiol. 33: 1225–1228.

Holm-Hansen, O., and C. R. Booth. 1966. The measurement of adenosine triphosphate in the ocean and its ecological significance. Limnol. Oceanogr. 11: 510–519.

Huettel, M., and G. Gust. 1992. Impact of bioroughness on interfacial solute exchange in permeable sediments. Mar. Ecol. Prog. Ser. 89: 253–267.

Hunt, M. J., and C. G. Alexander. 1991. Feeding mechanisms in the barnacle *Tetraclita squamosa* (Brugière). J. Exp. Mar. Biol. Ecol. 154: 1–28.

Hunter, T. 1989. Suspension feeding in oscillating flow: the effect of colony morphology and flow regime on plankton capture by the hydroid *Obelia longissima*. Biol. Bull. 176: 41–49.

Johnson, A. S. 1988. Hydrodynamic study of the functional morphology of the benthic suspension feeder *Phoronopsis viridis* (Phoronida). Mar. Biol. 100: 117–126.

Jokiel, P. L., and J. I. Morrissey. 1993. Water motion on coral reefs: evaluation of the "clod card" technique. Mar. Ecol. Prog. Ser. 93: 175–181.

Jones, D. A., J. G. Mumford, and P. A. Gabbott. 1974. Micro capsules as artificial food particles for aquatic filter feeders. Nature 247: 233–235.

Jones, H. D., and J. R. Allen. 1986. Inhalant and exhalant pressures in *Mytilus edulis* L. and *Cerastoderma edule* (L.). J. Exp. Mar. Biol. Ecol. 98: 231–240.

Jørgensen, C. B. 1990. Bivalve filter feeding: hydrodynamics, bioenergetics, physiology and ecology. Olsen and Olsen, Fredensborg, Denmark.

Judge, M. L., L. D. Coen, and K. L. Heck, Jr. 1992. The effect of long-term alteration of in situ currents on the growth of *Mercenaria mercenaria* in the northern Gulf of Mexico. Limnol. Oceanogr. 37: 1550–1559.

1993. Does *Mercenaria mercenaria* encounter elevated food levels in seagrass beds? Results from a novel technique to collect suspended food resources. Mar. Ecol. Prog. Ser. 92: 141–150.

Klein Breteler, W. C. M., and M. Laan. 1993. An apparatus for automatic counting and controlling density of pelagic food particles in culture of marine organisms. Mar. Biol. 116: 169–174.

Koehl, A. R. M., and R. S. Alberte. 1988. Flow, flapping and photosynthesis of *Nereocystis luetkeana*: a functional comparison of undulate and flat blade morphologies. Mar. Biol. 99: 435–444.

Kolber, Z., and P. G. Falkowski. 1993. Use of active fluorescence to estimate phytoplankton photosynthesis *in situ*. Limnol. Oceanogr. 38: 1646–1665.

Koochesfahani, M. M., and P. E. Dimolakis. 1985. Laser-induced fluorescence measurements of mixed fluid concentration in a liquid plane shear layer. J. Am. Inst. Aeronaut. Astrophys. 23: 1700–1707.

Krauss, R. M., P. Morrel-Samuels, and J. Hochberg. 1988. VIDEOLOGGER: a computerized multichannel event recorder for analyzing video tapes. Behav. Res. Methods Instrum. Comput. 20: 37–40.

LaBarbera, M. 1977. Brachiopod orientation to water movement. I. Theory, laboratory behaviour and field orientations. Paleobiology 3: 270–287.

Langdon, C. J., D. M. Levine, and D. A. Jones. 1985. Microparticulate feeds for marine suspension feeders. J. Microencapsulation 2: 1–11.

Leonard, A. B. 1989. Functional response in *Antedon mediterranea* (Lamarck) (Echinodermata: Crinoidea): the interaction of prey concentration and current velocity on a passive suspension feeder. J. Exp. Mar. Biol. Ecol. 127: 81–103.

Leversee, G. J. 1976. Flow and feeding in fan-shaped colonies of the gorgonian coral, *Leptogorgia*. Biol. Bull. 151: 344–356.

Lindegarth, M., P. R. Jonsson, and C. André. 1991. Fluorescent microparticles: a new way of visualizing sedimentation and larval settlement. Limnol. Oceanogr. 36: 1471–1475.

Lofquist, K. E. B. 1977. A positive displacement oscillatory water tunnel. Misc. Rep. 77-1, U.S. Army Corps of Engineers, CERC, Fort Belvoir, Va.

Loftus, M. E., and H. H. Seliger. 1975. Some limitations of the *in vivo* fluorescence technique. Chesapeake Sci. 16: 79–92.

Lorenzen, C. J. 1966. A method for the continuous measurement of *in vivo* chlorophyll concentration. Deep Sea Res. 13: 223–227.

Lowndes, A. G. 1942. The displacement method of weighing living aquatic organisms. J. Mar. Biol. Assoc. U.K. 25: 555–574.
Luckenbach, M. W. 1986. Sediment stability around animal tubes: the roles of hydrodynamic processes and biotic activity. Limnol. Oceanogr. 31: 779–787.
Lui, S. T., and G. H. Nancollas. 1971. The kinetics of dissolution of calcium sulphate dihydrate. J. Inorg. Nucl. Chem. 33: 2311–2316.
Magorrian, B. H., M. Service, and W. Clarke. 1995. An acoustic bottom classification survey of Strangford Lough, Northern Ireland. J. Mar. Biol. Assoc. U.K. 75: 987–992.
Malan, D. E., and A. Mclachlan. 1991. *In situ* benthic oxygen fluxes in a nearshore coastal marine system: a new approach to quantify the effect of wave action. Mar. Ecol. Prog. Ser. 73: 69–81.
McLatchie, S. 1992. Time series measurement of grazing rates of zooplankton and bivalves. J. Plankton Res. 14: 183–200.
Meyer, D. L., C. A. Lattaye, N. D. Holland, A. C. Arneson, and J. R. Strickland. 1984. Time-lapse cinematography of feather stars (Echinodermata: Crinoidea) on the Great Barrier Reef, Australia: demonstrations of posture changes, locomotion, spawning and possible predation by fish. Mar. Biol. 78: 179–184.
Miller, D. C., M. J. Bock, and E. J. Turner. 1992. Deposit and suspension feeding in oscillatory flows and sediment fluxes. J. Mar. Res. 50: 489–520.
Mills, E. L., and R. O. Fournier. 1979. Fish production and the marine ecosystems of the Scotian Shelf, eastern Canada. Mar. Biol. 54: 101–108.
Millward, A., and M. A. Whyte. 1992. The hydrodynamic characteristics of six scallops of the super family Pectinacea, class Bivalvia. J. Zool. London 227: 547–566.
Minchinton, T. E., and R. E. Scheibling. 1991. The influence of larval supply and settlement on the population structure of barnacles. Ecology 72: 1867–1879.
Miron, G., P. Pelletier, and E. Bourget. 1995. Optimizing the design of giant scallop (*Placopecten magellanicus*) spat collectors: flume experiments. Mar. Biol. 123: 285–292.
Møhlenberg, F., and H. U. Riisgard. 1978. Efficiency of particle retention in 13 species of suspension feeding bivalves. Ophelia 17: 239–246.
Monismith, S. G., J. R. Koseff, J. K. Thompson, C. A. O'Riordan, and H. M. Nepf. 1990. A study of model bivalve siphonal currents. Limnol. Oceanogr. 35: 680–696.
Moore, J. D., and E. R. Trueman. 1971. Swimming of the scallop *Chlamys opercularis* (L.). J. Exp. Mar. Biol. Ecol. 6: 179–185.
Mullineaux, L. S., and C. A. Butman. 1990. Recruitment of encrusting benthic invertebrates in boundary-layer flows: a deep-water experiment on Cross Seamount. Limnol. Oceanogr. 35: 409–423.
Mullineaux, L. S., and E. D. Garland. 1993. Larval recruitment in response to manipulated field flows. Mar. Biol. 116: 667–684.
Muschenheim, D. K. 1987. The dynamics of near-bed seston flux and suspension-feeding benthos. J. Mar. Res. 45: 473–496.
Muschenheim, D. K., J. Grant, and E. L. Mills. 1986. Flumes for benthic ecologists: theory, construction and practice. Mar. Ecol. Prog. Ser. 28: 185–196.
Navarro, J. M., and R. J. Thompson. 1995. Seasonal fluctuations in the size spectra, biochemical composition and nutritive value of the seston

available to a suspension-feeding bivalve in a subarctic environment. Mar. Ecol. Prog. Ser. 125: 95–106.
Nixon, S. W., C. A. Oviatt, C. Rogers, and K. Taylor. 1971. Mass and metabolism of a mussel bed. Oecologia 8: 21–30.
Noldus, L. P. J. J. 1991. The Observer: a software system for collection and analysis of observational data. Behav. Res. Methods Instrum. Comput. 23: 415–429.
Nowell, A. R. M., and P. A. Jumars. 1987. Flumes: theoretical and experimental considerations for simulation of benthic environments. Oceanogr. Mar. Biol. Annu. Rev. 25: 91–112.
Okamura, B. 1984. The effects of ambient flow velocity, colony size, and upstream colonies on the feeding success of Bryozoa. I. *Bugula stolonifera* Ryland, an arborescent species. J. Exp. Mar. Biol. Ecol. 83: 179–193.
Olson, R. J., and E. R. Zettler. 1995. Potential of flow cytometry for "pump and probe" fluorescence measurements of phytoplankton photosynthetic characteristics. Limnol. Oceanogr. 40: 816–820.
O'Riordan, C. A., S. G. Monismith, and J. R. Koseff. 1993. A study of concentration boundary layer formation over a bed of model bivalves. Limnol. Oceanogr. 38: 1712–1729.
Patterson, M. R. 1991. The effects of flow on polyp-level prey capture in an octocoral, *Alcyonium siderium*. Biol. Bull. 180: 93–102.
Penner, S. S., and T. Jerskey. 1973. Use of lasers for local measurement of velocity components, species densities and temperatures. Annu. Rev. Fluid Mech. 5: 9–30.
Petticrew, E. L., and J. Kalff. 1991. Calibration of a gypsum source for freshwater flow measurements. Can. J. Fish. Aquat. Sci. 48: 1244–1249.
Revsbech, N. P., and B. B. Jørgensen. 1983. Photosynthesis of benthic microflora measured with high spatial resolution by the oxygen microprofile method: capabilities and limitations of the method. Limnol. Oceanogr. 28: 749–756.
Revsbech, N. P., J. Sorensen, T. H. Blackburn, and J. P. Lomholt. 1980. Distribution of oxygen in marine sediments measured with microelectrodes. Limnol. Oceanogr. 25: 403–411.
Riisgård, H. U. 1977. On measurements of the filtration rates of suspension feeding bivalves in a flow system. Ophelia 16: 167–173.
Riisgård, H. U., and F. Møhlenberg. 1979. An improved automatic recording apparatus for determining the filtration rate of *Mytilus edulis* as a function of size and algal concentration. Mar. Biol. 52: 61–67.
Robertson, A. I. 1979. The relationship between annual production: biomass ratios and lifespans for marine macrobenthos. Oecologia 38: 193–202.
Saunders, H. E., and C. W. Hubbard. 1944. The circulating water channel of the David W. Taylor model basin. Soc. Naval Architects Mar. Engin. Trans. 52: 325–364.
Schlichting, H. 1979. Boundary-layer theory. 7th ed. McGraw-Hill, New York.
Schraub, F. A., S. J. Kline, J. Henry, P. W. Runstadler, and A. Littell. 1965. Use of hydrogen bubbles for quantitative determination of time-dependent velocity fields in low speed water flows. J. Basic Eng. 87: 429–444.
Schwinghamer, P. 1981. Characteristic size distributions of integral benthic communities. Can. J. Fish. Aquat. Sci. 38: 1255–1263.
Scoffin, T. P. 1968. An underwater flume. J. Sediment. Petrol. 38: 244–247.
Shumway, S. E., T. L. Cucci, R. C. Newell, and C. M. Yentsch. 1985. Particle

selection, ingestion, and absorption in filter-feeding bivalves. J. Exp. Mar. Biol. Ecol. 91: 77–92.
Snelgrove, P. V. R., C. A. Butman, and J. F. Grassle. 1995. Potential flow artifacts associated with benthic experimental gear: deep-sea mud box examples. J. Mar. Res. 53: 821–845.
Sorensen, R. M. 1993. Basic wave mechanisms for coastal and ocean engineers. John Wiley & Sons, New York.
Southgate, P. C., P. S. Lee, and J. A. Nell. 1992. Preliminary assessment of a microencapsulated diet for larval culture of the Sydney rock oyster *Saccostrea commercialis* (Iredale and Roughly). Aquaculture 105: 345–352.
Sponaugle, S., and LaBarbera, M. 1991. Drag-induced deformation: a functional feeding strategy in two species of gorgonians. J. Exp. Mar. Biol. Ecol. 148: 121–134.
Steele, J. H. 1974. The structure of marine ecosystems. Harvard University Press, Cambridge, Mass.
Strathman, R. R., T. L. Jahn, and J. R. C. Fontsia. 1972. Suspension feeding by marine invertebrate larvae: clearance of particles by ciliated bands of a rotifer, pluteus, and trocophore. Biol. Bull. 142: 508–519.
Strathman, R. R., and E. Leise. 1979. On feeding mechanisms and clearance rates of molluscan veligers. Biol. Bull. 157: 524–535.
Strickland, J. D. H., and T. R. Parsons. 1968. A practical handbook of seawater analysis. Bull. Fish. Res. Board Can. 167: 311.
Strickler, J. R. 1985. Feeding currents in calanoid copepods: two new hypotheses. Symp. Soc. Exp. Biol. 39: 459–485.
Sulkin, S. D. 1990. Larval orientation mechanisms: the power of controlled experiments. Ophelia 32: 49–62.
Svoboda, A. 1970. Simulation of oscillating water movement in the laboratory for cultivation of shallow water sedentary organisms. Helgol. Wiss. Meeresunters. 20: 676–684.
Taghon, G. L., A. R. M. Nowell, and P. A. Jumars. 1984. Transport and breakdown of fecal pellets: biological and sedimentological consequences. Limnol. Oceanogr. 29: 64–72.
Thompson, T. L., and E. P. Glenn. 1994. Plaster standards to measure water motion. Limnol. Oceanogr. 39: 1768–1779.
Thorburn, I. W., and L. D. Gruffyd. 1979. Studies of the behaviour of the scallop *Chlamys opercularis* (L.) and its shell in flowing seawater. J. Mar. Biol. Assoc. U.K. 59: 1003–1023.
Topping, J. 1957. Errors of observation and their treatment. Chapman & Hall, London.
Trager, G. C., J.-S. Hwang, and J. R. Strickler. 1990. Barnacle suspension-feeding in variable flow. Mar. Biol. 105: 117–129.
Turner, E. J., and D. C. Miller. 1991. Behaviour of a passive suspension feeder (*Spiochaetopterus oculatus*) (Webster) under oscillatory flow. J. Exp. Mar. Biol. Ecol. 149: 123–137.
Vogel, S. 1981. Life in moving fluids: the physical biology of flow. Willard Grant Press, Boston.
1985. Flow-assisted shell reopening in swimming scallops. Biol. Bull. 169: 624–630.
1994. Life in moving fluids: the physical biology of flow. 2nd rev. ed. Princeton University Press, Princeton, N. J.

Vogel, S., and N. Feder. 1966. Visualization of low-speed flow using suspended particles. Nature 209: 186–187.
Vogel, S., and M. LaBarbera. 1978. Simple flow tanks for research and teaching. Bioscience 28: 638–643.
Wainwright, S. C. 1990. Sediment-to-water fluxes of particulate material and microbes by resuspension and their contribution to the planktonic food web. Mar. Ecol. Prog. Ser. 62: 271–281.
Walne, P. R. 1972. The influence of current speed, body size and temperature on the filtration rate of five species of bivalves. J. Mar. Biol. Assoc. U.K. 52: 345–374.
Ward, J. E., B. A. MacDonald, and R. J. Thompson. 1993. Mechanisms of suspension feeding in bivalves: resolution of current controversies by means of endoscopy. Limnol. Oceanogr. 38: 265–272.
Wildish, D. J., and D. D. Kristmanson. 1979. Tidal energy and sublittoral macrobenthic animals in estuaries. J. Fish. Res. Board Can. 36: 1197–1206.
1984. Importance to mussels of the benthic boundary layer. Can. J. Fish. Aquat. Sci. 41: 1618–1625.
1985. Control of suspension-feeding bivalve production by current speed. Helgol. Wiss. Meeresunters. 39: 237–243.
1988. Growth response of giant scallops to periodicity of flow. Mar. Ecol. Prog. Ser. 42: 163–169.
Wildish, D. J., and U. Lobsiger. 1987. Three dimensional photography of soft sediment benthos, S.W. Bay of Fundy. Biol. Oceanogr. 4: 227–241.
Wildish, D. J., and M. P. Miyares. 1990. Filtration rate of blue mussels as a function of flow velocity: preliminary experiments. J. Exp. Mar. Biol. Ecol. 142: 213–219.
Wildish, D. J., and D. Peer. 1981. Method for estimating secondary production in marine Amphipoda. Can. J. Fish. Aquat. Sci. 38: 1019–1026.
Wildish, D. J., D. L. Peer, and D. A. Greenberg. 1986. Benthic macrofaunal production in the Bay of Fundy and the possible effects of a tidal power barrage at Economy Point-Cape Tenny. Can. J. Fish. Aquat. Sci. 43: 2410–2417.
Wildish, D. J., and A. M. Saulnier. 1993. Hydrodynamic control of filtration in *Placopecten magellanicus*. J. Exp. Mar. Biol. Ecol. 174: 65–82.
Wildish, D. J., J. D. Trynor, and H. M. Akagi. 1997. Measurement of water movement by gypsum dissolution method in the Bay of Fundy. Can. Tech. Rep. Fish. Aquat. Sci. (in press).
Winter, J. E. 1973. The filtration rate of *Mytilus edulis* and its dependence on algal concentration, measured by a continuous automatic recording apparatus. Mar. Biol. 22: 317–328.
Wolaver, T., G. Whiting, B. Kjerfve, J. Spurrier, H. McKellar, R. Dame, T. Chrzanowski, R. Zinmark, and T. Williams. 1985. The flume design: a methodology for evaluating material fluxes between a vegetated salt marsh and the adjacent tidal creek. J. Exp. Mar. Biol. Ecol. 91: 281–291.
Wright, R. T., R. B. Coffin, C. P. Ersing, and D. Pearson. 1982. Field and laboratory measurements of bivalve filtration of natural marine bacterioplankton. Limnol. Oceanogr. 27: 91–98.
Yentsch, C. M., P. K. Horan, K. Muirhead, Q. Dortch, E. Haugen, L. Legendre, L. S. Murphy, M. J. Perry, D. A. Phinney, S. A. Pomponi, R. W. Spinrad, M. Wood, C. S. Yentsch, and B. J. Zahuranic. 1983. Flow

cytometry and cell sorting: a technique for analysis and sorting of aquatic particles. Limnol. Oceanogr. 28: 1275–1280.
Young, R. A. 1977. Seaflume: a device for *in situ* studies of threshold erosion velocity and erosional behaviour of undisturbed marine muds. Mar. Geol. 23: M11–M18.
Yund, P. O., S. D. Gaines, and M. D. Bertness. 1991. Cylindrical tube traps for larval sampling. Limnol. Oceanogr. 36: 1167–1177.

3
Dispersal and settlement

Study of larval biology has been under way for ~150 years and a large literature is available, causing Young (1990) to suggest that ideas are often rediscovered by each new generation of scientists interested in this field. This literature has been repeatedly reviewed; the most useful and recent include Crisp (1974), Chia, Buckland-Nicks, and Young (1984), Scheltema (1986), Butman (1987), Pawlik (1992), Sammarco and Heron (1994), and McEdwards (1995). The overview presented here relies heavily on these publications and on selected references which have appeared in the decade subsequent to 1985. The basis for selection of references in this chapter is that only those with flow-related content are included.

Consistent with the aims outlined in Chapter 1, we emphasize hydrodynamic mechanisms in larval biology where these are known. Consequently, our concern here is limited to larval and post-larval dispersal at spatial scales >0.1 km and on the environmental factors, notably hydrodynamic ones, which influence larval settlement/recruitment at spatial scales <0.1 km on both soft and hard substrates. We begin with background coverage of the typical benthic suspension feeder life cycle and dispersal in general.

Benthic suspension feeder life cycle

The life cycle of many sessile benthic suspension-feeding animals includes a pelagic *larval stage* important in dispersal (Fig. 3.1). In the barnacle *Semibalanus balanoides*, for example, a number of larval stages can be recognized morphologically, culminating in the cyprid, which is the competent stage for settlement on a suitable hard substrate. Once settled on such a surface, the cyprid undergoes a metamorphosis to

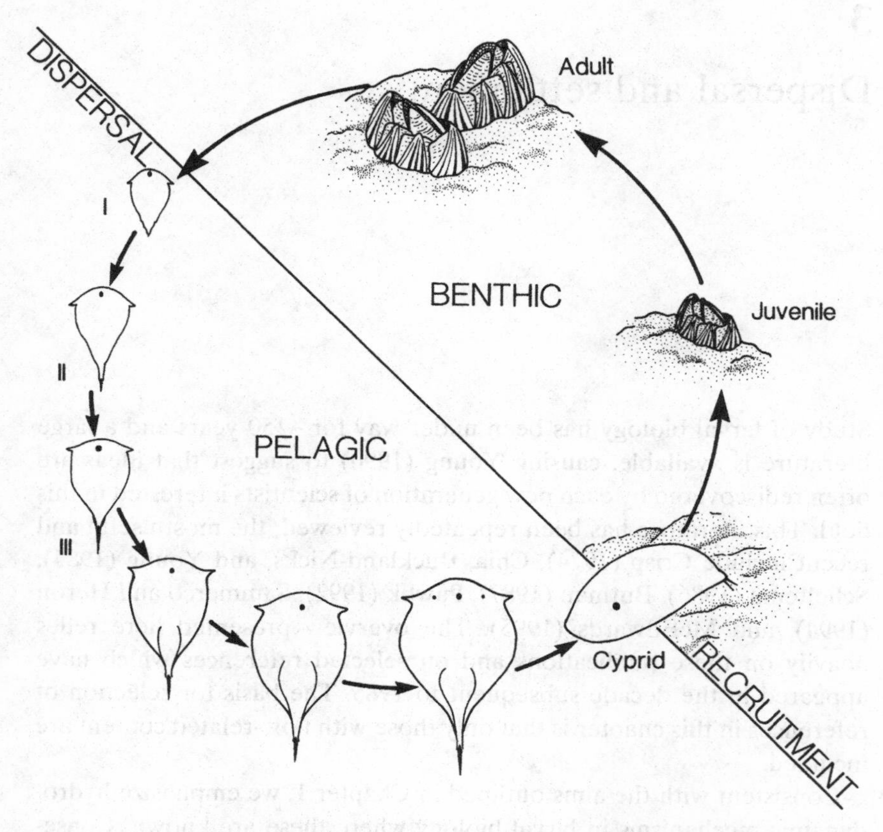

Figure 3.1 Life cycle diagram of the acorn barnacle, *Semibalanus balanoides* (based on Pyefinch 1948).

produce the *post-larval stage* or juvenile and, eventually, the adult barnacle. Barnacles are hermaphroditic and cross fertilization, by means of an intromittent organ, involves internal fertilization, followed by release of the first larval stage to begin the pelagic phase of life. There is a great range in shapes and sizes of larvae among suspension feeding invertebrates, although not all, e.g. polychaete larvae, have recognizably distinct stages as, for example, do barnacles. Some polychaetes simply add segments to the body as larval life in the plankton proceeds.

The duration of pelagic life for larval suspension feeders is dependent on both genetic factors and environmental variables. A characterization

Table 3.1. *Type of larvae.*

Type	Feeding behavior	Examples
1. Lecithotropic	Do not feed	Spirorbinidae Many Bryozoa Ascidiacea
2. Planktotrophic	Feed only in the pre-competent stage	Cirripede barnacles, where the cyprid is the non-feeding competent stage
3. Planktotrophic	Feed in both pre- and post-competent stages, but do not grow further in the latter period	Some gastropod molluscs
4. Planktotrophic	Feed in both pre- and post-competent stages and continue to grow in the latter period	Sand dollars (Highsmith and Emlet 1986) Some bivalve molluscs, e.g. blue mussels (Bayne 1965)

Source: Scheltema (1986).

based on whether they feed during planktonic life may indicate the approximate duration of larval life (Table 3.1). Thus, *lecithotrophic larvae* are short-lived (0.1–2 d) and do not feed, whereas *planktotrophic larvae* have the potential for a longer life (e.g. some teleplanic larvae ~2 years) and may feed at various periods during the larval stage (Sheltema 1986). The length of larval life in the plankton generally increases from type 1 through type 4 in Table 3.1.

Competent larvae are those with the ability to settle on a soft or hard substrate in the general way shown in Fig. 3.2. According to this scheme, the competent larva may make several contacts with the substrate, by either passive sinking or active swimming, before temporarily or permanently burrowing or attaching to it. Whether attachment or burrowing is involved after initial contact depends on the competent larva receiving ecologically relevant cues at sensory receptors. If appropriate environmental cues are present, the larvae undergo a hormonally coordinated physiological change, or *metamorphosis*, to produce the *post-larval stage*, which may precede, coincide with, or follow settlement (Scheltema 1974; Butman 1987). *Settlement* simply refers to the change from planktonic to benthic life of the competent larva; it may involve repeated explorations

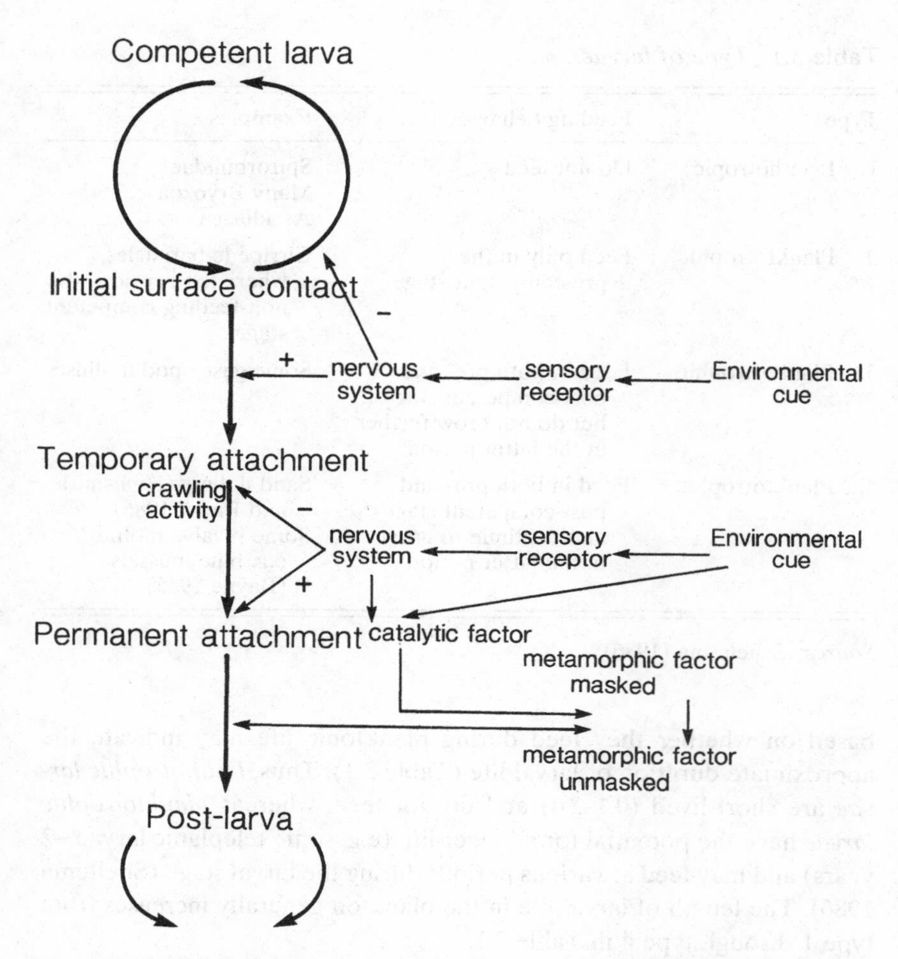

Figure 3.2 General model of settlement and metamorphosis of a planktotrophic larva, e.g. a barnacle cyprid (based on Chia 1978; Rittschof et al. 1984).

of the substrata, nervous coordination, and appropriate environmental cues (Scheltema 1974). As pointed out by Butman (1987), this definition is functionally inadequate in some cases as all benthic stages of soft sediment benthos, inclusive of post-larva, juvenile, and adult, may undertake pelagic excursions. Thus, if the post-larva cannot be morphologically distinguished from the competent larva, as in some polychaetes, confusion can occur. In those other cases where metamorphosis is coincident with settlement, as in barnacles, there will be no difficulty in distinguishing between competent and post-larval stages. The term *recruitment* is considered to be independent of the life history of the suspension feeder and is defined for investigator sampling convenience. Hence, the number of juveniles recruited to a settlement plate represents those retained on the sieve after removing them at some period after settlement (Keough and Downes 1982). This is not necessarily the same as the *initial settlement* of cyprid larvae (Butman 1987), since it excludes those that may have died after settlement, but before capture, or are too small to be retained by the sieve.

Dispersal

Dispersal is the general process by which organisms become distributed spatially, at scales generally >0.1 km, to potentially hospitable niches. Scheltema (1986) outlined four main types of dispersal common among benthic invertebrates:

- *Rafting dispersal*, generally involving passive transport of post-larval stages on a floating or submerged raft
- *Synanthropic dispersal*, involving passive transport by human intervention which includes all life history stages, particularly larvae
- *Larval dispersal*
- *Post-larval dispersal*

Most of this chapter deals with larval and post-larval dispersal, although we first briefly consider rafting and synanthropic dispersal. The purpose of this is to ensure for the reader that these forms of dispersal are not forgotten when pertinent questions regarding it are considered and dispersal hypotheses framed.

Dispersal by rafting occurs on a wide variety of living and dead floating substrates, which may carry all life history stages of suspension feeders wherever the surface currents dictate. Descriptions of the associated fauna with surface drifting macroalgal rafts stress the importance of a resident community, e.g. off the coast of Florida (Stoner and Livingston 1980). DeVantier (1992) described rafting of tropical marine animals, inclusive of reef and pearl oysters, as well as species of Bryozoa, on the buoyant skeleton of the reef coral *Symphylla agarica*, in which chambers in the coral were air-filled, giving lift during rafting. Other species of macroalgae may be submerged and moved by benthic currents along the seabed, rather like marine tumbleweeds. Holmquist (1994) reported that *Laurencia poiteau* does this and can therefore carry rafted fauna to new benthic niches.

The importance of synanthropic dispersal of a wide range of marine organisms is becoming better appreciated; Carlton and Geller (1993), for example, report on one important route for this. Since the 1880s, ships have pumped seawater into their holds as ballast when running empty between ports. In a study of 159 Asian ships which used Pacific coast ports of Oregon, Carlton and Geller (1993) found at least 367 taxa in ballast seawater – mostly larvae – inclusive of many suspension feeders. One was the larva of the Asian clam *Potamocorbula amurensis*. This clam, probably after discharge of ship ballast water (Carlton et al. 1990), rapidly spread through San Francisco Bay after 1986 and replaced the previously introduced exotic, the soft shell clam *Mya arenaria*, which in turn had replaced an indigenous bivalve *Macoma balthica* (Nichols and Thompson 1985). Another well studied example of synanthropic dispersal from ballast water is that of the zebra mussels *Dreissena polymorpha* and a second species of this genus *D. bugensis* introduced from Europe to the North American Great Lakes (Hebert, Mancaster, and Mackie 1989; May and Marsden 1992; Ricciardi, Serrouya, and Whoriskey 1995), where it has become a problem in blocking freshwater intake pipes.

These dramatic examples of the introduction of exotics and their sometimes rapid spread in the new location suggest the importance of this and similar processes in modifying nearshore habitats. Many other species introduced synanthropically may have replaced the indigenous fauna and flora without being recognized. It is often difficult to study rafting and synanthropic dispersal in the field because of the stochastic nature of these processes.

Larval dispersal

Planktotrophic larvae are generally <1000 μm in length and swim/sink at Reynolds numbers in the creeping flow range. Such larvae swim by means of beating cilia, although larvae which are longer than 1000 μm are unable to swim in this way as a result of inefficiencies linked to size, and are forced to use muscular contractions to swim (Chia et al. 1984). Examples include the later stages of some polychaete larvae or the lecithotrophic tadpole larvae of ascidians. Chia et al. (1984) showed that larvae using muscular power to swim increased their speed proportional to an increase in body size, although this was not the general case with ciliary swimming larvae.

Although data are readily available for larval size in the literature (for example, see the review of larval stages of bivalves presented by Ackerman et al. 1994), concomitant and reliable data on swimming speed and sinking rate are often lacking, explaining the few entries in Table 3.2.

The key factors which influence the final distances achieved by larvae during dispersal and before settlement of the competent stage are duration in the plankton, larval sinking rates, larval swimming speed, and direction and velocity of ambient currents. Environmental factors which influence the length of larval life, and therefore the potential dispersal distance, are physical, e.g. salinity and temperature, or biological, e.g. the availability and quality of sestonic food or predation.

The length of time that larvae are planktonic, both between species and within species, is highly variable. Hence, lecithotrophic larvae may spend a few hours or days in the plankton, with the outcome that they either settle or die because of predation or limiting energy resources (Scheltema 1986). Longer-lived planktotrophic larvae are affected by physical factors in various ways. As an example, the dispersal of the larval stages of the blue mussel, following the release of sperm and egg, is depicted in Fig. 3.3. After fertilization, the trocophore is followed by the veliger, the velichoncha, the eyed veliger, and the pediveliger, which is the competent settling stage, and this sequence can take from 26 to 60 d (Bayne 1964a). After settling on seaweed, the pediveliger metamorphoses to the first post-larval or spat stage, termed the plantigrade (not shown in Fig. 3.3). The plantigrade can actively migrate to a more suitable hard substrate nearby to grow into the juvenile and adult forms of the blue mussel. Bayne (1965) conducted growth experiments with blue mussel pediveligers in which optimum salinities of 30–33 parts per thousand (‰) and food concentrations of 1×10^5 cells $\cdot$ L^{-1} were offered while varying the temperature. It was found that mixed cultures of microalgae led to better growth rates than if a unialgal culture was offered as food and that the delay of metamorphosis achieved by pediveligers ranged from ~2 to 46 d, in inverse relation to temperature (Bayne 1965). Salinity also affected the time spent as a pediveliger, so that the delay was greater at the lower salinities tested of 12.5‰ (Bayne 1965).

From a knowledge of the duration of planktonic life it is possible to predict crudely the distances the larvae could disperse on the basis of the dispersion diagrams prepared by Okubo (1971). A three-dimensional mathematical model of ocean circulation over Georges Bank was constructed by Tremblay and Sinclair (1994) to predict the passive transport of larvae of the giant scallop *Placopecten magellanicus*. The model was reasonably successful in predicting the flow field and, consequently, in simulating the scallop larval drift over Georges Bank. Important vari-

Table 3.2. *Larval size, swimming speed, sinking rates, and duration of stay in the plankton for some benthic suspension feeders.*

Taxon	Size, μm		Sinking rate, $cm \cdot s^{-1}$	Swimming speed, $cm \cdot s^{-1}$			Duration, days	Reference
	Egg	Competent larva		Up	Down	Horizontal		
Cnidaria, Octocorallia								
Eunicella stricta			0.38	0.2	—	—		Theodor (1967)
Annelida, Polychaeta								
Phragmatopoma lapidosa californica	—	—	0.07–0.16	—	0.20–0.50	—	—	Pawlik et al. (1991)
Streblospio benedicti	—	300–1400	0.01–0.30	—	—	—	—	Butman (1989)
Mollusca, Bivalvia								
Crassostrea virginica	—	240–320	~0.83	0.13–0.27	0.13–0.27	0.01–0.1	—	Hidu and Haskins (1978)
Pecten maximus	66–70	240–260	—	0.10	—	0.12	32–41	Cragg (1980)
C. virginica	—	>250	—	0.037–0.102	—	—	>12–24	Mann (1988)
Spisula solidissima	—	196.1	0.22	0.03–0.04	—	—	24	Mann et al. (1991)
Mulina lateralis	—	159.7	0.13	0.03	0.02	—	19	Mann et al. (1991)
Rangia cuneata	—	168.5	0.17	0.02–0.05	0.06	—	12	Mann et al. (1991)
Crustacea, Cirripedia								
Balanus crenatus	—	—	0.17–0.55	—	—	—	12–20	DeWolf (1973)

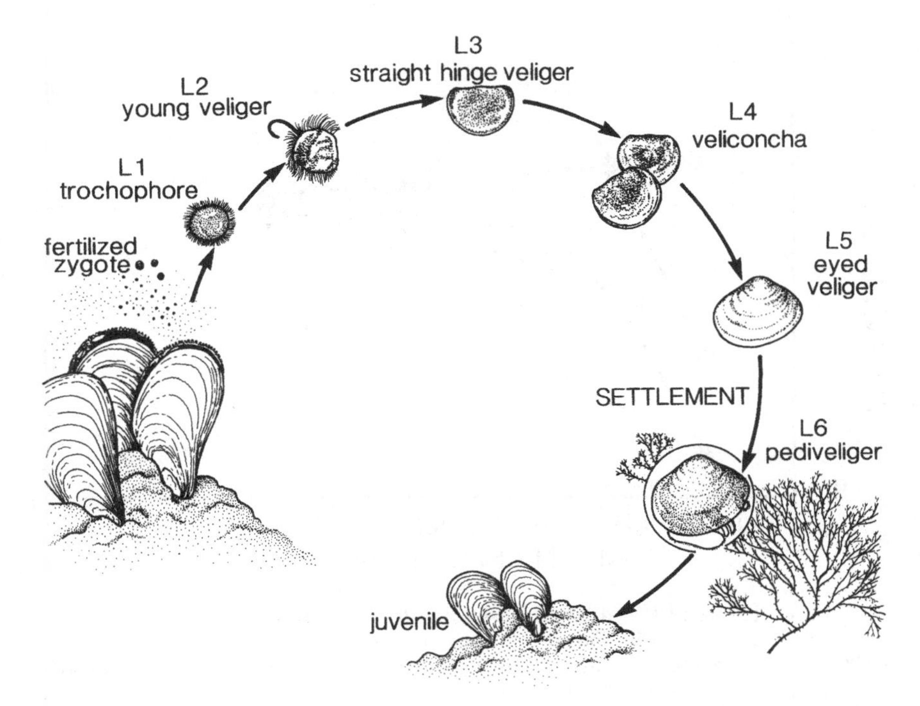

Figure 3.3 Diagram of larval dispersal and settlement of the blue mussel *Mytilus edulis* (based on photomicrographs and figures in Bayne 1976). The post-larval plantigrade is not shown.

ables in determining larval drift were thought to be larval growth rates, larval mortality rates, and depth maintained by the larvae during drifting.

Earlier attempts at measuring sinking rates and swimming speeds of dispersing larvae involved a measured distance and stopwatch, but were often made in a too small volume of seawater, e.g. on a microscope slide (Chia et al. 1984). As a consequence, "wall effects" (a sheared zone near the slide surface) may have influenced the results obtained in such a way as to underestimate the capability of the larva tested. Some examples thought to be free of this effect are shown in Table 3.2. DeWolf (1973) measured sinking rates of barnacle cyprids in a glass tube of 20 cm diameter and 1.25 m length. Hidu and Haskins (1978) used a sealed glass tube of 1.4 cm diameter and 15 cm length containing seawater to investigate both sinking rate and swimming speeds of oyster larvae. Cragg (1980) used various methods from a slide well up to a 4-L jar of 25 cm

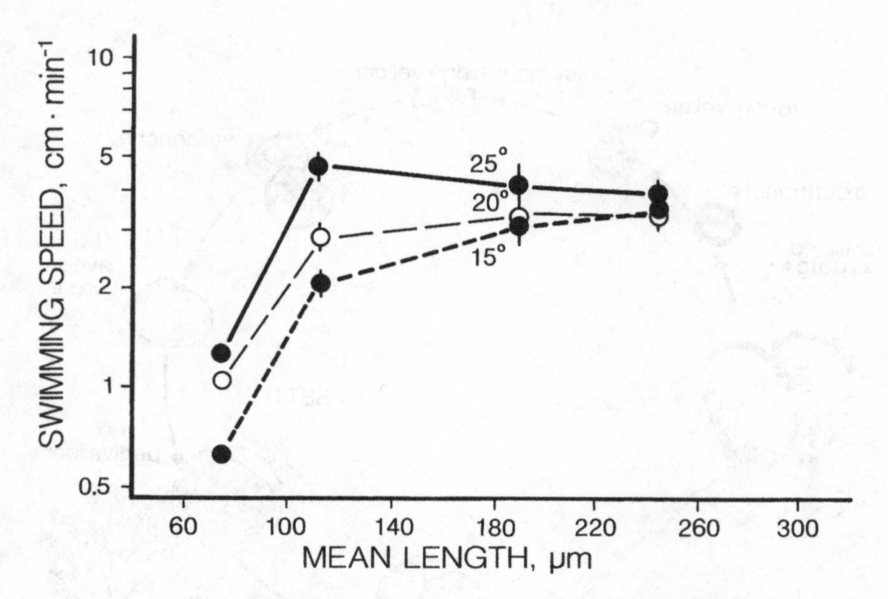

Figure 3.4 Effect of larval length of *Crassostrea virginica* on the log of swimming speed at $S = 25$ o/oo and three different temperatures (Hidu and Haskins 1978). To obtain the size range shown, trocophores, straight hinge, and eyed veligers were used. (From Hidu and Haskins 1978, reused by permission, Copyright Estuarine Research Federation.)

depth to determine larval swimming speeds of scallop. Butman (1989) and Pawlik, Butman, and Starczak (1991) measured sinking rates with narcotized or dead larvae and swimming speeds on live, suspension-feeding polychaete larvae in settling chambers (9.5 cm diameter × 122-cm length or 15 × 15 × 40 cm long) which were enclosed by a temperature-controlled water bath. Sinking rates shown in Table 3.2 vary from 0.01 to 0.83 $cm \cdot s^{-1}$, equivalent to the settling rates of inorganic sedimentary particles of <1 to 100 μm equivalent diameter in size. Swimming speeds in Table 3.2 vary from 0.01 to 0.50 $cm \cdot s^{-1}$, and if an average speed of 0.2 $cm \cdot s^{-1}$ was maintained throughout 1 d, the larva could swim a distance of 173 m. Larval ciliary swimming speeds are strongly influenced by environmental variables such as salinity and temperature because the efficiency of ciliary propulsion is influenced by seawater viscosity; thus, Hidu and Haskins (1978) were able to show that oyster larval swimming speed was inversely related to temperature (Fig. 3.4).

Finally, the dispersion of larvae is critically dependent on ambient currents, which are often of a larger magnitude than their swimming or sinking velocities. The scales of relevance, in both time and space, are very broad and may include the effects of molecular and turbulent diffusion, tides, storm-mixing events and wind-driven currents, Langmuir circulation, internal waves, mesoscale eddies, and large-scale general circulation (Okubo 1994).

A number of questions might be asked about the dispersion of larvae. For instance, how are they distributed downstream of a point or area from which they are released, either instantaneously or as a steady stream, from the bottom of a benthic boundary layer? There is advection of larvae along paths defined by stream lines, but they will also be subject to turbulent diffusion, which will move them from the mean streamlines in any direction. This problem can be analyzed in terms of the advection–diffusion equation, an application of which will be discussed in Chapter 7. Its solution will give the concentration of larvae at any point downstream. This equation can be made to include terms for larval mortality rate and for deposition on the bottom, if these are known.

In many circumstances, larval dispersion is dominated by more complex flows than in the benthic boundary layer. For example, it is well known that shoreward advection by currents driven by wind or waves may be an important transport mechanism (LeFèvre and Bourget 1992). Two-layer flows in salt wedge estuaries and partially mixed estuaries (Dyer 1979) will allow either landward or seaward transport of eggs or larvae, depending on their depth. Other important flows which may influence larval transport are tidally forced internal waves and bores, and coastal up-welling, as will be subsequently discussed in more detail. Topographically determined flows, such as observed around reefs and embayments (Fischer et al. 1979), are not easily analyzed by use of the advection–diffusion equation. Some of these topics are considered by Okubo (1994). The physical oceanography of larval dispersal has also been considered in conference proceedings edited by Sammarco and Heron (1994).

An interaction involving one of the physical oceanographic processes discussed and the swimming behavioral responses by the larvae of some suspension feeders is important in determining the outcome of the dispersal process. We have collected the most prominent hypotheses concerning such matters in Table 3.3.

Estuarine retention

In estuaries freshwater from a river is mixed with seawater from the ocean by the action of tides, wind effects at the surface, and river dis-

Table 3.3. *Physical oceanographic processes and larval behavior control the dispersal of larval benthic suspension feeders.*

No.	H_0/H_1	Description	Reference
1	H_0	Swimming by larvae in estuaries is not related to positioning in different water masses and dispersal is passive	Carriker (1951) Bousfield (1955) Mann (1988) Thiebaut et al. (1992)
	H_1	Swimming by larvae causes appropriate positioning in two layer flows which ensures estuarine retention	
2	H_0	Larvae are randomly dispersed in nearshore seawater	Hawkins and Hartnoll (1982) Shanks (1983, 1985, 1986, 1988)
	H_1	Shoreward transport of benthic larvae is caused by onshore winds	
3	H_0	Larvae are randomly dispersed in nearshore seawater	Shanks and Wright (1987)
	H_1	Shoreward transport of benthic larvae occurs after behavioral concentration in the surface over tidally forced internal waves	
4	H_0	Larvae are randomly dispersed in nearshore seawater	Pineda (1991)
	H_1	Shoreward transport of benthic larvae occurs in bottom waters where internal tidal bores are present	
5	H_0	Larvae are randomly dispersed in nearshore seawater	Farrel et al. (1991)
	H_1	Shoreward transport of benthic larvae results from shoreward advection which occurs if wind-induced upwelling is absent	
6	H_0	Plant canopies have no influence on suspension feeder recruitment and equal numbers of recruits occur on non-vegetated areas	Eckman (1983, 1987) Peterson (1986) Wilson (1990)
	H_1	Plant canopies of sufficient density entrain significantly more benthic recruits than non-vegetated areas	
	H_2	In plant canopies of sufficient density, suspension feeder recruitment depends linearly on volumetric flow rate	

charge moving seaward. Density differences between seawater and river water may also result in horizontal pressure gradients which affect flow patterns (Dyer 1979). On the basis of mixing processes and circulation, estuaries of temperate climates may be classified (Dyer 1979) as fjords, well mixed, partially mixed, or salt wedge. Only the two latter categories are of concern to us here. The salt wedge estuary is one in which there is a marked separation, at the halocline, between the bottom-more saline seawater and the surface, less saline river water. Seawater is mixed upwards by a process of entrainment across the halocline. In the partially mixed estuary, tidal action is dominant and tidal turbulence is the major mixing mechanism, causing seawater to be mixed upwards and river water downwards. Typically in the partially mixed estuary, the surface and bottom water are fresh toward the landward end and salt toward the seaward end, with the vertical salinity gradient slight. By contrast, in the salt wedge estuary, the seawater extends far landwards with the vertical salinity gradient sharp at the halocline.

Concerning larval dispersal in estuaries, the most important physical feature of both salt wedge and partially mixed estuaries is that the bottom water has a net movement landwards and surface water seaward when averaged over a number of tides. The basic estuarine retention hypothesis (number 1 in Table 3.3) is that by vertical swimming guided by appropriate environmental cues such as light, salinity, or gravity, the larva is able to position itself to be transported to a suitable site for settlement. This may involve a change in sign of the response to an environmental cue during larval ontogeny with larvae positively phototaxic early in life so that they are passively transported seawards. Later in larval life they become negatively phototaxic so that a competent larva is returned landwards by passive transport in bottom water, enabling it to settle near the adults producing it. An alternative mechanism, that greater larval size and/or sinking rate is achieved during ontogeny, accounts for depth position within the water column.

The estuarine retention hypothesis has been the subject of many field studies by fish and decapod larval biologists, but relatively few concern suspension feeders (Stancyk and Feller 1986). The field studies listed in Table 3.4 are of plankton sampling in estuaries where the authors consider that the temporal and spatial distributions of dispersing larvae of suspension feeders are consistent with the estuarine retention hypothesis. The field study in the Miramichi estuary, New Brunswick, Canada, by Bousfield (1955) of various barnacle larvae is probably the most comprehensive in support of an estuarine retention mechanism. This

Table 3.4. *Field studies involving plankton net sampling of larvae of suspension feeders within estuaries.*

Major taxa	Species	Reference
Cirripede barnacles	*Semibalanus balanoides*	Bousfield (1955)
	Balanus improvisus, *B. crenatus*	
Bivalve molluscs	*Crassostrea virginica*	Carriker (1951)
	C. virginica	Seliger et al. (1982), Boicourt (1982)
	C. virginica	Mann (1988)
Tubicolous polychaete	*Owenia fusiformis*	Thiebaut et al. (1992)

author showed that larval swimming depth increased with age so that competent cyprids were found at the greatest depths. Position of larvae within the water column varied with the tidal stage and current velocity. Bousfield (1955) calculated that approximately 18 d later, ~10% of those initially spawned remained as competent larvae within the estuary, and these were sufficient to maintain adult populations. By contrast, if the larvae of *Balanus improvisus* maintained a uniform distribution in the water column, his calculations showed that, as a result of seaward advection, insufficient numbers would remain to recruit the historic numbers of adult barnacles within this estuary.

In the study by Mann (1988) of the James River estuary, Chesapeake Bay, in the United States, oyster larvae were reported to migrate through a salinity discontinuity front present in the estuary and were then transported downwards, where they became entrained in the more saline bottom water which moved landwards. Laboratory behavioral experiments showed that intermediate stage oyster larvae could actively swim through a salinity gradient (Mann 1988) involving a salinity difference of 3‰.

As pointed out by Stancyk and Feller (1986), some authors using similar field observations consider that the observed larval distributions support the null hypothesis of number 1, Table 3.3. In all cases, though, the physical oceanographic conditions in the estuaries examined do not conform to salt wedge or partially mixed conditions, rendering the estuarine retention hypothesis invalid in these conditions. Thus, DeWolf

(1973), in a study of barnacle cyprids in the western Wadden Sea, the Netherlands, suggested that the larvae were transported passively by tidal currents, periodically sinking to the sediment at slack tide and being resuspended when currents were greater than 35–67 $cm \cdot s^{-1}$. Likewise, Korringa (1952) found that the larvae of *Ostrea edulis* in the well mixed Oosterschelde estuary, the Netherlands, were passively distributed in this tidally energetic system, where no salinity stratification was present. The larvae of the intertidal spionid polychaete *Pseudopolydora paucibranchiata* in the San Diego River estuary, California, were found by Levin (1983) to be mostly passively retained, with no evidence of adjustment of the vertical position of the larvae within the water column. This estuary was much impacted by human modification, with no evidence of a salt wedge or partially mixed conditions.

Support for the estuarine retention hypothesis is also to be found in laboratory measures of larval sinking or swimming rates, as shown in Table 3.2. The observed movements are well within the capabilities required to traverse the vertical distances for estuarine retention. Other support is to be found in the laboratory demonstration of positive or negative swimming responses directed to the source of the environmental cue, referred to as taxes, such as light and gravity. As an example, the earliest veliger larvae of blue mussels *Mytilus edulis* are briefly photonegative, but this condition rapidly changes so that no response to light is present, although during most of larval dispersion they are negatively geotaxic (Bayne 1964b). In the later stages of life, blue mussel pediveligers are negatively phototaxic and positively geotactic (Bayne 1964b). Other responses to light by larvae may be non-directional so that swimming continues until the larva reaches shade, a process termed *photokinesis* (Crisp 1974). In the review by Crisp (1974), only light, gravity, and flow direction were environmental cues considered to provide directional information for larvae. Other physical variables inclusive of light intensity, pressure, temperature, and salinity were scalar quantities lacking any directional information content.

Although not an estuarine study, but a study of blind ending coastal bays in the vicinity of Narragansett Bay, Rhode Island, United States, Gaines and Bertness (1992) showed that over a 10-year period, barnacle settling rates depended linearly on the average flushing time of each bay investigated. For *Semibalanus balanoides* and their particular habitat, passive dispersal based on the physical oceanographic conditions dominant at the site lead to good prediction of spatfall.

Shoreward advection

For open coastal environments such as wave pounded vertical rock faces, the question as to how planktotrophic larvae manage to recruit in a narrow zone on the shore has intrigued many investigators. Barnacle larvae have been a favorite in such studies because they are planktotrophic and feed only in the pre-competent stage. *Balanus crenatus*, for example, may spend from 12 to 20 d in the seawater of the Wadden Sea (DeWolf 1973) before settlement occurs. The question addressed here is, How does the competent cyprid regain the intertidal rock face required for the completion of the barnacle life cycle?

Working on the North Wales, United Kingdom, coast, Bennell (1981) suggested that *Semibalanus balanoides* settled on scraped natural rock surfaces which were cleaned daily when the wind direction was offshore and the sea state calm. In contrast, Hawkins and Hartnoll (1982), using similar field techniques but on the Isle of Man, United Kingdom, found a positive relationship between *S. balanoides* settlement density and timing with onshore winds (wind direction and velocity for the hour near high water) during May 1979 (see number 2, Table 3.3). In a California study by Shanks (1986), daily settlement, mainly of *Chthalamus* sp., was not correlated with wind direction for a 4-mo period in 1983 or wind direction at high water. Settlement rates of cyprids here were found not to be significantly different on days with onshore versus those with offshore winds.

In the study by Shanks (1986), three possible mechanisms were considered to explain how barnacle larvae reached suitable rock settlement sites in the intertidal:

- Cyprids swam ashore.
- Cyprids were randomly dispersed and deposited on the shore.
- Cyprids utilized onshore currents to transport them to shore.

The larval swimming capabilities were considered to be insufficient for the purpose, since cyprids are commonly found in abundance to ~10 km away from the coast (Shanks 1986) and up to 100 km offshore on the French Atlantic coast (LeFèvre and Bourget 1991). The second of the possibilities has been used as the null hypothesis for numbers 2–5 in Table 3.3, although field evidence, e.g. the neustonic concentration of barnacle cyprids of *Verruca stroemia* discovered by LeFèvre and Bourget (1991), suggests that, at least for some species, this is not true. The initial hypothesis of shoreward transport of larvae by internal waves was developed for crab megalopae (Shanks 1983), and an integral part of the

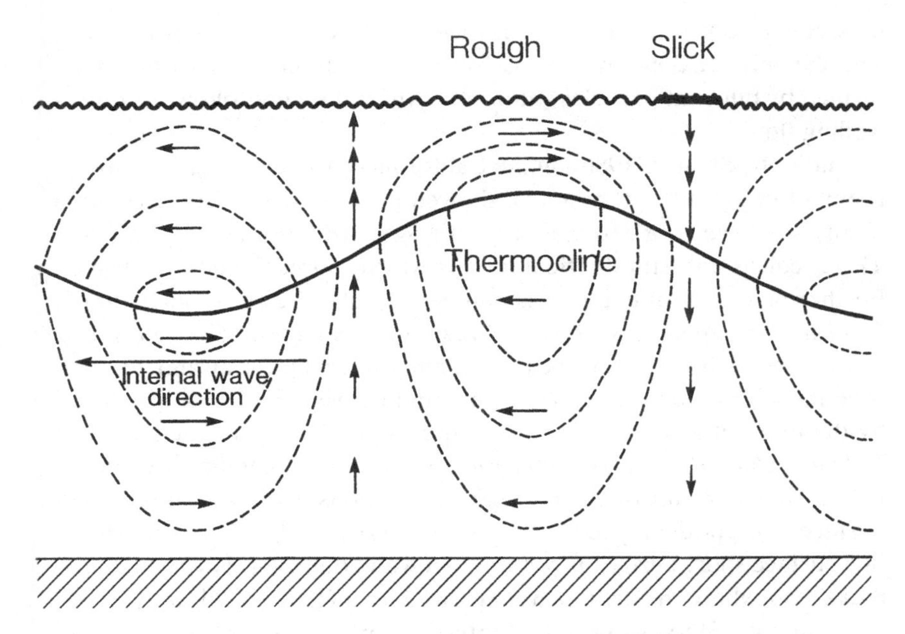

Figure 3.5 Diagram of currents over linear internal waves as postulated by Shanks and Wright (1987) for the Southern California coast. Larvae are transported shorewards in surface water by the internal wave.

mechanism involved a behavioral response by crab megalopae (Shanks 1985). Crab megalopae showed positive phototaxis, negative geotaxis, and kinetic responses to light and pressure in laboratory experiments and, in field experiments involving release of crab megalopae, a tendency to swim in surface waters (Shanks 1985). Whether the competent stages of suspension feeding larvae, such as barnacle cyprids, possess simple behavioral responses which keep them in surface seawater, as is required by the internal wave transport hypothesis, was not determined in Shanks's studies, although Crisp (1974) reports that many species of barnacle cyprids are positively phototaxic in laboratory conditions, and LeFèvre and Bourget (1991) demonstrated that cyprids of *V. stroemia* are concentrated in surface seawater.

The physical events caused by the passage of a linear internal wave are shown in Fig. 3.5. They are caused when the ebb tide slackens over a large-scale bottom feature such as the continental shelf break, causing a lee wave to be released, resulting in a series of large amplitude internal waves (Shanks and Wright 1987). During the internal waves passage, an up-welling occurs ahead and a down-welling behind the internal wave. The convergent down-welling is frequently marked by a surface slick formed by buoyant particles – flotsam – which does not follow the down-welling flow.

Studies by Shanks (1986) on the California coast, involving daily measurement of cyprid settling rates, showed that the rates were correlated tidally with peaks of settlement occurring 1–4 d before the spring tide. This is consistent with the tidal forcing which causes the internal waves. Further field tests of hypothesis number 3 (Table 3.3) were made in the San Juan Archipelago on the U.S. West Coast (Shanks and Wright 1987). These studies involved plankton sampling in surface seawater to compare the cyprid concentrations within and between slicks. As predicted by the internal wave shoreward transport hypothesis, the abundance of *Balanus glandula* and *Semibalanus cariosus* was significantly higher within the slicks on three of the four occasions when this was tested. Surface drifters deployed by Shanks and Wright (1987) also indicated that surface shoreward transport occurred and that cyprids settled on the rocks near where the drifters were carried. In a further field study of this type, Shanks (1988) showed that internal wave shoreward transport occurred, although few suspension-feeding larvae were included. This study was conducted near Beaufort Inlet, North Carolina, in the Atlantic Ocean, where the tidal range was lower (<2 m) and the continental shelf wider (>80 km) than in previous studies.

Two other shoreward advection hypotheses have been proposed (numbers 4 and 5, Table 3.3). Thus, Pineda (1991), working at La Jolla, California, found that invertebrate larvae, inclusive of barnacle cyprids, were concentrated in bottom water, contrary to the predictions of the internal wave transport hypothesis. To explain this he proposed that the shoreward advection of these larvae resulted from internal tidal bores associated with submarine canyons, which, in this part of the California coast, are very close to shore. One of the predictions from this hypothesis – that cyprid settlement peaks should correspond to low temperature events due to internal tidal bores – was confirmed by Pineda (1991). Thus, daily cyprid settlement rates of barnacles (*Chthamalus* sp. and *Pollicipes polymerus*) showed a statistically significant negative correlation with seawater temperatures. The shoreward advection of cold subsurface seawater by internal tidal bores is predictable because they are tidally related.

Another mechanism was proposed (number 5, Table 3.3) on the basis of field studies of the mid-California coast by Farrel, Bracher, and

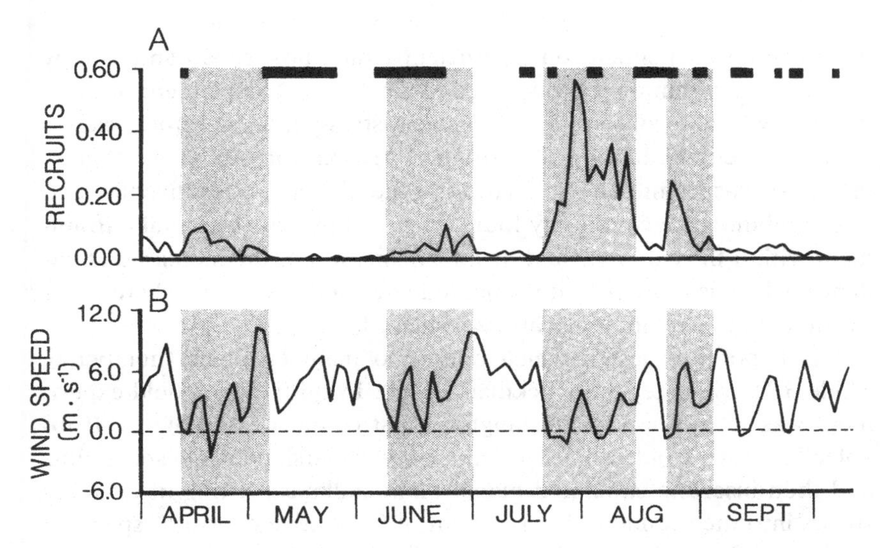

Figure 3.6 Recruitment density as mean number of recruits per settling plate ($n = 18$) against time for local barnacle settlement of *B. glandula* and *Chthalamalus* sp. on the mid-California coast (Farrel et al. 1991). Wind speed is shown (north > 0, south < 0). Bars at the top show when internal waves occurred.

Roughgarden (1991). Barnacle cyprid recruitment of *Balanus glandula* and *Chthamalus* sp. was monitored every other day and various physical oceanographic features coincidentally monitored. During the summer period of 1988, four major barnacle recruitment pulses (shaded area in Fig. 3.6) which correlated with the advection of warm, low salinity seawater shorewards were recorded. The pulses were not correlated with internal waves, onshore winds, or pulses of release of larvae by adults. Shoreward advection in this case appeared to be controlled by winds which caused up-welling at the narrow continental shelf edge. Coastal up-welling in association with along-shore winds caused net offshore transport. With the prolonged relaxation of winds, the transport reversed and became shorewards.

Plant canopies

Physical oceanographic interactions with seagrass meadows are discussed in Chapter 8 in the section, Marine plant canopies: seagrass meadows, and interactions with kelp forests in the section, Hydrodynamics. In general, they involve the development of a skimming flow over the plant canopy, dependent on a sufficiently high flow speed and canopy density, and within the canopy a reduction in speed, dependent on stem density, which results in reduced shear stress and, therefore, causes increased sedimentation. The original reason for studying seagrass meadows was to find an explanation of the higher species diversity and animal abundance commonly found there. Thus, passive larval entrainment within the canopy could be an explanation of the increased abundance, which is more fully documented later in this section, where other hypotheses to explain it are also considered.

The hypothesis tested which relates to plant canopies (number 6, Table 3.3) originated with Eckman (1983). In this study, a field experiment was set up on the Washington coast of the United States. It involved removing bulrush *Scirpus americanus* marsh plants in small plots and then inserting simulated marsh plant stalks consisting of drinking straws into the sediment. Recruitment of facultative deposit/suspension feeders such as a spionid and sabellarid polychaete, but mainly of meiofauna, was shown to have a significant positive correlation with the numerical density of straws, in agreement with a hydrodynamic mechanism of larval dispersal. Field observations on the North Carolina coast in a seagrass meadow consisting of 80% *Zostera marina* and 20% *Halodule wrightii* were conducted by Peterson (1986) to determine whether larval supply rates explained the higher density and biomass of hard clams within the meadow. The hard clam *Mercenaria mercenaria* was sampled in 0.25-m^2 cores with a sieving procedure that obtained only clams of >5 mm valve height. The size range of 5–25 mm sampled at annual intervals was considered to be a good indication of larval recruitment. The results, shown in Table 3.5, suggest that significantly more larval (0-year) recruits do appear within the seagrass beds than on the unvegetated sandflat. However, Peterson thought that passive hydrodynamic trapping of larvae within the seagrass meadow was only one part of the story since the older hard clams were present at a significantly greater density within seagrass than on the unvegetated sandflat. The numbers of older hard clams shown in Table 3.5 are higher than numbers of 0-year recruits, in part because several years are accumulated in calculating this mean. The larger differential within older hard clams between seagrass meadow and unvegetated sandflat was thought by Peterson (1986) to be due to differential survival within the two habitats, with more efficient post-settlement predation likely to be the cause of the

Table 3.5. Mercenaria mercenaria *densities per 0.25 m² in a North Carolina lagoon based on the mean ± standard error of 36 subsamples.*

	1980		1981	
	Plot 1	Plot 2	Plot 1	Plot 2
0-year recruits <25-mm valve height				
Seagrass	0.94 (±0.21)	0.72 (±0.15)	0.61 (±0.13)	0.27 (±0.09)
Sandflat	0.33 (±0.11)	0.36 (±0.11)	0.17 (±0.06)	0.05 (±0.04)
Older animals >25-mm valve height				
Seagrass	3.11 (±0.48)	1.78 (±0.33)	3.52 (±0.53)	2.20 (±0.24)
Sandflat	0.20 (±0.11)	0.36 (±0.11)	0.11 (±0.07)	0.33 (±0.09)

Source: Peterson (1986).

greater losses on the sandflat. That predation was more efficient on the sandflat and inhibited by seagrass roots and rhizomes had been demonstrated by Peterson (1982) and others.

In considering the second alternative hypothesis of number 6 in Table 3.3, Eckman (1987) mounted a field study in a *Zostera marina* meadow on Long Island, United States, in which the recruitment of two suspension-feeding bivalves – the bay scallop *Argopecten irradians* and the common jingle *Anomia simplex* – were determined at four sites within the local area. Both larval bivalves settle first on eelgrass blades by means of byssus threads. By increasing the numbers of available eelgrass blades with plastic mimics, recruitment as measured at approximately weekly intervals was dependent on the velocity of seawater through the meadow. Thus, high density eelgrass meadows had fewer recruits than those with fewer blades impeding the flow, as a result of differences of volumetric flow which passed through the meadow.

In all the plant canopy studies so far reviewed, *recruitment* was measured and the likelihood of post-settlement mortality at each site not independently assessed, and hence excluded as a possible confounding factor in determining whether the initial settlement rate was indicated by the recruitment rate. One such study in which settlement rates were determined directly was reported by Wilson (1990). Larval settlement rates were determined in settlement traps which were embedded in the sediment and the larval catch counted daily. The fieldwork, conducted in

Bogue Sound, North Carolina, showed that mean settlement of bivalve larvae within traps in a *Zostera marina–Halodule wrightii* meadow was 1.6 times greater than in nearby sandy areas which lacked seagrass. The between-habitat differences were reported by Wilson (1990) to be statistically significant and thus support H_1 of hypothesis number 6 (Table 3.3).

Duggins, Eckman, and Sewell (1990) reasoned that understorey kelps which share the same kind of interaction with flow (Chap. 8, the section, Marine plant canopies) as do seagrass meadows will also act to entrain larvae (number 6, Table 3.3). The field methods used in the San Juan Archipelago area of the U.S. West Coast involved acrylic plastic (Plexiglas) plates which were bolted to the rock bottom. Deployment of the plates was for 3- to 5-wk periods, although Duggins et al. (1990) claim that for the four major species which settled on the plates, they could measure initial settlement because each produced a calcareous secretion within hours of settling, which persisted and was counted even if the larva died soon after initial settlement. A complex field experiment was designed in which understorey kelp densities and substratum orientation – upward or downward facing plates – were manipulated. This allowed the following effects on settlement to be analyzed: kelp/no kelp, low flow/high flow, dark/light effect on microalgal growth, and sedimentation/no sedimentation. The results suggested that of the four species, all except *Membranipora membranacea* had higher recruitment within the kelp treatment, although for the spirorbid polychaetes which settled, the early season response changed to a preference for the no-kelp treatment later in the season. The bryozoan *Membranipora* preferred high-flow, low-sedimentation environments, suggesting that it is intolerant of deposition environments. These results therefore provide support for H_1 of hypothesis number 6 (Table 3.3) for some species but not others, and clearly demonstrate the importance of other factors such as flow, sedimentation, or growth of microalgal mats as involved in larval settlement within understorey kelps. As with seagrass meadows, kelp forests may also be subject to a key predator which affects larval settlement rates. For example, Roughgarden, Gaines, and Pacala (1987) showed that in central Californian kelp forests, the juvenile rock fish *Sebastes* sp. could significantly reduce any zooplankters passing through the kelp. Thus, barnacle cyprids within the kelp forest were 70 times less abundant than those outside as a result of rock fish predation, and when the rock fish left the kelp in the fall, the numbers of barnacle cyprids rapidly built up again.

Post-larval dispersal

Post-larval stages of many invertebrates frequently reappear in the water column after their settlement and metamorphosis. This was documented in a review by Butman (1987), which lists 33 field observational studies involving plankton tows, traps, or seawater pumping in which post-larvae of bivalves, polychaetes, plus a few gastropods and meiofauna were found in the samples.

A dispersal function for post-larval blue mussels *Mytilus edulis* in the Wadden Sea was proposed by Maas Geesteranus (1942). The young mussels were passively transported by currents until they detected a suitable substratum for their growth. Byssal detachment and reattachment of the young mussels could occur many times until a suitable substrate was found. Life cycle studies undertaken by Bayne (1964a) of blue mussels in the Menai Straits, North Wales, referred to earlier, suggest that the pediveliger larva first settles on a filamentous red alga. After settlement and metamorphosis to the post-larval or plantigrade stage, the young mussel may detach and be passively transported by seawater currents. Bayne (1964a) referred to primary and secondary settlement processes to describe this part of the blue mussel life history. However, McGrath, King, and Gosling (1988), in studies on the rocky shores of the west coast of Ireland, found that pediveligers of *M. edulis* settled directly onto the adult mussel reef without first settling on a filamentous red alga. Sigurdsson, Titman, and Davies (1976) presented field and laboratory observations designed to elucidate the mechanisms involved in post-larval dispersal in bivalves. They showed that slow currents induced post-larval bivalves to produce a long byssus thread which was functional in drifting dispersal. For example, juvenile tellinids *Abra alba* of 1 mm valve height could be induced to produce a mucopolysaccharide byssus thread which was 3 cm long and 2–4 μm in diameter. The byssus thread increases the surface area of the juvenile bivalve significantly. Because the settling velocity in creeping flow regimes is proportional to surface area, we can expect the settling velocity of the juvenile bivalve to be significantly less, thus enhancing dispersal. The post-larval bivalve was able to detach from its byssus thread, upon which it passively sank to the bottom. Sigurdsson et al. (1976) thought that byssus drifting was of importance for many bivalve species of up to 2.5 mm in valve height, inclusive of protobranchs, e.g. *Nucula tenuis*, and many lamellibranchs, e.g. *Mytilus edulis* and *Mya arenaria*.

Most of the studies described are observational and designed to aid in

Table 3.6. *Hydrodynamic hypotheses related to post-larval life of suspension feeders.*

No.	H_0/H_1	Description	Reference
1	H_0	Post-larval dispersal plays no part in determining density and production	Maas Geesteranus (1942) Bayne (1964a) Sigurdsson et al. (1976) Emerson and Grant (1991) Roegner et al. (1995)
	H_1	Post-larval dispersal is important in determining density and production	
2	H_0	Post-settlement mortality is not related to hydrodynamic–sediment conditions	Luckenbach (1984) Sammarco (1991)
	H_1	Post-settlement mortality is related to hydrodynamic–sediment conditions	

understanding the possible mechanisms of post-larval dispersal. But finding post-larvae within the water column does not necessarily mean that the dispersing individuals are successful in recolonizing or indicate the relative importance of this method of dispersal. In a field study based on the passive bedload transport of juvenile softshell clams *Mya arenaria*, Emerson and Grant (1991) were able to frame their hypothesis with an ecological emphasis, as shown in hypothesis 1 of Table 3.6. Specially designed subsediment bedload sampling traps, consisting of a 3-cm-diameter acrylic tube which could be removed from a supporting pipe, were placed so that the opening was flush with the sediment surface. Sedimentary bedload and macrofaunal trap catches were recorded on a daily basis from sandflats in Halifax Harbour, Canada, and compared to independent estimates of clam population density and spat settlement. The transport of juvenile clams >6 mm was commonly observed within Halifax Harbour, reaching maximum rates of 790 clams $\cdot$ $m^{-1} \cdot d^{-1}$ at a sheltered site and 2600 clams $\cdot$ $m^{-1} \cdot d^{-1}$ at a more exposed site. At both sites, clam transport was correlated with bedload sediment transport, with the maximum associated with storms when it reached 35 kg $\cdot$ $m^{-1} \cdot d^{-1}$. At the sheltered site, clam transport was negatively correlated with clam density, while at the exposed site, it accounted for a large jump in density in September. This suggested that the juvenile clams were able to reburrow and recolonize after transport. Clam transport at the exposed site caused a precipitous decline 2 mo later and the complete removal of any recently settled spat. These results clearly support H_1 of number 1 in Table 3.6. Roegner et al. (1995) have determined in flume experiments that 2-wk-old juvenile recruits of the softshell clam behave passively, essentially like sedimentary particles, since they avoid erosion until ambient velocities reach those necessary for the initiation of sediment transport. Comparing living and dead clams shows that burrowing behavior of young clams is one way of avoiding transport away from the site by flows sufficient to cause sediment erosion.

Early post-settlement mortality, as is implied in the wind–wave energetic sandflat in Halifax Harbour, is recognized to be a potentially important variable for soft sediment benthic community structure and function. The effect of environmental variables, particularly predation, on early recruit mortality patterns was studied, for example, by Luckenbach (1984) on coot clams, and Sammarco (1991) on coral reef recruitment (number 2, Table 3.6). Their results support H_0 of number 2, but do not consider flow as an alternate hypothesis. We could find no experimental tests which identify flow as a perturbing factor and assess the importance of early post-larval stage mortality caused by it.

Local hydrodynamics

The focus for the rest of this chapter shifts to a much smaller spatial scale – <0.1 km. Of most concern in benthic studies are flows near solid boundaries, whether relatively flat bottoms or more irregular shapes such as boulders and rocky surfaces. Except at very low velocities and at very small distances from the surface, these flows are turbulent. Creeping flows are seen very close to the solid boundaries, and, although these regions are thin, they may be large by comparison to typical sizes of larvae or eggs.

The hydrodynamically smooth benthic boundary layer is typically found over soft sediments and other fine-grained bottom materials. This flow is characterized by the following:

- The *viscous sublayer*, which is a viscosity-dominated layer near the bottom where the flow is mostly laminar
- The *logarithmic layer*, which is turbulent and where the mean velocity varies as the logarithm of the distance above the bottom
- The *outer layer*, which is also turbulent but with turbulent energy decreasing with height
- The *free stream flow*, which drives the flow below it (Soulsby 1983)

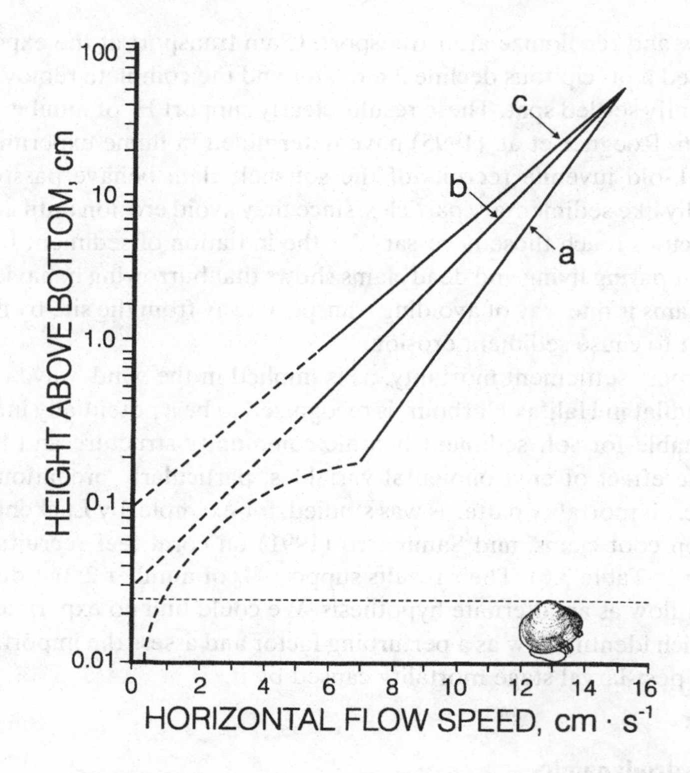

Figure 3.7 Turbulent velocity profiles at different shear stresses over a soft sediment: a, smooth, $U^* = 0.60\,\text{cm} \cdot \text{s}^{-1}$; b, rough, $U^* = 0.82\,\text{cm} \cdot \text{s}^{-1}$; c, rough, $U^* = 0.98\,\text{cm} \cdot \text{s}^{-1}$. The curved region of line a represents the viscous sublayer. A 240-μm pediveliger larva of the blue mussel is shown to scale (based on Butman 1986).

The viscous sublayer is typically of the order of millimetres thick, while the logarithmic layer is a few metres deep.

A larva which has to move between the bulk flow and the bottom must cross turbulent zones as well as the viscous sublayer, a flow dominated by viscosity. A typical velocity profile for a smooth boundary (from Butman 1986) is shown in Fig. 3.7. This profile (curve a) is based on a velocity of $15\,\text{cm} \cdot \text{s}^{-1}$ at a depth of 50 cm and a shear velocity of $0.60\,\text{cm} \cdot \text{s}^{-1}$. It is clear that the viscous sublayer is less than 2 mm thick under these circumstances. This dimension might be compared to the dimensions of some larvae and their sinking or swimming speeds (Table 3.2). A competent larva of the blue mussel pediveliger is about 0.024 cm in size and can be expected to have a typical bivalve swimming speed of ~0.2 $cm \cdot s^{-1}$. It is only in the viscous sublayer, however, that the larva can be said to control its own trajectory. In the logarithmic layer, the velocity fluctuations about the mean are typically about 10% of the mean value (Butman 1986), about 1 $cm \cdot s^{-1}$ in the case under discussion. These fluctuating velocities will dominate the movement of most larvae, although a high settling velocity or swimming in a preferred direction should result in a net movement across the stream lines.

The hydrodynamically rough boundary layer is characterized by very thin viscous sublayers on the upstream faces of roughness elements and a wake flow in their lee. The roughness elements are higher than the depth of the viscous sublayer which would be expected in their absence. Movement of larvae will be determined by the random motion in the flow a short distance above the tops of the roughness elements and by processes specific to the geometry of the roughness near the boundaries. Similarly, wave-swept rocky shores will be characterized by thin, time-dependent boundary layers and wakes which are difficult to generalize in terms of larval transport and diffusion.

Larval settlement/recruitment

Before we begin an overview of the rapidly expanding body of work which deals with the detailed hydrodynamics of settlement of competent suspension-feeding larvae, we should recall the general ecological features that are involved.

The life history of suspension feeders begins with spawning, and field studies have suggested that the timing of this event is seasonally adjusted for each species (Thorson 1950; Lacalli 1981; Hawkins and Hartnoll 1982). As an example, Lacalli (1981) showed that spawning periods in Passamaquoddy Bay, within the Bay of Fundy, Canada, e.g. of polynoid polychaetes (Fig. 3.8), are correlated with the timing of the spring diatom bloom in Passamaquoddy Bay. The mechanism which allows this linkage to be made was first suggested in the experimental work of Starr, Himmelman, and Therriault (1990). In this work, two species were examined, one of which was a suspension feeder, *Mytilus edulis*. Adult blue mussels were triggered to spawn by release of a chemically unidentified ectocrine produced by phytoplankton cells found in the Gulf of St. Lawrence, Canada. In the laboratory, phytoplankton such as *Skelatonema costatum*, *Phaeodactylum tricornutum*, and *Thallassosira nordenskioldii* produced sufficient ectocrine to induce mussel spawning.

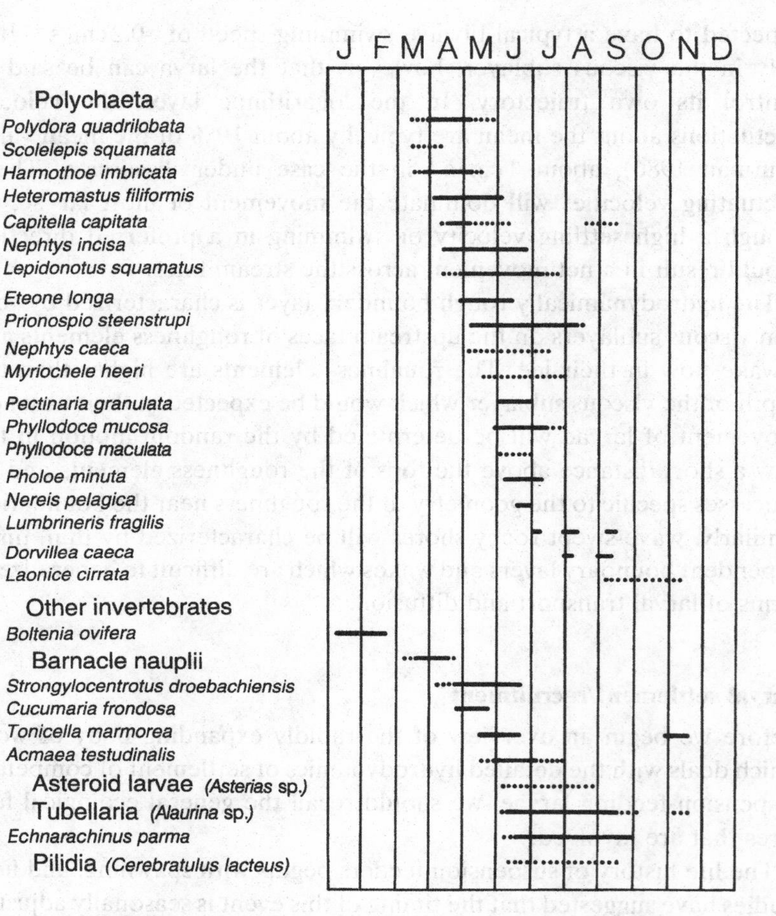

Figure 3.8 Seasonal presence/absence of planktotrophic larvae for some common benthic invertebrates in Passamaquoddy Bay, Bay of Fundy, Canada: solid lines, maximum abundance; dotted lines, lesser abundance in plankton tows (Lacalli 1981).

In *M. edulis*, the response was concentration dependent, requiring at least 40×10^7 cells $\cdot$ L^{-1} to induce spawning (Starr et al. 1990). As pointed out by the authors, such high phytoplankton densities would rarely occur in field conditions; however, continuous suspension feeding by the mussel from lower cell densities might be sufficient to concentrate enough phytoplankton in the mantle cavity to trigger spawning.

Because of the seasonal nature of spawning in cold temperate conditions, it is likely that larval recruitment will also show a strong seasonal signal. As far as we are aware, there is limited information (e.g. Menge 1975; Strathman 1985) regarding larval survivorship during dispersal, but before settlement. Larval settlement has been studied for a variety of reasons, one of which concerns the economic importance of fouling on man-made structures in the sea, such as saltwater intake pipes, ship hulls, oil rigs, and floating mariculture pens. Suspension feeding larvae in the period when they are near to, and just after recruitment to, the benthos are subject to a variety of variables – physical, chemical, and biological – which can influence the outcome of this process. In the following account, we are concerned primarily with how hydrodynamics at a scale of <0.1 km influences larval settlement.

A central question of benthic ecology is to determine ways to predict which characteristic macrofaunal assemblage will occur in a particular de-faunated patch of marine sediment (Gray 1974; Wildish 1977; Underwood and Fairweather 1989). One obvious and potentially important variable in determining the spatial pattern of soft sediment macrofaunal assemblages, as discussed in Chapter 7 (and see Chap. 6 for a discussion of aggregation), is the supply of colonizing larvae. Larval supply is affected by the proximity to the patch of adults producing fertilized zygotes, the length of planktonic life, the species composition of adults producing larvae, larval production rates, hydrodynamic effects which affect dispersal and settlement, as well as survival of larvae during their planktonic life. These considerations may be boiled down to the following four ecological parameters which can indicate recruitment success (see also Keough and Downes 1982; Luckenbach 1984):

- Survivorship of the larval stage from zygote to final settlement
- Success rate of larval settlement
- Losses associated with post-settlement dispersal (see this chapter, the section, Post-larval dispersal)
- Early post-settlement mortality (see Chap. 3, the section, Hard substrates).

Because there is limited information available for mortality during the planktonic phase of the life history of most suspension feeders, most of this section will be concerned with the factors which influence larval settlement.

Soft sediments

In a comprehensive review of invertebrate larval settlement in soft sediments, Butman (1987) considered two major hypotheses – passive deposition of competent larvae (number 1, Table 3.7) and active habitat

Table 3.7. *Hypotheses relating hydrodynamic factors which may influence larval settlement/recruitment of soft sediment macrofauna.*

No.	H_0/H_1	Description	Reference
1	H_0	Competent larval suspension and deposit feeders are not passively co-deposited like similarly sized sediment particles	Hannan (1984), Butman (1989), Pawlik et al. (1991), Pawlik and Butman (1993)
	H_1	Competent larval suspension and deposit feeders are passively co-deposited like similarly sized sediment particles	Snelgrove (1994), Harvey et al. (1995)
2	H_0	Larval swimming near the sediment surface has no influence on selection of a settlement site	Butman et al. (1988), Bachelet et al. (1992) Butman and Grassle (1992), Grassle et al. (1992a,b), Turner et al. (1994)
	H_1	Larval swimming near the sediment surface is guided by chemical cues and results in larval choice of a settlement site	
3	H_1	Recruitment enhancement does not occur near small-scale protuberances from the sediment	Eckman (1979, 1983, 1985), Gallagher et al. (1983)
	H_2	Recruitment enhancement occurs near small-scale protuberances, e.g. animal tubes, seagrass shoots	Ertman and Jumars (1988)

selection by competent larvae (number 2, Table 3.7). In the former, purely hydrodynamic forces are considered to deliver larvae which act like passive sedimentary particles with similar fall velocities. In the active selection hypothesis, competent larvae are considered to be able to swim (Table 3.2) to a preferred ecological niche. A third hypothesis – passive deposition near sediment protuberances – is really a special case of the passive deposition hypothesis (number 3, Table 3.7). Considering all possible types of initial settlement pattern – random; irregular, e.g. after incorporation of larvae in marine snow (Shanks and Edmondson 1990) or other rapidly sinking masses; that caused by active swimming; or that due to passive deposition – we believe that it is unnecessary to use active habitat selection as H_0 and passive deposition as H_1, or vice versa, and we have phrased the null and alternate hypotheses non-exclusively in Table 3.7. This table also lists those studies, both field and flume experimental, which test the three hypotheses and which, for the latter, pay adequate attention to scale and hydrodynamic similarity in simulating conditions in the field. Butman (1987) concluded that passive deposition and active habitat selection were not exclusive concepts because they often operated at very different scales of space and time. With regard to space, competent larvae settle at millimetre to metre scales, with passive deposition often operating at the higher end of the scale. Temporal scales of importance are tidal (hours) and the stochasticity associated with wind–wave motion (days to months).

As is frequently the case in marine biology, the origin of useful concepts such as the passive deposition hypothesis resulted from keen observation during fieldwork. Tyler and Banner (1977), for example, followed larval transport of the brittle star *Ophiura alba*, considering the larvae to be physically sorted by local currents and passively deposited along with fine sand in the Bristol Channel, United Kingdom. Even earlier field observations such as those by Pratt (1953), who studied bivalve distribution in Narragansett Bay, and Baggerman (1953), who studied oyster spatfall in the Wadden Sea, suggest the importance of passive larval transport and deposition.

Rigorous experimental testing of the passive deposition hypothesis for infauna originated with Hannan (1984) and Butman (1989). In these field experiments, sediment traps of various designs were set 0.4 m above the soft sediments of Buzzards Bay, Massachusetts, United States. The traps had previously been calibrated in a flume, which met dynamic similarity criteria with boundary layer–sediment conditions in the field, and in which the relative particle collection efficiencies of larval mimics (polystyrene or glass spheres of the same fall velocity) had been determined. The H_1 that planktotrophic larvae sink in seawater as passive particles in turbulent flow can be tested by determining the rank order of collection in traps compared with the a priori determined rank order of larval mimics. Differences in trap design (see Chap. 2) caused differences in local disturbances of flow which resulted in the differences observed. The results from field deployment of the traps 1 m above bottom showed that most larvae were collected according to a priori expectations as passive particles, including polychaetes, bivalve post-larvae, enteropneust, and gastropod larvae (Butman 1989). A few exceptions were found; thus the polychaete larva of *Pectinaria gouldii* and the larvae and post-larvae of a seastar (? *Asterias* sp.) did not settle passively. The results provide some support for the larval passive deposition hypothesis, although they do not provide the benthic boundary layer conditions that a larva would experience in reaching the sediment. Further field experiments in Buzzards Bay designed by Snelgrove (1994) overcame this problem. Two types of settlement tray were deployed at 20 m depth directly in a mud habitat. One tray had an 11.3-cm-diameter by 10-cm-deep depression filled with a 2-cm-deep layer of de-faunated muddy sediment, and the other had a 2-cm-deep cup of the same diameter to which de-faunated muddy sediment was added so that it was flush with the tray walls. Most of the naturally occurring settling larvae of bivalves, gastropods, juvenile *Mediomastus ambiseta*, and, on one occasion, spionid polychaetes and juvenile nemerteans preferred the depression trays, thus supporting H_1 of the passive deposition hypothesis (number 1, Table 3.7). Only the polychaete *Capitella* sp. did not prefer the depression trays. These experiments were adequately replicated and, since the trays were deployed for 3–4 d, the results suggest that initial larval settlement was involved.

A laboratory test of the passive deposition hypothesis comes from flume experiments with the aggregating tube worm *Phragatopoma lapidosa californica*. This sabellarid polychaete constructs tubes of sand grains which can form into extensive reefs (see Chap. 6, Fig. 6.2) on the hard substrates of the Southern California coast (Pawlik et al. 1991; Pawlik and Butman 1993). Because the larvae are easily cultured and delay settlement until they find sand originating from adult worm tubes (Pawlik 1986), they are ideal for larval settlement choice experiments in which unidirectional flume flows over sand are offered. The laminar sublayer and logarithmic layer can be characterized by the shear or friction velocity, U^*, which is related to the shear force exerted by the bottom. Six different flume flows of 5–30 $cm \cdot s^{-1}$ were tested, corresponding to U^* of 0.26–1.22 $cm \cdot s^{-1}$ calculated from the "full profile" method (Pawlik and Butman 1993). A long flume 17 m by 0.6 m wide which contained temperature-controlled seawater at 20°C was used. In preliminary tests, larval mimics were used to determine bedload shear velocities at which sediment transport was initiated. This was determined to be at $U^* = 0.47$–$0.86\ cm \cdot s^{-1}$ in the conditions of this experiment. The flume was filled to a 12-cm depth with filtered seawater and 2000 larvae added by plastic tubing and funnel at a point just upstream of a flow straightener. At $U^* > 1.03\ cm \cdot s^{-1}$, the larval mimics began to make brief excursions into the benthic boundary layer, and, as the shear velocities increased, so the number of spheres present there and distance travelled by each mimic also increased. Only one pass by live larvae of the acrylic settlement plate which contained four wells, each with a sand treatment, was permitted in the later experiments (Pawlik and Butman 1993). Two treatments were offered – either "filmed" sand, that is, aged sand coated with a microbial film but not processed by worms, or "tube" sand, prepared from the broken off tubes that had been cemented by older worms. Results showed that there was a unimodal response between larval settlement rates determined in 3-h-long experiments and flume bulk velocity flows. Peak settlement occurred at bulk velocities of 15–25 $cm \cdot s^{-1}$, and at these velocities, larvae tumbled end over end, sometimes briefly adhering to the floor by means of their tentacles. At $<10\,cm \cdot s^{-1}$, larvae swam near the release point, eventually swimming to the flume surface, where they were advected downstream in the bulk layer, so that they did not come into contact with the settlement plate. At $>30\,cm \cdot s^{-1}$, larval settlement was poor, either because the shear velocity was too high and/or because the increased turbulence meant that fewer larvae actually came into contact with the sand treatments. In field conditions, *Phragmatopoma* is much more likely to experience oscillatory wind–wave flows so it is not possible to place these results in a field context.

Observations on the free-swimming larvae of the suspension-feeding bivalve *Cerastoderma edule* in still water and a Vogel–LaBarbera flume, 3.5 m long × 0.5 wide × 0.4 m high, were made by Jonsson, André, and Lindegarth (1991). In still water, upward helical swimming of the competent pediveliger was observed and interpreted as geotaxis although statocysts were not involved. In unidirectional flow flume observations, four bulk velocities were offered: 2, 5, 10, and 15 $cm \cdot s^{-1}$. At 5–10 $cm \cdot s^{-1}$, the larvae became confined to the viscous sublayer, where they drifted in a streamwise direction at 0.45–1.60 $mm \cdot s^{-1}$, periodically touching the flume floor. At 15 $cm \cdot s^{-1}$, bedload transport of tumbling larvae occurred with a much higher probability of larval resuspension. Jonsson et al. (1991) suggested that boundary-shear forces created by $U > 15\,cm \cdot s^{-1}$ prevented the larvae from swimming properly. An hypothesis involving viscous drag torque on the larval body by benthic boundary layer forces, which can help explain settlement by *C. edule*, is proposed as a specific mechanism of viscous sublayer larval confinement by Jonsson et al. (1991).

Spat of the Iceland scallop *Chlamys islandica* preferentially settles on various filamentous red algae and on the perisarcs of dead hydroids in the Gulf of St. Lawrence (Harvey, Bourget, and Miron 1993). Model studies with three-dimensional branching structures made of plastic were designed by Harvey et al. (1995) to determine the factors influencing larval recruitment. The effect of branching patterns and branch diameter of the red alga models on scallop larval settlement was determined in field experiments. Concurrent flume laboratory experiments using similar algal models and simulated larvae, actually inert polyvinylchloride microparticles of the right size, became trapped on the branches in silicone grease. Flume experiments gave results similar to the field ones, suggesting that passive settlement was the mechanism at a scale of ~10 cm, although active processes at lower scales could not be excluded by this study (Harvey et al. 1995).

Butman (1987) pointed out that experimental tests of the active habitat hypothesis prior to her review were nearly all larval choice experiments in still water. Most experiments were conducted at inappropriate spatial scales of <10 cm so that they were not simulating field conditions. Nevertheless, this was the "ruling hypothesis" of this subject (number 2, Table 3.7) prior to 1987, with much compelling evidence that chemical cues were utilized during the settlement of competent larvae (Pawlik 1992).

The first experimental test in which realistic hydrodynamic conditions were simulated during larval choices to set on different sediment types was presented by Butman, Grassle, and Webb (1988). This work involved comparisons of a still water test and unidirectional flow in a recirculating flume. The annular flume had a 6-m-long working section × 0.5 × 0.3 m deep, total volumetric capacity of 1000 L, and flow induced by paddle wheel. Competent larvae introduced to the flume were given settlement choices in 4 × 4 × 1-cm-deep compartments filled with sediment. Two choices – a natural mud and an abiotic glass bead mixture similar in grain size (~70% < 63 μm) to the mud – were offered. Competent larvae tested included *Capitella* sp. 1, a subsurface deposit feeder associated with organic-rich mud, and the suspension-feeding hard clam *Mercenaria mercenaria*. Bachelet et al. (1992) discovered a confounding factor in those experiments with the hard clam that was due to decomposition of the larval shell on preservation in acidic conditions, which led to an underestimate of larval numbers in all mud treatments. These results will not be discussed further, although Bachelet et al. (1992) showed that the hard clam larvae showed no preference over mud and clean sand in still water conditions. For *Capitella* sp. 1, Butman et al. (1988) found that for the one flow tested, equivalent to $U^* = 0.30\,\mathrm{cm \cdot s^{-1}}$, the worm larvae were able to choose the ecologically appropriate muddy sediment. In a continuation of these studies with *Capitella*, Butman and Grassle (1992) and Grassle, Butman, and Mills (1992a), in choice experiments at flows of 0, 5, and $15\,\mathrm{cm \cdot s^{-1}}$ (equivalent to boundary shear velocities of 0.26 and $0.64\,\mathrm{cm \cdot s^{-1}}$), provided further evidence that the larvae of this worm can select highly organic sediments at up to the highest shear velocity tested. It is of interest that at $U^* = 0.64\,cm \cdot s^{-1}$, greater numbers of larvae settle in the 2-h flume test than in still water. This is at least in part due to the greater flux at higher flows, which allows more larval contacts with sediments in a limited test time.

Using the same annular paddle wheel flume, settlement experiments were conducted with pediveliger larvae of the coot clam *Mulina lateralis* by Grassle et al. (1992b). In 24-h experiments, the clam larvae chose an organically rich natural mud over abiotic glass beads of approximately the same particle size in a free-stream flow of $\sim 5\,cm \cdot s^{-1}$ and $U^* = 0.30\,cm \cdot s^{-1}$. At zero flow, the clams made similar significant choices to those in flow because of their helical swimming abilities. Grassle, Snelgrove, and Butman (1992b) conclude that the coot clam pediveliger can exercise habitat choice by swimming, but that older pediveligers which are less efficient swimmers require ambient flows to transport them between sediment patches.

Further experiments in the same annular flume with larvae of *M. lateralis* and *Capitella* sp. were designed by Snelgrove, Butman, and Grassle (1993) to test the relative importance of active selection versus passive deposition mechanisms. In these laboratory experiments, different size depressions in an acrylic plastic (Plexiglas) settlement plate were variably filled with organic-rich mud or glass beads of the same particle size. Both species generally chose mud over glass beads, although in *M. lateralis* in some treatments, larval settlement was greatest among glass beads. This "wrong" behavior was probably due to physical trapping in the depression, i.e. physically controlled or passive dispersal. Because *M. lateralis* pediveligers are poor swimmers, they were unable to swim out of the depressions and settle in the appropriate mud-filled depression. These results were shown to be scale dependent, and, hence, the relevance of the results in field conditions was not made clear.

One question of importance in regard to hypothesis 2 (Table 3.7) is whether the chemical cue is adsorbed on a suitable surface, as suggested in the work on *Capitella* and *Mercenaria* by Butman et al. (1988), or is waterborne, as proposed by Zimmer-Faust and Tamburri (1994) in work on oyster larvae of *Crassostrea virginica*. Other chemical inducers of larval settlement are reviewed in Pawlik (1992) and all are lipophilic and substrate bound. A laboratory test of the water-soluble chemical hypothesis was conducted by Turner et al. (1994) with *C. virginica* larvae in flows of 2 and $6\,cm \cdot s^{-1}$ at $U^* < 0.25\,cm \cdot s^{-1}$. The experiments were set up in a manner similar to those previously described: tests in both still water and an annular, recirculating flume of 3516-L capacity. The settlement choice area was an acrylic plastic (Plexiglas) plate in which circular wells of 6.98 cm diameter 1.27 cm deep were made with a small hole in the bottom so that solutions were continuously injected at $2\,ml \cdot s^{-1}$ into each well. The wells were filled with crushed oyster shell which had been cleaned and dried before use. A low molecular weight peptide at concentrations of 10^{-6} to 10^{-10} *M* glycl-glycl-L-arginine was the chemical tested because Zimmer-Faust and Tamburri (1994) had found in still water tests that it induced oyster larval settlement. The oyster larvae clearly preferred to settle in wells with water-soluble cues over those without the chemical. The chemical inducer caused rapid downward swimming or sinking ($>0.3\,cm \cdot s^{-1}$) of the larvae, resulting in concentration of the larvae in near-bottom water, and hence enhanced larval contact and settlement. Thus, the larval response does not appear to depend on a chemical gradient but to be triggered at low threshold concentrations of $<10^{-10}$ *M* glycl-gylcl-L-arginine, which caused more rapid swimming. The applicability of this result in the field, of course, depends on whether the chemical concentrations simulated are realistic and what happens in more energetic flows.

The study of how small-scale protuberances from the sediment surface affect larval recruitment was initiated by Eckman (1979). These studies utilized field experimentation and observation to investigate the distribution of recently settled recruits to an intertidal sandflat in Puget Sound, United States. Eckman showed that increased abundances of recently settled spat occurred within 1 cm of the tubes of polychaetes, amphipods, and tanaids. Recruitment enhancement experiments involving model tubes in natural sediments suggested that the resultant enhancement was controlled by localized flow–tube–sediment interactions. These relationships were investigated with small (in relation to benthic boundary layer thickness) tube mimics in flume studies by Eckman and Nowell (1984). Whether resuspension or deposition near the tube occurs depends on the tube height and the boundary shear stress imposed by the flow. Characteristic horseshoe patterns of sediment erosion occur at high boundary shear stress on the upstream side of the tube, with discrete depositional areas a few tube diameters downstream of the tube. Recruitment enhancement occurred in depositional areas downstream of the tube, although it was not possible with the experimental design to determine whether local transport of larvae, spat mortality, or movement subsequent to settlement was the cause. Eckman (1983) set up field experiments in which de-faunated sand cores were placed at control or treatment sites on a Puget Sound sandy beach. Each site had a core at the centre, with either a bare sand control or a treatment site surrounded by different densities of simulated marsh grass stalks. The latter were actually 0.6-cm-diameter plastic straws placed at three densities of 56–2500 straws $\cdot$ m^2. Coincident experiments in a recirculating flume where three velocities, equivalent to $U^* = 0.52$–$1.67\,\text{cm} \cdot \text{s}^{-1}$, and similar straw densities to the field manipulations were set up. Enhanced recruitment in the field was found for harpacticoids, amphipods, facultative deposit–suspension feeding, tube-living polychaetes, and meiofauna. The results closely follow those predicted from flume observations of boundary shear stresses around rigid cylinder mimics and thus support H_1 of hypothesis 3 (Table 3.7). A similar type of field experiment with tube mimics at densities less than that causing skimming flow showed that enhancement of bacterial colonization occurred in the downstream wakes (Eckman 1985). Bacterial colonization may be important for larval recruitment since a bacterial film may stimulate larval settlement (e.g., Crisp 1974).

Although the field experiments of Gallagher, Jumars, and Trueblood (1983) were designed to test the well known conceptual model of Connell and Slatyer (1977) of mechanisms of succession (inhibition, tolerance, or facilitation), the results provide support for the passive deposition hypothesis (number 3, Table 3.7). The experiments involved placing cores containing clean sand into 10-cm^2 patches to which live or simulated tubes of the common tube builders in the Puget Sound, soft-sediment intertidal area were added. Either live or simulated tubes enhanced the recruitment of other taxa to the de-faunated cores in comparison to that of controls without tubes, suggesting the physical nature of the mechanism involved.

The hydrodynamic characteristics of the exhalant jet of a cockle from Puget Sound, *Clinocardium nutalli*, were studied by Ertman and Jumars (1988) in a 2.5-m-long-working-section recirculating flume. Exhalant velocity in *Clinocardium* was estimated as 9–$11\,\text{cm} \cdot \text{s}^{-1}$ and patterns of flow around it varied with ambient velocity in much the same way as around the animal tube mimics studied by Eckman and Nowell (1984). Polystyrene beads with the same settling velocity as that of narcotized larvae were injected into the exhalant flow of an actively pumping cockle and the sediment near it sampled with 0.6-cm-diameter plastic drinking straws as corers. The free stream velocities tested ranged from 3.3 to $16.0\,\text{cm} \cdot \text{s}^{-1}$, which was equivalent to $U^* = 0.19$–$0.78\,\text{cm} \cdot \text{s}^{-1}$. The results were consistent with the passive deposition hypothesis (number 3, Table

3.7) because the downstream distribution of polystyrene beads showed localized patterns in the downstream wake of the flow, as indicated by the high variance of the measured densities of recruits. Similar results were found by field sampling downstream of the exhalant of resident soft-shell clams *Mya arenaria* in Puget Sound.

Hard substrates

Just as for soft sediments, for benthic ecologists interested in hard substrates an important question is the prediction of the type of macrofaunal assemblage which will develop on a particular de-faunated area of rock or coral. Like those of larvae settling on soft sediments, the ecological parameters of primary concern for hard substrate recruitment are planktonic survivorship, larval settlement rate, and post-settlement mortality. One way that larval settlement of suspension feeders of hard substrates appears to differ from that of soft sediment is the absence of post-settlement dispersal in the former. Hypotheses which have concerned researchers in this field are similar to passive deposition (number 1, Table 3.8) and active habitat selection (numbers 2 and 3, Table 3.8) of soft sediments. An additional hypothesis, which is more ecological in outlook, has focused on larval supply and settlement rates, which are proposed to control the final densities achieved (number 4, Table 3.8). This approach has been referred to as "supply side ecology" (Roughgarden et al. 1987; Underwood and Fairweather 1989), although many of the studies in support of it have not adequately separated larval settlement rates from early post-settlement mortality rates.

Much of the field and laboratory experimental work on hard substrate settlement has been done with the competent settling stage, or cyprid, of cirripede barnacles. Crisp (1955) conducted settling experiments with cyprids in 1.5-m-long, smooth-walled, glass tubes of from 2- to 10-mm internal diameter. Water flow rates were varied by adjusting a constant-head device and recording the volumetric flow rate. All of the flows were laminar. The velocity gradient at the wall of the tube can be related to the average velocity and the tube diameter. The friction velocity, U^*, is related to the velocity gradient at the wall by

$$\left(\frac{du}{dr}\right)_0 = U^{*2}\upsilon^{-1}$$

Friction velocities have been calculated for the data of Crisp (1955); they are shown in Table 3.9. The experiments involved introducing individual

Table 3.8. *Hypotheses relating hydrodynamic factors which may control larval recruitment of hard substrate suspension feeders.*

No.	H_0/H_1	Description	Reference
1	H_0	Larval settlement is independent of benthic boundary layer flow	Crisp (1955), Mullineaux and Butman (1990, 1991) Mullineaux and Garland (1993), Abelson et al. (1994)
	H_1	Larval settlement is passively controlled by benthic boundary layer flow	
2	H_0	Competent larvae settle indiscriminately	Wethey (1986), LeTourneux and Bourget (1988) Walters (1992)
	H_1	Competent larvae behaviorally choose particular hard substrates on the basis of surface roughness	
3	H_0	Competent larvae settle indiscriminately	Rittschoff et al. (1984) Neal and Yule (1994)
	H_1	Competent larvae behaviorally choose particular hard substrates on the basis of chemical cues which are either waterborne or bound to surfaces	
4	H_0	Larval supply rates are independent of the subsequent density of rocky shore recruits	Keough and Downes (1982), Bushek (1988) Bertness et al. (1992), Miron et al. (1995)
	H_1	Larval supply rates control the subsequent density of rocky shore recruits	

cyprids into the tube in a pipette and then observing their passage in a 1-m section of the tube. Results were expressed as a percentage of those tested which made contact with the tube walls as a function of flow velocity. Larval initial contacts were unimodally related to a range of tube flow velocities tried, equivalent to velocity gradients up to $1000\,s^{-1}$ for two species of barnacle, as shown in Table 3.9. Maximum attachment was at velocity gradients up to, but less than, that in which the swimming cyprid could maintain its position by swimming against the current. Peak settling, b, occurred at ~$75\,s^{-1}$ for *Semibalanus balanoides* and ~$40\,s^{-1}$ for *Elminius modestus*. In stage a response, the larvae did not swim and were moved passively. Flows greater than $50\,s^{-1}$ caused the cyprids to swim and attach. Presumably in the c stage, increasing shear made it harder for the cyprid to attach, until at >$400\,cm^{-1}$ for *S. balanoides* and >$700\,cm^{-1}$ for *E.*

Table 3.9. *Summary of initial settling responses of barnacles in laminar tube flow experiments.*

Species	Response stage	Velocity gradient (s^{-1})	U^* ($cm \cdot s^{-1}$)
Semibalanus balanoides	a	<75	<0.87
	b	75	0.87
	c	75–400	0.87–2.00
	d	>400	>2.00
Elminius modestus	a	<40	<0.63
	b	40	0.63
	c	40–700	0.63–2.65
	d	>700	>2.65

Source: Crisp (1955).

modestus no further larval attachment was possible. The hydrodynamic conditions in the experimental tube in these experiments do not simulate natural benthic boundary layer conditions very well, although they do show the importance of the velocity gradient to initial contact in line with H_1 of hypothesis 1 (Table 3.8). Crisp's experiments are among the few discussed for larvae that adequately consider a wide enough velocity range.

The technique of using thin and bluff plates to create particular hydrodynamic conditions on surfaces without changing the plate surface roughness was introduced by Mullineaux and Butman (1990) in field recruitment experiments at a deepwater seamount of 410 m depth in the central North Pacific. Circular plates were used in this work, and the recruitment results suggested that settlement had been influenced by small-scale boundary layer flows over the plates. Unidirectional flume flows at free-stream velocities of 5 or 10 $cm \cdot s^{-1}$ were used for visualization of the flow over rectangular settlement plates and for behavioral studies of settling by barnacle cyprids of *Balanus amphitrite* by Mullineaux and Butman (1991). The objectives of this study were to determine whether and how larval settlement was influenced by passive benthic boundary layer phenomena, and whether active behavioral processes were involved after initial larval contact. Three types of polycarbonate plate, 10 × 26 cm long, were oriented so that the flat surface was parallel to the flume flow. Each type of plate produced characteristically different streamlines dependent on its shape, as shown in Fig. 3.9. About 20,000 competent cyprids were

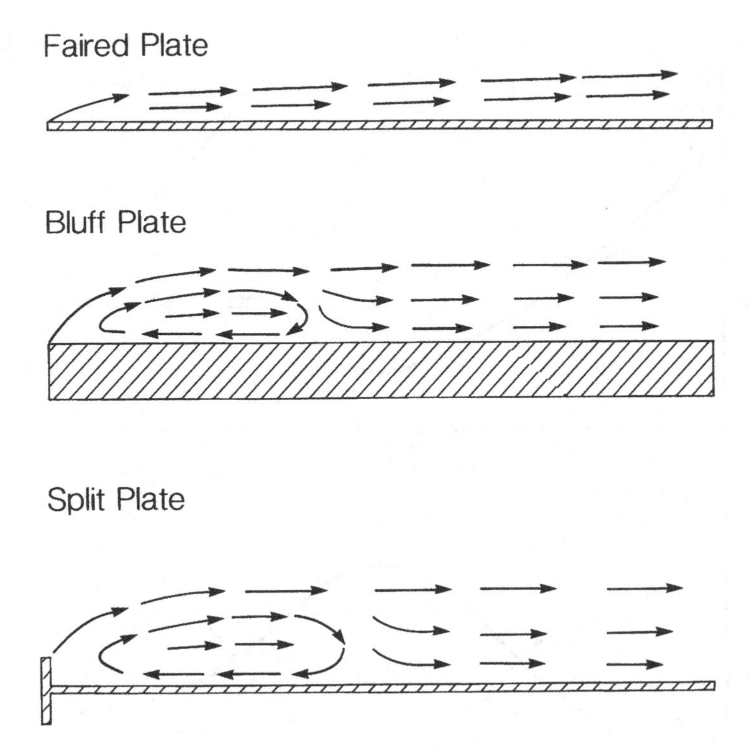

Figure 3.9 Streamlines over three types of settlement plate placed horizontal with respect to flume flow (Mullineaux and Butman 1991).

placed in the ~1000-L-capacity flume and allowed to recirculate past the plates which had been pre-soaked with adult barnacle extract, known to induce settlement chemically (Crisp and Meadows 1963). Cyprid settlement was recorded by video viewing of an area 10 × 10 cm during a 48-h experimental period. Typical results for the bluff plate are shown in Fig. 3.10. Initial cyprid contacts were found to be negatively correlated with shear stress and vertical advection in support of H_1 (hypothesis 1, Table 3.8). After initial contacts, many cyprids began crawling with the aid of their antennules, and videophotography showed that downstream of the re-attachment point they moved downstream, whereas above this point they moved upstream with respect to the bulk flow. As Fig. 3.9 indicates, near-bed streamlines are in the same direction as larval movement, and this behavior results in two peaks of settlement (Fig. 3.10B), although cyprids could, and did, actively swim off the plate. This was not due to

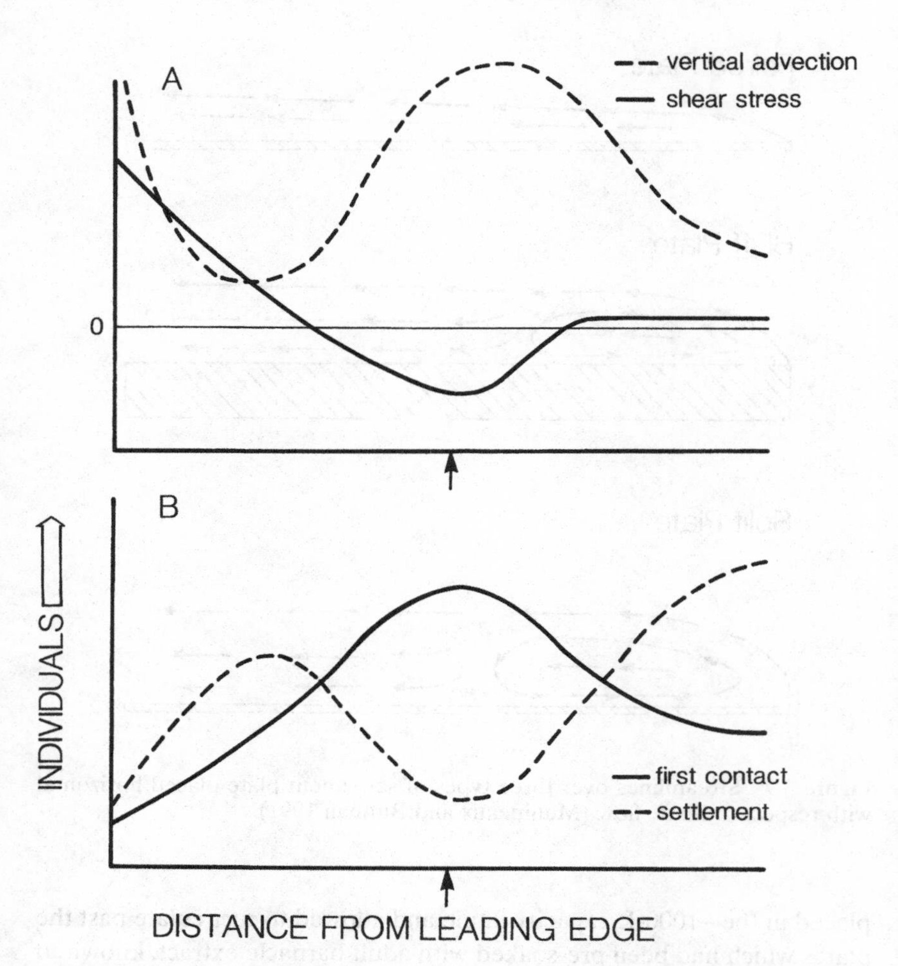

Figure 3.10 Diagram of boundary layer flow conditions over a bluff plate: A, vertical advection (dashed line) and shear stress (solid line); B, initial contact points (solid line) and final settlement location (dashed line) of barnacle cyprid larvae; arrow, reattachment point of the eddy; see also Fig. 3.9 (redrawn from Mullineaux and Butman 1991).

erosion of the cyprids from the substrate because Eckman et al. (1990) have shown that the critical drag force required to dislodge cyprids is not reached at bulk flows of $10\,cm \cdot s^{-1}$. The flume studies of Eckman et al. (1990) showed that the force required to detach *B. amphitrite* cyprids was positively correlated with the duration of attachment in periods of up to 3h.

Mullineaux and Garland (1993) reported on flume and field work with rectangular settling plates of similar design to that used by Mullineaux and Butman (1991), but with the addition of a 30° angled plate type. Hydrodynamic characteristics of each type of settlement plate were determined in flume flows followed by deployment on polyvinylchloride (PVC) poles so that the plates were raised 1.2 m above the sediment at sites near Woods Hole, Massachusetts, United States. The settlement plates were placed into a swivel by SCUBA divers. Flows of ca. $>2\,cm \cdot s^{-1}$ were enough to cause the swivel to turn so the leading edge of each plate opposed the flow. A hydrodynamic model of passive settling by planktonic larvae onto hard substrates presented by Eckman (1990) could be used to analyze the field results obtained by Mullineaux and Garland (1993). The model considers a depth limited boundary layer, boundary shear velocities, larval fall velocities, as well as roughness features of the substrate. Settlement responses of the locally abundant larvae were characteristic for each species. Thus, the hydroid *Tubularia croeca* settled in regions of high turbulence and strong shear stress, whereas the bryozoan *Schizoporella unicornis* was associated with regions of high shear only. Another bryozoan, *Bugula turrita*, settled in a region of reduced shear. The results obtained by Mullineaux and Garland (1993) are consistent with H_1 of hypotheses 1 and 2 of Table 3.8.

Abelson, Weihs, and Loya (1994) reported some interesting work on the settlement choices made by four species of coral larvae found at Eilat, Israel. Experiments were conducted in a special recirculating flume, termed a nozzle-diffuser flow tank, in which an expansion section caused deceleration and acceleration flow patterns. Two of the species, *Seriatopora caliendrum* and *Alveopora daedalea*, settled on the flume walls, in either non-accelerating or decelerating zones, e.g. in the downstream wakes of tubes, and two species, *Litophyton arboreum* and *Dendronephthya hemprichi*, preferentially settled only in the accelerating part of the flume expansion section or on protruding tubes. The latter species are found in natural conditions on protruding body substrates so the flume results are consistent with field observations. During the experiments, all four of the larval species were passively deposited by decelerating flows, although the two species found on protruding tubes actively swam back into the flow and delayed settling until an appropriate settlement site was found. The mechanism of settlement on protruding bodies involved secretion of aboral, adhesive mucous threads up to 100 times the larval body length, which allowed nearly instantaneous attachment to the substrate.

Shown in Table 3.10 are some examples of settlement or recruitment

Table 3.10. *Selected larval settlement or recruitment experiments on hard substrate where flow is simulated (S) or ambient flow (F) is available.*[a]

Species	Hypotheses tested	Flow conditions	Reference
BARNACLES			
Semibalanus balanoides, Elminius modestus	1	S	Crisp (1955)
Balanus amphitrite	1, 3	S	Rittschof et al. (1984)
S. balanoides	1, 2, 3	F	Wethey (1984)
S. balanoides, Chthamalus fragilis	1, 2	S, F	Wethey (1986)
S. balanoides	1, 2, 3	F	LeTournex and Bourget (1988)
S. balanoides	1, 2, 3	F	Chabot and Bourget (1988)
B. amphitrite	1	S	Eckman et al. (1990)
B. amphitrite	1, 2	S	Mullineaux and Butman (1991)
B. amphitrite (Bugula neritana)	1, 2, 3	F	Walters (1992)
B. perforatus, E. modestus	1, 3	S	Neal and Yule (1994)
Balanus sp.	1, 2	F, S	Lemire and Bourget (1996)
B. crenatus	1, 2, 3	F, S	Miron et al. (1996)
CORALS			
Local coral spat	1, 2, 3	F	Carleton and Sammarco (1987)
Local coral spat	1, 2, 3 + light	F	Maida et al. (1994)
Senatopora caliendrum	1	S	Ableson et al. (1994)
Alveopora daedalea	1	S	Ableson et al. (1994)
Litophyton arboreum	1	S	Ableson et al. (1994)
Dendronephthya hemprichi	1	S	Ableson et al. (1994)

[a] Refer to Table 3.8 for hypotheses tested.

experiments where flow was adequately simulated in flume flows and field experiments where ambient water movement was utilized. The rugophilic behavior of barnacle cyprids was demonstrated by Crisp and Barnes (1954) in still seawater experiments. Some tests of responses to surface roughness have involved field experiments in ambient flow. A difficulty inherent in field experiments where most factors influencing the outcome are not controlled is in identifying the causal factor(s). To overcome this, Wethey (1986) used plastic models made from casts of adult and juvenile barnacles on rock surfaces which were cemented to natural rock surfaces on Long Island. The models were examined daily by photography and ~30% of *S. balanoides* and *Chthamalus fragilis* cyprids settled at the same microsites, low in shear stress, on three replicated plates, and – at least initially – ecologically relevant chemical cues were absent. Because of the nature of these experiments, it was impossible to determine whether passive or active processes in settlement were involved (hypotheses 1 and 2, Table 3.8). The experiments designed by Walters (1992) in a low energy environment off the coast of North Carolina were an attempt to overcome the difficulty of distinguishing passive and active settlement processes. The plastic settlement plates were 8 × 8 cm with 3-mm-high, 4.5-mm-diameter bumps which were separated by 3-mm spaces. Half of the plates were treated with a thin film of silicone vacuum grease, thereby trapping settling larvae at the place of initial passive contact; half were untreated, thereby indicating subsequent active movement during the 4-d experimental period. The metamorphosis of two species selected for study, *Balanus amphitrite* and *Bugula neritina*, occurred most frequently around the bases of each bump on untreated plates and was significantly different from the passive deposition observed on greased plates. Active larval crawling on a millimetre to centimetre scale after settlement explained these results since differential post-settlement mortality or passive erosion could be ruled out as explanations. These results were confirmed by using similar methods for *Balanus* sp. cyprids and by comparing field experiments on the North Carolina coast and flume experiments with larval mimics to separate active and passive processes (Lemire and Bourget 1996).

The field settling study of LeTourneux and Bourget (1988) involved plastic settling plates of 100 × 30 cm and observations at metre, millimetre, and micrometre scales. The settling of *S. balanoides* was observed at two locations: in the Bay of Fundy and in the St. Lawrence estuary. For the smallest scale observed, surface microheterogeneity was significantly greater at the site occupied by the cyprid in comparison with vacant sites. The results suggest that larvae can discriminate microheterogeneity down to a length of 35 μm, which is approximately the size of the anntenular disc used by the cyprid both to explore and to attach to the substrate. Again, it is not possible to decide from these field observations whether the results are due to a passive or active process.

In common with earlier writers who used non-flowing larval bioassays (e.g. Crisp and Meadows 1963), Rittschof, Sansford Branscomb, and Costlow (1984) assumed that the chemical cues associated with settling were adsorbed onto surfaces. The method they used to demonstrate chemical attractiveness was similar to the glass tube bioassay for cyprid settling in flow of Crisp (1955). *B. amphitrite* larvae were used in these tests with a constant flow rate of $6.8\,\mathrm{ml \cdot min^{-1}}$ in a 3.25-mm-internal-diameter tube, which gave a velocity gradient 0.5 mm from the wall of $39.4\,\mathrm{s^{-1}}$. The experiments carried out by Rittschof et al. (1984) showed that rinsing the glass tube with $2.5\,\mu\mathrm{g \cdot ml^{-1}}$ of settlement factor, consisting of ground-up, whole *B. amphitrite* and adjusted to $50\,\mu\mathrm{g}$ protein $\cdot\,\mathrm{ml^{-1}}$, significantly increased settlement in young cyprids compared to a non-rinsed or control tube rinsed with bovine serum of the same protein content. Besides attractant chemicals from conspecifics of poorly defined identity, biofilms on surfaces have been identified as a settlement cue. Thus, enhancement of settlement of *B. amphitrite* cyprids (Maki et al. 1990) or inhibition (Maki et al. 1988) by biofilms may occur.

Neal and Yule (1994) designed an experiment in which *B. perforatus* and *E. modestus* cyprids, after initial contact on two contrasting biofilms – high ($83\,\mathrm{s^{-1}}$) and low ($15\,\mathrm{s^{-1}}$) shear stress – were tested for the tenacity of their attachment to each substrate. The films were developed on cover slips in flume flows over a 2-mo. period. The high shear stress biofilm was thinner and denser than the other and resulted in increased cyprid tenacity. The low shear stress substrate was relatively thicker and looser and caused cyprids to attach with lower tenacity. In these experiments, it was not clear how the benthic boundary layer flow would influence the results, although both cyprid species tested can obviously choose between the substrates offered.

In combined flume and field experiments in the St. Lawrence estuary, Miron et al. (1996) were able to examine both passive and active controls of initial settlement of barnacle cyprids. To distinguish between hydrodynamic and chemical settlement cues, responses by *Balanus crenatus* cyprids to either live or ceramic model adult barnacles were compared. It was found that spat were absent on control model barnacle panels but present on panels with live barnacles, suggesting the importance of a chemical cue to final settlement location. Flume tests suggested that initial settlement was passive, but the cyprids could swim away if appropriate cues were absent. Live cyprids actively moved after settlement, using chemical and roughness cues to accept or reject a settlement site (Miron et al. 1996).

An important question, to which we will return later in Chapter 6, is, How do groups of suspension feeders become aggregated? One mechanism – gregarious settlement – will be considered here in more detail. We use the term *gregarious* in a general sense to apply to aggregated larval settlement that results from any active or passive aggregating mechanism. It is thus used here differently than by Havenhand and Svane (1991), who consider gregarious larval settlement to imply active movement and a specific chemical cue. Reviews by Meadows and Campbell (1972) and Burke (1986) also present the view that gregarious larval settlement must involve chemicals as pheromones from conspecifics.

A further field study in the Bay of Fundy and Gulf of St. Lawrence concerning settling of *B. balanoides* was presented by Chabot and Bourget (1988). One aim of this study was to determine the effect of post-settlement conspecifics on the settlement rate. In the Bay of Fundy, settling *S. balanoides* larvae preferred open vertical rock faces, the reverse of that found in the Gulf of St. Lawrence and elsewhere. At smaller spatial scales, of cracks <1.5 cm deep, settlement increased as adult density increased up to a maximum when 22% to 30% of the rock was covered with adult barnacles; at greater densities, settlement rates decreased. Crisp (1974) considered that secretions containing arthropodin were the chemical cues for this behavior after first contact and subsequent crawling.

It was suggested by Keough (1984) that larvae of the bryozoan *Bugula neritina* settled gregariously. He devised settlement experiments to determine the mechanism in static glass dishes and used aggregation statistics to show that sibling larvae settled in an aggregated pattern, while unrelated larvae settled randomly. These tests showed that settling larvae responded only to conspecific larvae and not to conspecific juveniles. Many colonial suspension feeders and some sponges, cnidarians, bryozoans, and tunicates are known to have self-/non-self-recognition mechanisms associated with local competition and the ability to fuse if the adjacent colonies are genetically related (Keough 1984), so a kin recognition system has already been identified in these species.

In field experiments, Havenhand and Svane (1991) were the first to show that an hydrodynamic cue could be involved in gregarious settlement, leading to the broadening of the meaning of the term that we have adopted. In laboratory static bioassays, these authors showed that larvae of the solitary ascidian *Ciona intestinalis* settled randomly on a smooth plastic surface and that extracts of adult body tissues could not induce gregarious settlement. The field recruitment experiments were conducted in the Gullmarsfjorden, Sweden, with 16 × 16-cm plastic settlement plates to which were attached real and mimic adult *C. intestinalis* at one or five adult *Ciona* per plate. Recruitment around either real or mimic *Ciona* at five per plate was significantly higher than that on a bare plate. The results of Havenhand and Svane (1991) are consistent with the hypothesis that hydrodynamic rather than chemical cues, operating either at the time of initial settlement or at some point following this event and involving post-settlement movement, are the primary determinants of gregarious settlement in *Ciona intestinalis*.

The final hypothesis, number 4 in Table 3.8, we consider quite different from the mechanistic nature of the first three because it is ecological in character and provides a link to some later chapters (Chaps. 6, 7, 8). The ecological parameters of importance during early life history stages of suspension feeders are larval supply, settlement, and early post-settlement mortality rates. Methods for determining these rates for a few available species are presented in Chapter 2. As suggested by Bertness et al. (1992), proper consideration of "supply side ecology," a concept originating with Thorson (see Olafsson, Petersen, and Ambrose 1994), requires independent and coincident measurement of all three parameters for a given species and at a number of different sites.

Keough and Downes (1982) were among the first to distinguish the possible importance of larval supply and settlement and post-settlement mortality rates in establishing the spatial structure of the macrofaunal association of a hard substratum. Settlement plates were employed on the southern California coast, coupled with predator exclusion cages over half the panels, to indicate post-settlement mortality, but their work did not coincidentally monitor larval supply of locally dominant bryozoans and polychaetes. The study of Gaines and Roughgarden (1985) on central California coasts included settlement and post-settlement mortality rates of *Balanus glandula*, but not a measure of larval supply rates in the coastal waters.

At various sites in Galveston Bay, Texas, United States, Bushek (1988) measured settlement on ceramic tiles and larval supply by quantitative plankton sampling. The species involved were the oyster

Crassostrea virginica and the barnacle *Balanus eburneus*. The pattern of zonation on pilings in Galveston Bay was barnacles high and oysters lower in the intertidal zone. The zonation was suggested by Bushek (1988) to be controlled by larval supply rates and preferences for larval settlement in high and low water motion, respectively. This author also considered juvenile growth with respect to water motion measured by gypsum dissolution, showing that there is a positive linear relationship for the barnacle, but no correlation with oyster growth. Two populations of *Semibalanus balanoides* found on the rocky shore of Rhode Island were compared by Bertness et al. (1992). Larval sampling was by especially designed sediment traps in which larvae collected were preserved in formalin. Larval settlement was determined on a 5 × 5-cm cleared rock quadrat, some cleaned daily and others allowed to accumulate recruits, thereby giving an estimate of post-settlement mortality rates. This study provides a clear appreciation of the interplay of all of the three potentially important early life history parameters for these two Rhode Island locations. Thus, at the high settlement rate site, saturation of settlement sites meant that cleared sites (such as experimentally cleared quadrats) received 15 times more settlers than those at the low settlement site. However, recruitment of juveniles to the high settlement site was comparable to that at the other site because of a significantly greater density-dependent, post-settlement mortality rate at the former. Both field results provide support for H_1 in hypothesis 4 of Table 3.8.

The correlation between supply of competent larvae of *Semibalanus balanoides* and daily settlement of cyprids was tested daily on a wooden substrate fixed vertically to a wharf at St. Andrews, New Brunswick, Canada, by Miron, Boudreau, and Bourget (1995). The pattern of settlement was assessed daily, followed by removal of the spat. Water samples at five different depths, including the neustonic surface, were pumped at $400\,L \cdot min^{-1}$ so that 2000 L was filtered through a 64-μm plankton net for each depth. The best correlation was found to be with bottom-collected larvae in the plankton and settled larvae in the low intertidal, supporting H_1 of number 4 in Table 3.8. Increased variability was found by Miron et al. (1995) if integrated larval abundance at all depths was used to correlate with settlement numbers.

Summary

From the foregoing overview of larval research of benthic suspension feeders, we conclude that larvae, rather than post-larvae, are the most important means of dispersal and recruitment. We identify two other passive dispersal methods, rafting and synantrophic transport, which may also be important and complementary in dispersal of some species or at particular times in their ontogeny.

During dispersal, a larva a few hundred micrometres in size may be transported distances >1000 km. Yet, during settlement, larval crawling over distances of a few millimetres may be crucial in allowing the larva to recruit successfully to the benthos, rather than being swept away by the flow. Because of these considerations, temporal and spatial scale are of primary importance in defining a field or laboratory study concerning larval biology.

The importance of physical oceanographic processes at scales >0.1 km during larval dispersal has been demonstrated by the work discussed herein. We can conclude that each of the alternate hypotheses in Table 3.3 is supported by some evidence, although the applicability of the result may be limited and localized in extent. The evidence for estuarine retention and shoreward advection – both of which presuppose appropriate directed behavioral responses by swimming larvae – is circumstantial. To render the evidence more direct, it is necessary to set up a field experiment with live larvae to determine their swimming patterns. Shanks (1995) conducted one such field experiment by releasing large decapod megalopa larvae, which were observed directly by SCUBA divers. He found that some species swam in surface seawater in the direction of the sun's bearing, while others swam near the surface and parallel to the prevailing flow direction. These swimming patterns of the decapod larvae helped them to remain in surface seawater, where they were transported shorewards by surface currents. It should be possible to extend this type of field experiment to smaller suspension-feeding larvae using live or dead individuals identified by vital staining or other means, so that they could act as models during dispersion studies.

Because of limitations imposed by the methods used, most larval studies reported from within plant canopies measured recruitment rather than settlement rates. In such studies, it is not possible to determine the cause of recruitment as settlement or early post-settlement mortality. Only the study by Wilson (1990), who measured larval catches on a daily basis, could be used to indicate larval settlement rates. The results suggest that it was indeed that larval settlement occurred at a higher rate within the eelgrass meadow than on nearby bare sand.

The possibility that small benthic juvenile post-larval stages are important in further dispersal has been confirmed in some bivalves but has not been adequately investigated in other species. Its importance is indicated by the degree to which it affects the productivity of populations, e.g. in mussels and clams (number 1, Table 3.6). Whether flow–sediment interaction influences juvenile mortality rate during dispersal (number 2, Table 3.6) does not appear to have been adequately considered, and this possibility remains to be tested.

The central question of both fundamental and applied research in larval biology is focussed on the environmental factors which influence initial settlement and recruitment at spatial scales of <0.1 km. The importance of both passive deposition and active habitat selection by soft-sediment benthic animals (Table 3.7) has been confirmed by both field and laboratory experiments. These experiments have mainly been done with bivalve and polychaete larvae. Both mechanisms are operating simultaneously but at different spatial scales (Butman 1987), with active habitat selection common at the lower end of the scale (micrometres to centimetres). The response of competent larvae to chemical inducers – either adsorbed on surfaces or free in seawater – has also been confirmed in laboratory experiments which include flow among the variables. With regard to small-scale protuberances, field and laboratory experiments strongly suggest that settlement is passively controlled over millimetre distances downstream from either tubes or bivalve exhalant jets. Thus, passive deposition mechanisms may also include very small-scale phenomena within the benthic boundary layer.

For hard substrate suspension feeders, the equivalent hypotheses (Table 3.8) to passive deposition (1) and active habitat selection include choice on the basis of surface roughness (2) and on the basis of chemical cues (3). Both field and laboratory experiments to test the passive deposition and active habitat selection hypotheses have been undertaken mainly with barnacle cyprids, although in some cases, coral reef larvae were the experimental subjects. Evidence supporting the alternate in the first three hypotheses of Table 3.8 is presented, and the importance of chemical and surface roughness cues confirmed in the presence of natural or simulated flows.

Evidence for a kin recognition mechanism operating among some bryozoan larvae at settlement, given by Keough (1984), explained aggregation for this species. There is further evidence from species which use chemicals or pheromones from conspecific larvae to aid in gregarious larval settlement. For solitary ascidians, Havenhand and Svane (1991) showed that hydrodynamic cues were important at settlement. The relative importance of these active and passive mechanisms among a wider range of suspension feeders which aggregate remains to be assessed. The subject of aggregation is further considered in Chapter 6, where Table 6.8 lists some hypotheses in which the mechanisms, enhanced settlement or differential post-settlement mortality rate, are formally proposed.

Considering the currently popular question concerning supply side ecology (number 4, Table 3.8), a number of authors have confirmed the alternative hypothesis that absolute larval supply rates control the subsequent density of barnacles, although we may expect species-specific responses here. The coincident measurement of larval supply, settlement, and early post-settlement mortality rates perhaps heralds a new and welcome quantitative approach to the ecology of early life history stages of suspension feeders.

References

Abelson, A., D. Weihs, and Y. Loya. 1994. Hydrodynamic impediments to settlement of marine propagules, and adhesive filament solutions. Limnol. Oceanogr. 39: 164–169.

Ackerman, J. D., B. Sim, S. J. Nichols, and R. Claud. 1994. A review of the early life history of zebra mussels (*Dreissena polymorpha*): comparisons with marine bivalves. Can. J. Zool. 72: 1169–1179.

Bachelet, G., C. A. Butman, C. M. Webb, V. R. Starczak, and P. V. R. Snelgrove. 1992. Non-selective settlement of *Mercenaria mercenaria* (L.) larvae in short-term, still-water laboratory experiments. J. Exp. Mar. Biol. Ecol. 161: 241–280.

Baggerman, B. 1953. Spatfall and transport of *Cardium edule* L. Arch. Neerland. Zool. 10: 315–342.

Bayne, B. L. 1964a. Primary and secondary settlement in *Mytilus edulis* L. (Mollusca). J. Anim. Ecol. 33: 513–523.

1964b. The responses of the larvae of *Mytilus edulis* L. to light and gravity. Oikos 15: 162–174.

1965. Growth and delay of metamorphosis of the larvae of *Mytilus edulis* (L.). Ophelia 2: 1–47.

1976. The biology of mussel larvae, p. 81–120. *In* B. L. Bayne (ed.) Marine mussels: their ecology and physiology. Cambridge University Press, Cambridge.

Bennell, S. J. 1981. Some observations on the littoral barnacle populations of North Wales. Mar. Environ. Res. 5: 227–240.

Bertness, M. D., S. D. Gaines, E. G. Stephens, and P. O. Yund. 1992. Components of recruitment in populations of the acorn barnacle *Semibalanus balanoides* (Linnaeus). J. Exp. Mar. Biol. Ecol. 156: 199–215.

Boicourt, W. C. 1982. Estuarine larval retention mechanisms on two scales, p. 445–457. *In* V. S. Kennedy (ed.) Estuarine comparisons. Academic Press, New York.

Bousfield, E. L. 1955. Ecological control of the occurrence of barnacles in the Miramichi estuary. Bull. Nat. Mus. Can. 137: 69.

Burke, R. D. 1986. Pheromones and the gregarious settlement of marine invertebrate larvae. Bull. Mar. Sci. 39: 323–331.

Bushek, D. 1988. Settlement as a major determinant of intertidal oyster and barnacle distributions along a horizontal gradient. J. Exp. Mar. Biol. Ecol. 122: 1–18.

Butman, C. A. 1986. Larval settlement of soft-sediment invertebrates: some predictions based on an analysis of near bottom velocity profiles, p. 487–513. *In* J. C. J. Nihoul (ed.) Marine interfaces ecohydrodynamics. Elsevier, Amsterdam.

1987. Larval settlement of soft-sediment invertebrates: the spatial scales of pattern explained by active habitat selection and the emerging role of hydrodynamical processes. Oceanogr. Mar. Biol. Annu. Rev. 25: 113–165.

1989. Sediment-trap experiments on the importance of hydrodynamical processes in distributing settling invertebrate larvae in near-bottom waters. J. Exp. Mar. Biol. Ecol. 134: 37–88.

Butman, C. A., and J. P. Grassle. 1992. Active habitat selection by *Capitella* sp. 1 larvae. I. Two-choice experiments in still water and flume flows. J. Mar. Res. 50: 669–715.

Butman, C. A., J. P. Grassle, and C. M. Webb. 1988. Substrate choices made by marine larvae settling in still water and in a flume flow. Nature 333: 771–773.

Carleton, J. C., and P. W. Sammarco. 1987. Effects of substratum irregularity on success of coral settlement: quantification of comparative geomorphological techniques. Bull. Mar. Sci. 40: 85–98.

Carlton, J. T., and J. B. Geller. 1993. Ecological roulette: the global transport of non-indigenous marine organisms. Science 261: 78–82.

Carlton, J. T., J. K. Thompson, L. E. Schemel, and F. H. Nichols. 1990. Remarkable invasion of San Francisco Bay (California, USA) by the Asian clam. I. Introduction and dispersal. Mar. Ecol. Prog. Ser. 66: 81–94.

Carriker, M. R. 1951. Ecological observations on the distribution of oyster larvae in New Jersey estuaries. Ecol. Monogr. 21: 19–38.

Chabot, R., and E. Bourget. 1988. Influence of substratum heterogeneity and settled barnacle density on the settlement of cypris larvae. Mar. Biol. 97: 45–56.

Chia, F-S. 1978. Perspectives: settlement and metamorphosis of marine invertebrate larvae, p. 283–285. *In* F-S. Chia and M. E. Rice (eds.) Settlement and metamorphosis of marine invertebrate larvae. Elsevier, New York.

Chia, F.-S., J. Buckland-Nicks, and C. M. Young. 1984. Locomotion of invertebrate larvae: a review. Can. J. Zool. 62: 1205–1222.

Connell, J. H., and R. O. Slatyer. 1977. Mechanisms of succession in natural communities and their role in community stability and organization. Am. Nat. 111: 1119–1144.

Cragg, S. M. 1980. Swimming behaviour of the larvae of *Pecten maximus* (L.) (Bivalvia). J. Mar. Biol. Assoc. U. K. 60: 551–564.

Crisp, D. J. 1955. The behaviour of barnacle cyprids in relation to water movement over a surface. J. Exp. Biol. 32: 569–590.

1974. Factors influencing the settlement of marine invertebrate larvae, p. 177–265. *In* P. T. Grant and A. M. Mackie (ed.) Chemoreception in marine organisms. Academic Press, London.

Crisp, D. J., and H. Barnes. 1954. The orientation and distribution of barnacles at settlement with particular reference to surface contour. J. Anim. Ecol. 23: 142–162.

Crisp, D. J., and P. S. Meadows. 1963. Adsorbed layers: the stimulus to settlement in barnacles. Proc. Roy. Soc. Ser. B. 156: 500–520.

DeVantier, L. M. 1992. Rafting of tropical marine organisms on buoyant coralla. Mar. Ecol. Prog. Ser. 86: 301–302.
DeWolf, P. 1973. Ecological observations on the mechanisms of dispersal of barnacle larvae during planktonic life and settling. Neth. J. Sea Res. 6: 1–129.
Duggins, D. O., J. E. Eckman, and A. T. Sewell. 1990. Ecology of understorey kelp environments. II. Effects of kelp on recruitment of benthic invertebrates. J. Exp. Mar. Biol. Ecol. 143: 27–45.
Dyer, K. R. (Editor). 1979. Estuarine hydrography and sedimentation: a handbook. Cambridge University Press, Cambridge.
Eckman, J. E. 1979. Small-scale patterns and processes in a soft-substratum, intertidal community. J. Mar. Res. 37: 437–457.
1983. Hydrodynamic processes affecting benthic recruitment. Limnol. Oceanogr. 28: 241–257.
1985. Flow disruption by an animal-tube mimic affects sediment bacterial colonization. J. Mar. Res. 43: 419–435.
1987. The role of hydrodynamics in recruitment, growth, and survival of *Argopecten irradians* (L.) and *Anomia simplex* (D'Orbigny) within eelgrass meadows. J. Exp. Mar. Biol. Ecol. 106: 165–191.
1990. A model of passive settlement by planktonic larvae onto bottoms of differing roughness. Limnol. Oceanogr. 35: 887–901.
Eckman, J. E., and A. R. M. Nowell. 1984. Boundary skin friction and sediment transport about an animal-tube mimic. Sedimentology 31: 851–862.
Eckman, J. E., W. B. Savidge, and T. F. Gross. 1990. Relationships between duration of cyprid attachment and drag forces associated with detachment of *Balanus amphitrite* cyprids. Mar. Biol. 107: 111–118.
Emerson, C. W., and J. Grant. 1991. The control of soft-shell clam (*Mya arenaria*) recruitment on intertidal sandflats by bedload sediment transport. Limnol. Oceanogr. 36: 1288–1300.
Ertman, S. C., and P. A. Jumars. 1988. Effects of bivalve siphon currents on the settlement of inert particles and larvae. J. Mar. Res. 46: 797–813.
Farrell, T. M., D. Bracher, and J. Roughgarden. 1991. Cross-shelf transport causes recruitment to intertidal populations in central California. Limnol. Oceanogr. 36: 279–288.
Fischer, H. B., J. Imberger, E. J. List, R. C. Y. Koh, and N. H. Brooks. 1979. Mixing in inland and coastal waters. Academic Press, New York.
Gaines, S. D., and M. D. Bertness. 1992. Dispersal of juveniles and variable recruitment in sessile marine species. Nature 360: 579–580.
Gaines, S., and J. Roughgarden. 1985. Larval settlement rate: a leading determinant of structure in an ecological community of the marine intertidal zone. Proc. Nat. Acad. Sci. U.S.A. 82: 3707–3711.
Gallagher, E. D., P. A. Jumars, and D. D. Trueblood. 1983. Facilitation of soft-bottom benthic succession by tube builders. Ecology 64: 1200–1216.
Grassle, J. P., C. A. Butman, and S. W. Mills. 1992a. Active habitat selection by *Capitella* sp. 1 larvae. II. Multiple-choice experiments in still water and flume flows. J. Mar. Res. 50: 717–743.
Grassle, J. P., P. V. R. Snelgrove, and C. A. Butman. 1992b. Larval habitat choice in still water and flume flows by the opportunistic bivalve *Mulina lateralis*. Neth. J. Sea Res. 30: 33–44.
Gray, J. S. 1974. Animal–sediment relationships. Oceanogr. Mar. Biol. Annu. Rev. 12: 223–261.

Hannan, C. A. 1984. Planktonic larvae may act like passive particles in turbulent near-bottom flows. Limnol. Oceanogr. 29: 1108–1116.
Harvey, M., E. Bourget, and R. G. Ingram. 1995. Experimental evidence of passive accumulation of marine bivalve larvae on filamentous epibenthic structures. Limnol. Oceanogr. 40: 94–104.
Harvey, M., E. Bourget, and G. Miron. 1993. Settlement of Iceland scallop spat (*Chlamys islandica*) in response to hydroids and filamentous red algae: field observations and laboratory experiments. Mar. Ecol. Prog. Ser. 99: 283–292.
Havenhand, J. N., and I. Svane. 1991. Roles of hydrodynamics and larval behaviour in determining spatial aggregation in the tunicate *Ciona intestinalis*. Mar. Ecol. Prog. Ser. 68: 271–276.
Hawkins, S. J., and R. G. Hartnoll. 1982. Settlement patterns of *Semibalanus balanoides* (L.) in the Isle of Man (1977–1981). J. Exp. Mar. Biol. Ecol. 62: 271–283.
Hebert, P. D. N., B. W. Mancaster, and G. L. Mackie. 1989. Ecological and genetic studies on *Dreissena polymorpha* (Pallas) a new mollusc in the Great Lakes. Can. J. Fish. Aquat. Sci. 46: 1587–1591.
Hidu, H., and H. H. Haskins. 1978. Swimming speeds of oyster larvae of *Crassostrea virginica* in different salinities and temperatures. Estuaries 1: 252–255.
Highsmith, R. C., and R. Emlet. 1986. Delayed metamorphosis: effect on growth and survival of juvenile sand dollars (Echinoidea: Clypeasteroida). Bull. Mar. Sci. 39: 347–361.
Holmquist, J. G. 1994. Benthic macroalgae as a dispersal mechanism for fauna: influence of a marine tumbleweed. J. Exp. Mar. Biol. Ecol. 180: 235–251.
Jonsson, P. R., C. André, and M. Lindegarth. 1991. Swimming behaviour of marine bivalve larvae in a flume boundary-layer flow: evidence for near-bottom confinement. Mar. Ecol. Prog. Ser. 79: 67–76.
Keough, M. J. 1984. Kin-recognition and the spatial distribution of larvae of the bryozoan *Bugula neritina* (L.). Evolution 38: 142–147.
Keough, M. J., and B. J. Downes. 1982. Recruitment of marine invertebrates: the role of active larval choices and early mortality. Oecologia 54: 348–352.
Korringa, P. 1952. Recent advances in oyster biology. Quart. Rev. Biol. 27: 266–308, 339–369.
Lacalli. T. 1981. Annual spawning cycles and planktonic larvae of benthic invertebrates from Passamaquoddy Bay, New Brunswick. Can. J. Zool. 59: 433–440.
LeFèvre, J., and E. Bourget. 1991. Neustonic niche for cirripede larvae as a possible adaptation to long-range dispersal. Mar. Ecol. Prog. Ser. 74: 185–194.
1992. Hydrodynamics and behaviour: transport processes in marine invertebrate larvae. Trends Ecol. Evol. 7: 288–289.
Lemire, M., and E. Bourget. 1996. Substratum heterogeneity and complexity influence micro-habitat selection of *Balanus* sp. and *Tubularia croeca* larvae. Mar. Ecol. Prog. Ser. 135: 77–87.
LeTourneux, F., and E. Bourget. 1988. Importance of physical and biological settlement cues used at different spatial scales by the larvae of *Semibalanus balanoides*. Mar. Biol. 97: 57–66.
Levin, L. A. 1983. Drift tube studies of bay–ocean water exchange and implications for larval dispersal. Estuaries 6: 364–371.

Luckenbach, M. 1984. Settlement and early post-settlement survival in the recruitment of *Mulinia lateralis* (Bivalvia). Mar. Ecol. Prog. Ser. 17: 245–250.

Maas Geesteranus, R. A. 1942. On the formation of banks by *Mytilus edulis* L. Arch. neerl. Zool. 6: 283–325.

Maida, M., J. C. Coll, and P. W. Sammarco. 1994. Shedding new light on scleractinian coral recruitment. J. Exp. Mar. Biol. Ecol. 180: 189–202.

Maki, J. S., D. Rittschof, J. D. Costlow, and R. Mitchell. 1988. Inhibition of attachment of larval barnacles, *Balanus amphitrite*, by bacterial surface films. Mar. Biol. 97: 199–206.

Maki, J. S., D. Rittschof, M.-O. Samuelsson, U. Szewzyk, A. B. Yule, S. Kjeuebert, J. D. Costlow, and R. Mitchell. 1990. Effects of marine bacteria and their exopolymers on the attachment of barnacle cypris larvae. Bull. Mar. Sci. 46: 499–511.

Mann, R. 1988. Distribution of bivalve larvae at a frontal system in the James River, Virginia. Mar. Ecol. Prog. Ser. 50: 29–44.

May, B., and J. E. Marsden. 1992. Genetic identification and implications of another invasive species of dressenid mussel in the Great Lakes. Can. J. Fish. Aquat. Sci. 49: 1501–1506.

McEdwards, L. (Editor). 1995. Ecology of marine invertebrate larvae. CRC Press, Boca Raton, Florida.

McGrath, D., P. A. King, and E. M. Gosling. 1988. Evidence for the direct settlement of *Mytilus edulis* larvae on adult mussel beds. Mar. Ecol. Prog. Ser. 47: 103–106.

Meadows, P. S., and J. I. Campbell. 1972. Habitat selection by aquatic invertebrates. Adv. Mar. Biol. 10: 271–382.

Menge, B. 1975. Brood or broadcast? The adaptive significance of different reproductive strategies in the intertidal seastars *Lepasterias hexactis* and *Piaster ochraceus*. Mar. Biol. 31: 87–100.

Miron, G., B. Boudreau, and E. Bourget. 1995. Use of larval supply in benthic ecology: testing correlations between larval supply and larval settlement. Mar. Ecol. Prog. Ser. 124: 301–305.

Miron, G., E. Bourget, and P. Archambault. 1996. Scale of observation and distribution of adult conspecifics: their influence in assessing passive and active settlement mechanisms in the barnacle *Balanus crenatus* (Brugière). J. Exp. Mar. Biol. Ecol. 201: 137–158.

Mullineaux, L. S., and C. A. Butman. 1990. Recruitment of encrusting benthic invertebrates in boundary layer flows: a deep-water experiment on Cross Seamount. Limnol. Oceanogr. 35: 409–423.

1991. Initial contact, exploration and attachment of barnacle (*Balanus amphitrite*) cyprids settling in flow. Mar. Biol. 110: 93–103.

Mullineaux, L. S., and E. D. Garland. 1993. Larval recruitment in response to manipulated field flows. Mar. Biol. 116: 667–683.

Neal, A. L., and A. B. Yule. 1994. The tenacity of *Elminius modestus* and *Balanus perforatus* cyprids to bacterial films grown under different shear regimes. J. Mar. Biol. Assoc. U.K. 74: 251–257.

Nicholls, F. H., and J. K. Thompson. 1985. Time scales of change in the San Francisco Bay benthos. Hydrobiologia 129: 121–138.

Okubo, A. 1971. Oceanic diffusion diagrams. Deep Sea Res. 18: 789–802.

1994. The role of diffusion and related physical processes in dispersal and recruitment of marine populations, p. 5–32. *In* P. W. Sammarco and M. L.

Heron (ed.) Coastal and estuarine studies. Vol. 45. The bio-physics of marine larval dispersal. American Geophysical Union, Washington, D.C.

Olafsson, E. B., C. H. Petersen, and W. G. Ambrose. 1994. Does recruitment limitation structure populations and communities of macro-invertebrates in marine soft sediments: the relative significance of pre- and post-settlement processes. Oceanogr. Mar. Biol. Annu. Rev. 32: 65–109.

Pawlik, J. R. 1986. Chemical induction of larval settlement and metamorphosis in the reef building tube worm *Phragmatopoma californica* (Polychaeta: Sabellaridae). Mar. Biol. 91: 59–68.

1992. Chemical ecology of the settlement of benthic marine invertebrates. Oceanogr. Mar. Biol. Annu. Rev. 30: 273–335.

Pawlik, J. R., and C. A. Butman. 1993. Settlement of a marine tube worm as a function of current velocity: Interacting effects of hydrodynamics and behaviour. Limnol. Oceanogr. 38: 1730–1740.

Pawlik, J. R., C. A. Butman, and V. R. Starczak. 1991. Hydrodynamic facilitation of gregarious settlement of a reef-building tube worm. Science 251: 421–424.

Peterson, C. H. 1982. Clam predation by whelks (*Busycon* spp.): experimental tests of the importance of prey sizes, prey density, and seagrass cover. Mar. Biol. 66: 159–170.

1986. Enhancement of *Mercenaria mercenaria* densities in seagrass beds: is pattern fixed during settlement season or altered by subsequent differential survival? Limnol. Oceanogr. 31: 200–205.

Pineda, J. 1991. Predictable upwelling and the shoreward transport of planktonic larvae by internal tidal bores. Science 253: 548–551.

Pratt, D. M. 1953. Abundance and growth of *Venus mercenaria* and *Callocardia morrhiana* in relation to the character of bottom sediments. J. Mar. Res. 12: 60–74.

Pyefinch, K. A. 1948. Methods of identification of the larvae of *Balanus balanoides* (L.), *B. crenatus* Brug. and *Verruca stoemia* O. F. Muller. J. Mar. Biol. Assoc. U.K. 27: 451–463.

Ricciardi, A., R. Serrouya, and F. G. Whoriskey. 1995. Aerial exposure tolerance of zebra and quagga mussels (Bilvalvia: Dreissenidae): implications for overland dispersal. Can. J. Fish. Aquat. Sci. 52: 470–477.

Rittschof, D., E. Sansford Branscomb, and J. D. Costlow. 1984. Settlement and behavior in relation to flow and surface in larval barnacles, *Balanus amphitrite* Darwin. J. Exp. Mar. Biol. Ecol. 82: 131–146.

Roegner, C., C. André, M. Lindegarth, J. E. Eckman, and J. Grant. 1995. Transport of recently settled soft-shell clams (*Mya arenaria* L.) in laboratory flume flow. J. Exp. Mar. Biol. Ecol. 187: 13–26.

Roughgarden, J., S. D. Gaines, and S. W. Pacala. 1987. Supply side ecology: the role of physical transport processes, p. 491–518. *In* J. H. R. Gee and P. S. Giller (ed.) Organization of communities past and present. Blackwell Scientific Press, Oxford.

Sammarco, P. W. 1991. Geographically specific recruitment and post-settlement mortality as influences on coral communities: the cross continental shelf transplant experiment. Limnol. Oceanogr. 36: 496–514.

Sammarco, P. W., and M. L. Heron (Editors). 1994. The bio-physics of marine larval dispersal. American Geophysical Union, Washington, D.C.

Scheltema, R. S. 1974. Biological interactions determining larval settlement of marine invertebrates. Thal. Jugosl. 10: 263–296.

1986. On dispersal and planktonic larvae of benthic invertebrates: an eclectic overview and summary of problems. Bull. Mar. Sci. 39: 290–322.

Seliger, H. H., J. A. Boggs, R. B. Rivkin, W. H. Biggley, and K. R. H. Aspden. 1982. The transport of oyster larvae in an estuary. Mar. Biol. 71: 57–72.

Shanks, A. L. 1983. Surface slicks associated with tidally forced internal waves may transport pelagic larvae of benthic invertebrates and fishes shoreward. Mar. Ecol. Prog. Ser. 13: 311–315.

1985. The behavioural basis of internal wave-induced shoreward transport of the megalopae of *Pachygrapsus crassipes*. Mar. Ecol. Prog. Ser. 24: 289–295.

1986. Tidal periodicity in the daily settlement of intertidal barnacle larvae and an hypothesized mechanism for the cross shelf transport of cyprids. Biol. Bull. 170: 429–440.

1988. Further support for the hypothesis that internal waves can cause shoreward transport of larval invertebrates and fish. Fish. Bull. 86: 703–714.

1995. Orientated swimming by megalopae of several eastern North Pacific crab species and its potential role in their onshore migration. J. Exp. Mar. Biol. Ecol. 186: 1–16.

Shanks, A. L., and E. W. Edmondson. 1990. The vertical flux of metazoans (holoplankton, meiofauna, and larval invertebrates) due to their association with marine snow. Limnol. Oceanogr. 35: 455–463.

Shanks, A. L., and W. G. Wright. 1987. Internal-wave mediated shoreward transport of cyprids, megalopae, gammarids and correlated longshore differences in the settling rate of intertidal barnacles. J. Exp. Mar. Biol. Ecol. 114: 1–13.

Sigurdsson, J. B., C. W. Titman, and P. A. Davies. 1976. The dispersal of young post-larval bivalve molluscs by byssus threads. Nature 262: 386–387.

Snelgrove, P. V. R. 1994. Hydrodynamic enhancement of invertebrate larval settlement in microdeposition environments: colonization tray experiments in a muddy habitat. J. Exp. Mar. Biol. Ecol. 176: 149–166.

Snelgrove, P. V. R., C. A. Butman, and J. P. Grassle. 1993. Hydrodynamic enhancement of larval settlement in the bivalve *Mulina lateralis* (Say) and the polychaete *Capitella* sp. 1 in microdepositional environments. J. Exp. Mar. Biol. Ecol. 168: 71–109.

Stancyk, S. E., and R. J. Feller. 1986. Transport of non-decapod invertebrate larvae in estuaries: an overview. Bull. Mar. Sci. 39: 257–268.

Starr, M., J. H. Himmelman, and J.-C. Therriault. 1990. Marine invertebrate spawning induced by phytoplankton. Science 247: 1071–1074.

Stoner, A. W., and R. J. Livingston. 1989. Distributional ecology and food habits of the banded blenny *Parclinus fasciatus* (Clinidae): a resident in a mobile habitat. Mar. Ecol. 56: 239–246.

Strathmann, R. R. 1985. Feeding and non-feeding larval development and life history evolution in marine invertebrates. Annu. Rev. Evol. Syst. 16: 339–361.

Theodor, J. 1967. Contributions a l'étude des gorones (VII): écologie et comportament de la planula. Vie Milieu Ser. A 18: 291–301.

Thiébaut, E., J. C. Dauvin, and Y. Lagadeuc. 1992. Transport of *Owenia fusiformis* larvae (Annelida: Polychaeta) in the Bay of Seine. I. Vertical distribution in relation to water column stratification and ontogenic vertical migration. Mar. Ecol. Prog. Ser. 80: 29–39.

Thorson, G. 1950. Reproduction and larval ecology of marine bottom invertebrates. Biol. Rev. Cambridge Philos. Soc. 25: 1–45.

Tremblay, M. J., and M. Sinclair. 1994. Drift of sea scallop larvae *Placopecten magellanicus* on Georges Bank: a model study of the roles of mean advection, larval behaviour and larval origin. Deep Sea Res. Part II 41: 7–50.

Turner, E. J., R. K. Zimmer-Faust, M. A. Palmer, M. Luckenbach, and N. D. Peutcheff. 1994. Settlement of oyster (*Crassostrea virginica*) larvae: effects of water flow and a water soluble chemical cue. Limnol. Oceanogr. 39: 1579–1593.

Tyler, P. A., and F. T. Banner. 1977. The effect of coastal hydrodynamics on the echinoderm distribution in the sublittoral of Oxwich Bay, Bristol Channel. Est. Coastal Mar. Sci. 5: 293–308.

Underwood, A. J., and P. C. Fairweather. 1989. Supply side ecology and benthic marine assemblages. Trends Ecol. Evol. 4: 16–20.

Walters, L. J. 1992. Field settlement locations on subtidal marine hard substrata: Is active larval exploration involved? Limnol. Oceanogr. 37: 1101–1107.

Wethey, D. S. 1984. Spatial patterns in barnacle settlement: day to day changes during the settlement season. J. Mar. Biol. Assoc. U.K. 64: 687–698.

1986. Ranking of settlement cues by barnacle larvae: influence of surface contour. Bull. Mar. Sci. 39: 393–400.

Wildish, D. J. 1977. Factors controlling marine and estuarine sublittoral macrofauna. Helgolander. Wiss. Meeresunters. 30: 445–454.

Wilson, F. S. 1990. Temporal and spatial patterns of settlement: a field study of molluscs in Boque Sound, North Carolina. J. Exp. Mar. Biol. Ecol. 139: 201–220.

Young, C. M. 1990. Larval ecology of marine invertebrates: a sesquicentennial history. Ophelia 32: 1–48.

Zimmer-Faust, R. K., and M. N. Tamburri. 1994. Chemical identity and ecological implications of a waterborne, larval settlement cue. Limnol. Oceanogr. 35: 1075–1087.

4

Flow and the physiology of filtration

In this chapter, our concern is primarily with the first filtration stage of feeding as defined later. In the next chapter, we deal with the second stage of suspension feeding, which involves the mechanisms of particle capture on the appropriate surface. Because these two stages of feeding are the only ones directly affected by flow, they are the only ones considered in physiological detail in this book.

Background

In suspension-feeding animals, feeding and growth involves a series of coordinated steps. Each of these steps is dependent on the preceding one (Table 4.1). We will refer to suspension feeding as inclusive of the first four steps shown and reserve the term *filtration* for the first two only. Filtration or clearance rates (see Chap. 2, the section, Initial feeding responses) need not, therefore, equal feeding rates if particle sorting leads to rejection as pseudofaeces. Growth encompasses the remaining catabolic and anabolic steps. Growth rates need not be directly related to the feeding rate because of differential digestion of sestonic particles. For example, in the giant scallop, some microalgae pass through the gut undigested (Shumway et al. 1985b). Residence times of food in the gut may be increased or decreased (Bricelj, Bass, and Lopez 1984) and thus change the contact time between ingested food, and digestive and absorptive surfaces. Thus, filtration rate is not necessarily equal to feeding rate and feeding rate need not always be directly proportional to the growth rate.

Two major categories of suspension feeding benthic animals can be recognized by the extent to which the filtration process is dependent on external flow (Vogel 1981; LaBarbera 1984):

Table 4.1. *Stages in the physiology of feeding, assimilation, growth, and elimination in suspension feeding benthic animals.*

Stage	Stage number	Description
Feeding	1	– Transport of seawater past the filtration surface
	2	– Capture of seston at the filtration surface
	3	– Transport of particles to the mouth involving sorting and rejection, e.g. as pseudofaeces in bivalves
	4	– Ingestion
Assimilation	5	– Transport through the gut
	6	– Uptake across specific gut surfaces
Growth	7	– Anabolic metabolism
	8	– Catabolic metabolism
Elimination	9	– Waste solids voided as true faeces
	10	– Metabolic wastes voided as urine

- *Passive suspension feeders* – the individual is solely dependent on the external ambient flow to bring seston close enough to the filtration surface for its capture
- *Active suspension feeders* – the individual must supply its own energy in the form of ciliary or muscular pump power to transport seawater with its load of seston across the filtration surface where particle capture is effected

In addition, there are *facultative active suspension feeders*, where individuals are able to switch from passive to active feeding and back again, depending on the ambient flow conditions. These individuals behave as passive suspension feeders at high velocities and as active suspension feeders at low velocities. In another category, the *combined passive–active suspension feeders*, the individual has two parallel mechanisms, one passive and one active, for creating a flow through the body for feeding purposes. The latter group differs from the facultative active suspension feeder in that there is no switching at critical velocities and both active and passive filtration may occur simultaneously. The passive mechanism involves ambient velocity induced flow, and the active mechanism involves a ciliary or flagellar pump. This is a new category, not explicitly considered by Vogel and LaBarbera, and tentatively includes some ascidians, sponges, and, possibly, brachiopods. Finally, there are the *deposit-suspension feeders*, individuals that ingest already deposited sedimentary particles at low velocities, but switch to suspension feeding at a higher threshold velocity. Examples include spionid polychaetes and some infaunal bivalve molluscs.

We view these generalizations as helpful in analyzing the wide diversity of methods of suspension feeding to be found among benthic animals. As further detailed knowledge of suspension-feeding mechanisms accumulates from a wider range of benthic animals, we can expect further modification of the categories presented. We feel that creating too many categories is counter-productive because, eventually, the process will be competing with taxonomy. It is not the artificial categories that developed new suspension-feeding mechanisms, but populations from individual species (Chap. 9).

Expected filtration responses of suspension feeders to increases in unidirectional flows are shown in Fig. 4.1. In this representation, it is assumed that the suspension feeder is optimally positioned with respect to flow direction. The filtration responses appear to be some form of unimodal function of velocity. In earlier studies this response was described as a "reverse ramp" for part of b and c (Wildish and Saulnier 1993). Since this engineering term is unfamiliar to biologists, we have replaced it with the more general term *continuous unimodal function*. Its use does not imply a precise mathematical relationship between filtration and velocity. For filtration responses and velocity, parabolic, orthogonal, sharp peaked, or ramp functions may be obtained, depending on the species and environmental conditions considered. Given constant seston concentrations, increments in velocity enhance filtration (Fig. 4.1a) at very low velocities up to a maximum (Fig. 4.1b) beyond which velocity increase has no further effect on filtration because other processes (e.g. stage 3 or 4; see Table 4.1) limit it. An example of suspension feeding in Protozoa is given by Fenchel (1980); in it particle transport after capture and the rate of ingestion may be the rate-limiting factors. Bivalves have a pseudofaecal rejection mechanism that avoids overloading ingestion (stage 4) and assimilation (stages 5 and 6). At higher velocities (Fig. 4.1c), flow conditions begin to have a negative effect on the filtration process and finally filtration (Fig. 4.1d) is completely suppressed, e.g. by mantle edge closure or withdrawal of siphons in infaunal bivalve molluscs. The precise mechanisms by which the filtration responses are governed depend on the species of suspension feeder, as will be discussed in some examples in later sections of this chapter. Thus, in active suspension feeders such as siphonate bivalves, in which the inhalant is normal to the flow, flow-induced inhibition may be absent, while in non-siphonate bivalves, ambient flow inhibits seston capture (Fig. 4.1). In

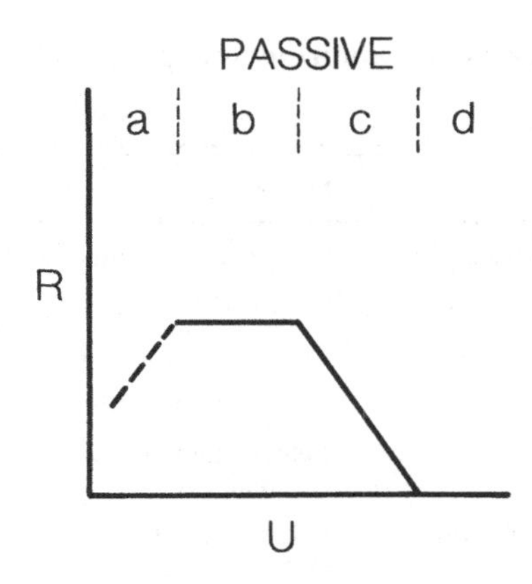

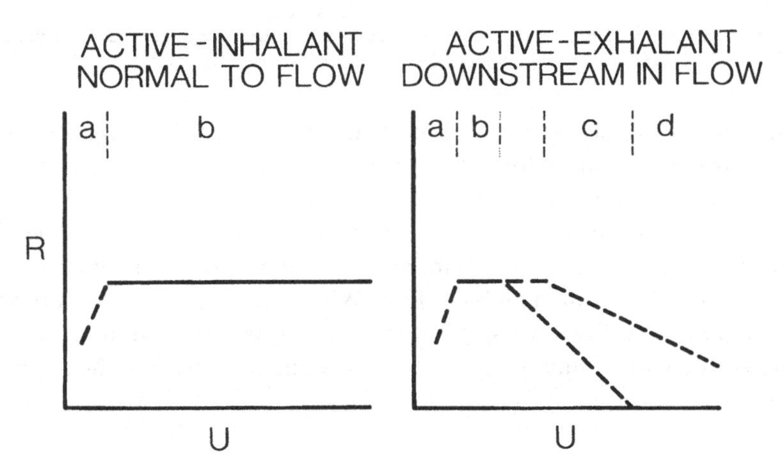

Figure 4.1 Effects of velocity, *U*, on filtration rate, *R*: a, increasing velocity enhances filtration; b, optimum filtration rates; c, velocity inhibition of filtration; d, filtration suppressed by velocity.

general, in stage a, filtration is limited by seston supply as a result of seston dilution and limited transport; in b, filtration is not limited by seston dilution and transport; and in c, filtration is inhibited by excessive ambient flow forces that interfere with seston capture processes.

Seston concentration and quality may also have an influence on the filtration rate, which may be either independent of or interactively linked to velocity; this is the case, for example, for bivalve molluscs (Wildish, Kristmanson, and Saulnier 1992). These responses permit active suspen-

Table 4.2. *Representative list of macrobenthic passive suspension feeders.*

Common name	Scientific name	Major taxa	Reference
Sea pen	*Ptilosarcus guerneyi*	Cnidaria: Pennatulacea	Best (1988)
Hydroid	*Plumilaria setacea*	Cnidaria: Hydrozoa	Warner (1977)
Sea whip	*Leptogorgia virgulata*	Cnidaria: Anthozoa	Leversee (1976)
Black coral	*Cirrhipathes lutkeni*	Cnidaria: Antipatharia	Warner (1977)
Feather star	*Oligometra serripinnia*	Echinodermata: Crinoidea	Leonard et al. (1988)
Brittle star	*Ophiothrix fragilis*	Echinodermata: Ophiuroidea	Warner and Woodley (1975)
Sea cucumber	*Cucumaria curata*	Echinodermata: Holothuroidea	Jørgensen (1966)

sion feeders rapidly to increase both the time spent filtering and the volumetric throughput for filtration to take advantage of temporally discrete pulses of good quality sestonic food.

For active suspension feeders, the filtration response also depends critically on whether the exhalant siphon flow follows or is opposed to the principal direction of ambient flow. Where the exhalant flow opposes the ambient flow, the adverse pressure field causes a reduction in filtered flow, as extraciliary pump work cannot be done to overcome the applied external flow force.

Passive suspension feeders

Passive suspension feeders are found in many of the phyla which have benthic representatives. Table 4.2 gives an idea of the diversity of the taxonomic forms represented. A detailed hydrodynamic study of filtration responses to velocity and the mechanism of seston capture is available for only a few passive suspension feeders.

Cnidarians

The sea pen *Ptilosarcus gurneyi* is a common cnidarian of soft, sandy-mud sediments in Puget Sound and the San Juan Archipelago of the U.S. West Coast (Best 1988). Its body consists of a central supporting peduncle inserted in the sediment, a joint, and a rachis. The sea pen is oriented so that its flat surface faces the major prevailing flow. A parallel series of semi-circular leaves carries the filtering elements. These filtering elements are polyps with radiating tentacles bearing pinnules which are located on the downstream part of each leaf. The rachis is flexible and bends downstream as ambient velocities increase.

Volume flow rates through *P. gurneyi* have been determined by estimating the rate of dye movement to determine velocity and a measure of the rachis area from the dye-stained part of the sea pen (Best 1988). Volumetric flow rates were related to the square of the height since the filter area increased linearly with size. The maximum volumetric flow through the rachis occurred at an ambient velocity which was a function of sea pen size: thus 6.5–8.5 $cm \cdot s^{-1}$ for small, 12–14 $cm \cdot s^{-1}$ for medium, and 14–18 $cm \cdot s^{-1}$ for large sea pens.

Velocity measurements between the leaves and near the polyps made with a thermistor flow probe showed that near the polyps velocities were much reduced at the highest free-stream flows tested (25 $cm \cdot s^{-1}$) as a result of drag near the filtration surfaces.

Filtration efficiency (*RE*; see Chap. 2) was examined in flume experiments by taking seawater samples upstream and downstream of the sea pen (with a suction sampler so that sampling was possibly not isokinetic). The sample was examined in a particle counter and the efficiency of capture of each particle size determined. Smaller sea pens were significantly less efficient than medium or large individuals. Free stream velocities ranging from 1.5 to 6.0 $cm \cdot s^{-1}$ significantly reduced RE.

Filtration and seston uptake rates, determined from independent estimates of volume flow rate and filtering efficiency as outlined, were equal in *P. gurneyi* and unimodal functions of velocity (Fig. 4.2). For the sea pen, increasing velocity in low flow conditions enhances the filtration seston uptake rate because of the increase in volumetric flow through the filter. The plateau reached around 7 $cm \cdot s^{-1}$ (Fig. 4.2) occurs as ambient velocity just begins to deform the sea pen. This process eventually causes both volume flows and filtering efficiency to decline in the c stage because pinnules and tentacles are swept back as the flow increases and, consequently, less filtering surface is presented to the flow. For small sea pens, the c stage begins at 13–16 $cm \cdot s^{-1}$ and, in medium and larger ones, at 18–21 $cm \cdot s^{-1}$. Filtration/seston uptake rate inhibition in *P. gurneyi* can be seen to begin in Fig. 4.2 but, because the highest velocities tested were only 25 $cm \cdot s^{-1}$, the full unimodal function responses were not obtained.

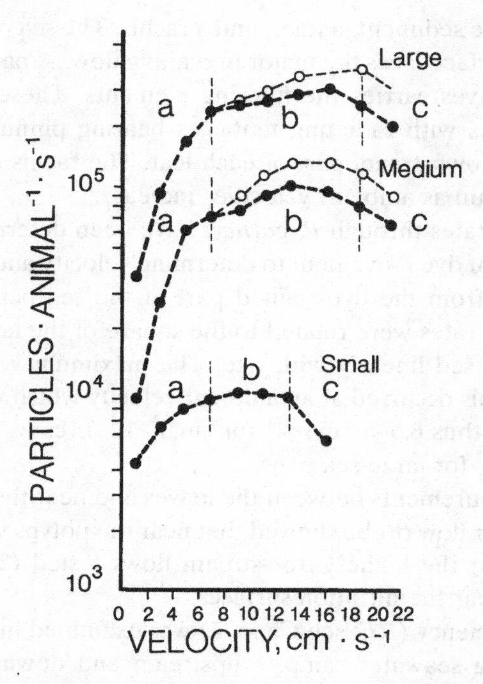

Figure 4.2 Seston uptake rate in *Ptilosarcus gurneyi* as a function of ambient velocity: open circles, retention efficiency at 30%; closed circles, at 20%. The unimodal response limits are suggested (modified from Best 1988).

No test was made of the effect of seston concentration on capture rates during this study (Best 1988).

The gorgonian corals exhibit whole colony flexibility where water movement forces reach moderate to high levels. Two species, *Pseudopterogorgia acerosa* and *P. americana*, found off Jamaica, were studied by Sponaugle and LaBarbera (1991). Coral feeding was studied in a recirculating flume in unidirectional flows measured within 2 mm of the polyp using a thermistor flow probe with a sensing head of 0.5 mm. *Artemia* nauplli were used as food at concentrations of 2.5–4 nauplii $\cdot L^{-1}$. Captured rhodamine-stained nauplii were easily observed in the individual polyp gastrovascular cavity. Results showed that the capture rate of *Artemia* was a unimodal function of velocity with a peak at intermediate velocities (~15 $cm \cdot s^{-1}$).

Model studies of the individual polyp of the tropical gorgonian *Pseudopterogorgia acerosa* were made by Sponaugle (1991). She noticed that, in addition to colony flexion, which keeps flow velocities near the polyps low, individual polyps are deformed by increased velocities. Two effects were noticed: the first that polyp flexion helped maintain reduced and constant velocities near the polyp with the effect most marked at intermediate flows; the second that flow patterns downstream of the polyp at intermediate ambient flows suggested an upward circulation, similar to that in tube-living suspension feeders discussed in Chapter 6 (see the section, Tube-living epifauna) which enhanced particle capture. Both mechanisms contribute to the unimodal feeding response of this gorgonian to flow described by Sponaugle and LaBarbera (1991).

Studies of particle capture by individual polyps of the octocoral *Alcyonium siderium* collected from either the Massachusetts or the California coast were made in a cooled, recirculating flume at three flows: 2.7, 12.2, or 19.8 $cm \cdot s^{-1}$ (Patterson 1991). These flows were measured with a thermistor flowmeter probe positioned 1 cm above the coral polyp. It was shown at the lowest flow tested that *Artemia* cysts were captured on the upstream tentacles of the *Alcyonium* polyp, but that at flows of 12.2 $cm \cdot s^{-1}$ or greater, capture was transferred to the downstream tentacles. At the highest flow tested, the number of cysts captured per polyp per unit time was significantly less than at the lower flows. These observations are consistent with the unimodal feeding response of other suspension feeders to flow, although there are insufficient data for *A. siderium* to determine the precise velocity limits. At the 19.8-$cm \cdot s^{-1}$ flow the polyps are bent downstream and eddies form over the tentacular surfaces. At these higher flows, the cysts are captured all over the tentacles in a radially symmetrical pattern and move towards the tentacle tip, although capture efficiency decreases with increasing Reynolds number.

Filtration efficiency of three species of gorgonian corals were shown by Dai and Lin (1993) in recirculating flume experiments to be a unimodal response to velocity with a maximum efficiency near 8 $cm \cdot s^{-1}$ (Fig. 4.3). A wide range of velocities was tested during these experiments with *Subergorgia suberosa* feeding limited to the range 7–9 $cm \cdot s^{-1}$, *Melthaea ochracea* to 4–40 $cm \cdot s^{-1}$, and *Acanthogorgia vegae* to 2–22 $cm \cdot s^{-1}$. Dai and Lin (1993) found that the typical feeding range for a species was related to the polyp size and its deformability in flow.

Echinoderms

The stalkless crinoid *Oligometra serripinnia* from the Great Barrier Reef was used in experimental flume tests by Leonard, Strickler, and Holland (1988). This crinoid lives at 10–15 cm above the sediment–water interface

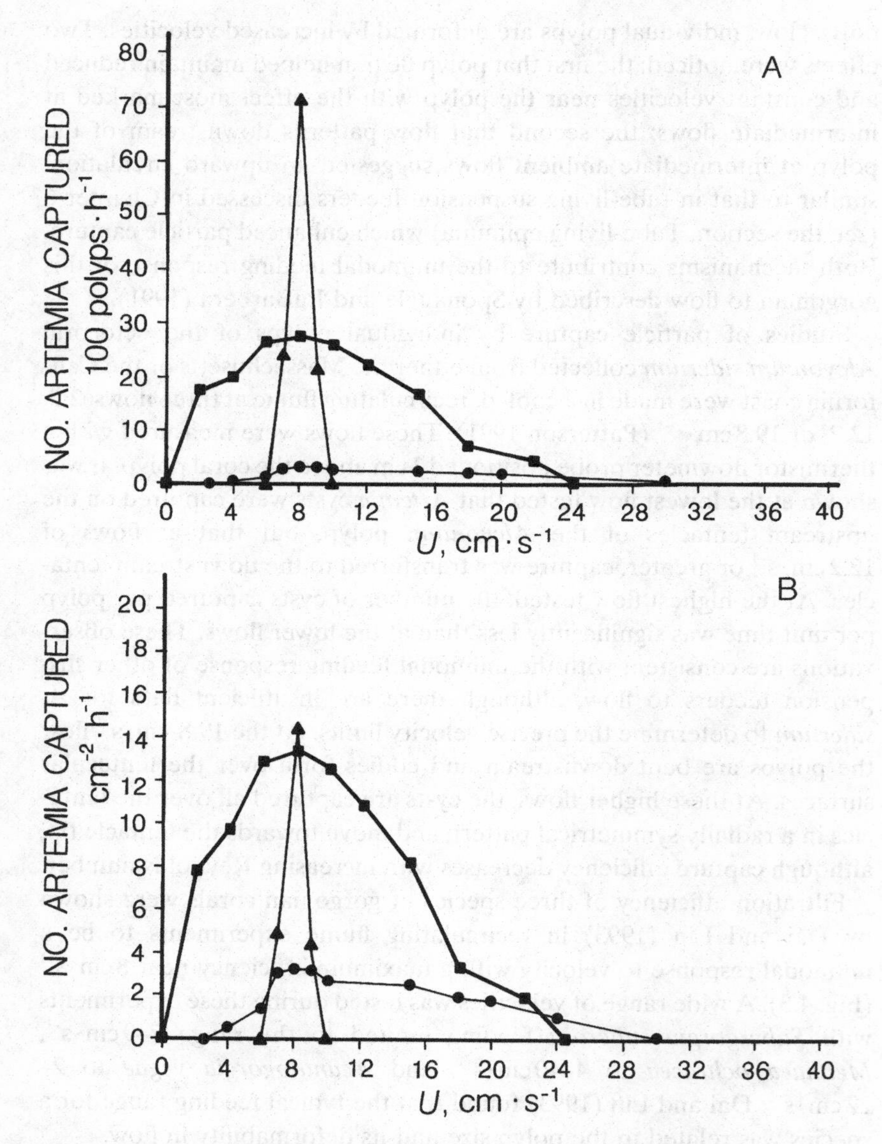

Figure 4.3 Normalized feeding effectiveness with respect to ambient velocity: ■, *Acanthogorgia vegae*; •, *Melithaea ochracea*; and ▲, *Subergorgia suberosa*; A, polyp feeding rates; B, feeding rate per unit surface area (Dai and Lin 1993).

as an ectocommensal on sea fans, where it occupies the downstream side of its host. The adult animals used in the experiments described here were star-shaped with 10 radiating arms, each ~3 cm in length. Each arm had 55–80 pinnules with 40–60 tube feet per pinnule for feeding. There was a relatively wide gap between arms but smaller distance (up to 0.69 mm) between pinnules. At higher flows, the distance between pinnules was a function of velocity due to flow deformation.

Velocity was determined by Leonard et al. (1988) to affect filtration in the following ways:

- By changing the rate of encounters with seston particles
- By affecting the size and quality of available seston
- By changing the fine-scale flow regime near the filter
- By changing the shape of the filter
- By changing the efficiency of capture by tube feet and pinnules

Capture efficiency rates were determined independently in this study by tracking individual particles by dark field photography so that seston approach, capture, and escape rates as a function of velocity (Fig. 4.4) could be determined. The results suggest that for the adult size crinoids used ($n = 4$), the following unimodal responses occur:

a. 0.9–4.8 $cm \cdot s^{-1}$: rate of encounter with seston low but capture efficiency high
b. 4.8–6.4 $cm \cdot s^{-1}$: optimum flow rate with rate of encounter with seston higher and capture efficiencies still high
c. >6.4 $cm \cdot s^{-1}$: highest rate of encounter with seston but high flows causing deformation of the filter and the capture efficiency lower

The maximum velocity tested by Leonard et al. (1988) was 13.3 $cm \cdot s^{-1}$, at which the absolute approach rate of seston particles for a whole crinoid had reached an asymptotic level.

Studies by Leonard (1989) on another stalkless crinoid, *Antedon mediterranea*, found in the Mediterranean Sea off Monaco, included measurement of both velocity and seston concentration to determine the effect on feeding responses. Feeding activity was recorded by the dark field illumination technique (see Chap. 2), focused on the pinnules bearing approximately 70 tube feet. The feeding rate was assumed to be directly proportional to the number of single or multiple tube foot flicks made when a sestonic particle was captured by the filter. The crinoids in a small capacity Vogel–LaBarbera recirculating flume were subjected to a cycle of increasing step changes of velocity (0.4–3.7 $cm \cdot s^{-1}$) followed by decreasing step changes of velocity (3.7–0.4 $cm \cdot s^{-1}$) for a wide range of seston concentrations (25–10,000 cells of *Hymenomonas elongata*/ml). This microalga is a spherical coccolithophore of ~11 μm diameter. The

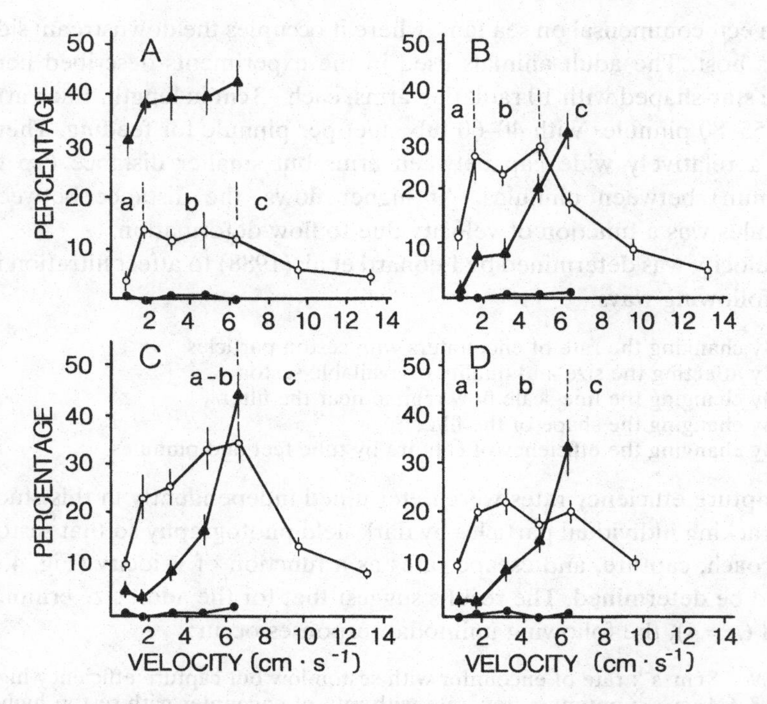

Figure 4.4 *Oligometra serripinnia* capturing pollen grains at different ambient velocities in four individuals (A–D): ▲, passes; ○, captures; •, escapes as a percentage of those possible. The unimodal function limits are suggested (modified from Leonard et al. 1988).

optimum concentration for filtration was 2000 cells·ml^{-1}, beyond which further increases caused a progressive decrease in feeding activity (flick responses). For each concentration tested, there was a maximum flick activity at 2.3 cm·s^{-1} (a). Beyond 2.3 cm·s^{-1}, feeding rate declined (c), suggesting a limited b region. Statistical analysis of the data by analysis of variance suggested that flick activity depends on velocity and particle concentration, but not on the rate at which particles approach the filter. Thus, one cannot assume that filtration rate is the product of velocity and seston concentration.

Active suspension feeders

Bivalve molluscs are perhaps best described as active suspension feeders, although some are deposit feeders and others are facultative suspension–

Table 4.3. *Representative macrobenthic active suspension feeders.*

Common name	Scientific name	Major taxa	Reference
Scallop	*Placopecten magellanicus*	Bivalvia	Wildish et al. (1987)
Mussel	*Mytilus edulis*	Bivalvia	Walne (1972)
Winkle	*Serpulorbis natalensis*	Gastropoda	Hughes (1978)
Bryozoan	*Flustrellidra hispida*	Bryozoa	Best and Thorpe (1983)
Sea squirt	*Ciona intestinalis*	Ascidiacea	Flood and Fialo-Médioni (1981)
Worm	*Chaetopterus variopedatus*	Polychaeta	Flood and Fialo-Médioni (1982)
Sand dollar	*Dendrastus excentricus*	Echinodermata	Timko (1976)

deposit feeders. Examples of taxa which are thought to be active suspension feeders are listed in Table 4.3. Besides bivalves, they include Bryozoa and sea squirts, plus a few species of echinoderms, polychaetes, and gastropods.

In most cases, the pump which moves seawater in the trophic fluid transport system is a ciliary one. Only in a few polychaete species, such as *Chaetopterus variopedatus*, is a muscular piston pump employed.

Bivalves

Although many early bivalve biologists (e.g. Kerswill 1949) recognized the importance of flow in bivalve production, the precise mechanisms by which this was achieved were not addressed until relatively recently. Kirby-Smith (1972) was the first to study the effect of flow on bay scallops (*Argopecten irradians*) in a growth tube apparatus (Chap. 2). Flow was measured as the bulk velocity flow rate. Kirby-Smith (1972) was also the first to demonstrate that higher velocities could inhibit bay scallop growth. Wildish et al. (1987) have reinterpreted his growth results as a unimodal function of velocity, in general agreement with later studies made in a flume with this bay scallop (Cahalan, Siddall, and Luckenbach 1989; Eckman, Peterson, and Cahalan 1989). Growth results with the giant scallop (*Placopecten magellanicus*) in flume experiments also show that the growth of this species has a unimodal function response to velocity. Flume studies (Wildish and Saulnier 1993) where a wide range of velocities were tested confirm that filtration by the giant scallop is a unimodal function of velocity, suggesting that it is the relationship underlying the growth responses.

Considering the possible mechanisms involved in the giant scallop filtration response to velocity and using the labelling of Fig. 4.1, the following observations can be made:

a. $U = 0$ to ~$3\,cm \cdot s^{-1}$. At zero velocity, the filtration rate is time dependent and, because of mixing limitations, usually affected by the geometry of the experimental apparatus. The exhalant jet itself, consisting as it does of seawater cleared of seston, and in otherwise static conditions will cause some mixing and result in seston dilution near the inhalant. As the seston concentration declines, inhibition of filtration may result. The effect of small increments in velocity is to help remove the localized buildup of seston-diluted seawater around the scallop.
b. Beyond the point where the ambient flow is sufficient to remove completely any localized seston dilution around the scallop, flow has no further influence on filtration. The flat-topped unimodal response seen and lack of a continuing increase in filtration rate with increasing velocity argue against the latter being used for enhancement of filtration rates, as is suggested by LaBarbera (1977) in brachiopods.
c. As ambient velocity further increases, the pressure field around the animal will be altered. The difference between ambient pressure at the inhalant (P_A) and the exhalant (P_B) may become significant with respect to the pressure difference generated by the ciliary pump. If $P_A \gg P_B$, seawater will tend to be forced through the system at a rate greater than can be handled by the ciliary pump filter, as described by Jørgensen et al. (1986a). This causes a behavioral response involving valve and mantle wall partial closure, thereby reducing the volume that can be processed. The internal homeorheostat (Wildish and Saulnier 1993) can to some extent overcome the closing response when seston concentration is sufficiently high (Fig. 4.5).
d. At high ambient velocities, the siphons retract and close because they cannot filter. This velocity is not fixed but variable for the scallop because seston concentration and other environmental variables, such as temperature, can influence the endpoint.

Behavioral responses of a bivalve to external pressure fields which oppose the internal pressure field created by the ciliary pump can be expected in phases c and d. The ciliary pump in some bivalve molluscs can generate pressure differences of the order of 0.1–2.0mm of water (Foster-Smith 1976). If the external pressure differences are of the same order of magnitude, it can be predicted that the flow through the pump, in the absence of behavioral response, will be either much reduced or considerably enhanced, depending on the sign of the pressure differential.

If the scallop is oriented with the inhalant pointing upstream, the pressure difference between the inhalant and the exhalant is given approximately by the stagnation pressure. The streamlined shape of the

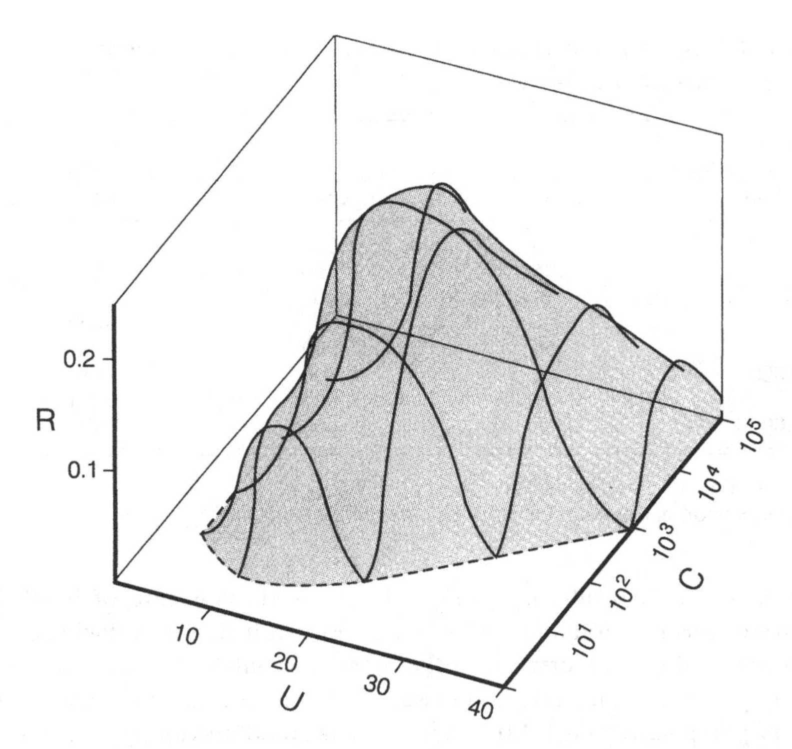

Figure 4.5 The seston uptake rate, *R*, as arbitrary units, is a function of velocity, *U*, in $cm \cdot s^{-1}$, and seston concentration, *C*, as cell number$\cdot ml^{-1}$ of *Chroomonas salinus* (Wildish et al. 1992).

valves would suggest that the pressure defect would be relatively small in the vicinity of the exhalant, which will lie in the turbulent wake of the scallop. The bulk stream velocity required to give a stagnation pressure of 1.0 mm of water is $14 \, cm \cdot s^{-1}$. Because the flow through the trophic fluid transport system is laminar, the volumetric flow is linearly related to the external pressure difference (see Jørgensen et al. 1986a). If the scallop's ciliary pump can generate a pressure difference of 1.0 mm of water, the volumetric flow through the animal can be doubled by an external pressure difference of the same magnitude. How the ciliary pumping system of other bivalve molluscs would respond to extreme pressure differences caused by ambient flow is not clear. As has been demonstrated by Wildish and Saulnier (1993), by direct video viewing of the exhalant siphon, the live scallop reacts behaviorally by partially closing the mantle and valves in order to resist the external pressure

Table 4.4. *Scallop seston uptake rates as affected by flux, seston concentration, and velocity.*[a]

Seston concentration $cells \cdot ml^{-1}$	Velocity $cm \cdot s^{-1}$	Flux rate $cells \cdot cm^2 \cdot s^{-1}$	Seston uptake rate μg Chl a $\cdot$ g wt $\cdot h^{-1}$
10	50	500	0
100	5	500	0.018
	50	5,000	0
1,000	5	5,000	0.09
	50	50,000	0
10,000	5	50,000	0.30
	50	500,000	0.24
100,000	5	500,000	15.0

Source: Extrapolated from Wildish et al. (1992).
[a] *Note*: Seston is a unialgal culture of *Chroomonas salinus.*

difference. This results in a reduced volumetric flow rate of inhaled seawater for filtration and hence feeding rate at increased velocities.

If the scallop is reversed with the inhalant pointing downstream, the velocity at which the ambient pressure differential is just balanced by a ciliary pump delivering a 1.0-mm head is, as noted previously, $14 cm \cdot s^{-1}$. For a head of 2.0 mm, the required velocity is $20 cm \cdot s^{-1}$. The reversed live scallop must maintain its ciliary pumping work rate but, in the unfavorable pressure conditions, cannot maintain its filtration throughput, as suggested by Jørgensen (1990) in mussels. Evidence that growth rates of live scallops forced to oppose unimodal flume velocities $>12 cm \cdot s^{-1}$ is reduced is presented by Wildish and Saulnier (1992).

In addition to velocity, seston concentration itself has been shown by Wildish and Saulnier (1993) to interact to influence the degree of valve and mantle edge opening, as indicated by seston uptake (Table 4.4). For a unialgal culture of *Chroomonas salinus*, filtration rates by individual, starved scallops were maximized at a cell concentration of $\sim 1 \times 10^3$ $cells \cdot ml^{-1}$ and at a constant ambient velocity of $8 cm \cdot s^{-1}$. Flume experiments in which velocities and seston concentrations were widely varied (Wildish et al. 1992) and giant scallop filtration rates determined show that the maximum uptake rate occurs at $\sim 25 cm \cdot s^{-1}$ when the optimum seston concentration ($\sim 10^3$ $cells \cdot ml^{-1}$ of *C. salinus*) is offered. Presumably, at this optimum point, giant scallops are filtering near the maximum intrinsic capacity of their pump. The ambient velocity at which scallops

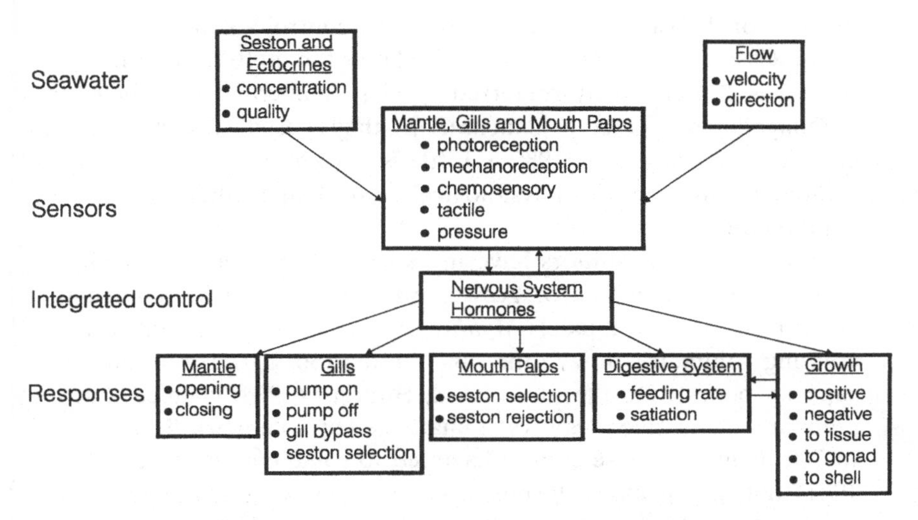

Figure 4.6 Environmental/physiological model of giant scallop feeding and growth (Wildish and Saulnier 1993).

stop filtering is directly proportional to seston concentration (Fig. 4.5). At seston concentrations greater than the optimum, sufficient ration can be obtained at lower clearance rates, and at both 10^4 and 10^5 cells $\cdot$ ml^{-1} of *C. salinus*, filtration was still occurring at up to 45 cm $\cdot$ s^{-1}. This was the maximum velocity achievable with the recirculating Vogel–LaBarbera type flume used in these experiments (Wildish et al. 1992).

The demonstration that giant scallop filtration rates are controlled by at least two independent environmental variables, velocity and seston concentration (Fig. 4.5), suggests an integrated environmental–physiological model for feeding and growth (Wildish and Saulnier 1993). This model (Fig. 4.6) depends on a constant output ciliary pump whose filtration rate can be controlled by closing/opening valves and mantle edge according to changes in these environmental variables. The evidence also indicates that velocity and seston concentration are interactive so that pumping and filtration only continue if it is nutritionally advantageous for the scallop. This implies central physiological control, perhaps by the nervous system, in which the bioenergetic profit and loss can be optimized.

The model incorporates evidence that ectocrines (substances released from cells which elicit chemosensory responses in target organisms) associated with seston alter the filtration response by causing closing/

opening of the valves, which automatically control volume throughput. Ward, Cassel, and MacDonald (1992) showed that filtration of the giant scallop is stimulated by preferred species of microalgae. When toxic dinoflagellates were fed to giant scallops, they caused valve clapping and closure (Shumway et al. 1985a). Little is known about the structure of sensors involved in the physiological control of feeding in the giant scallop (Beninger 1991).

Surprisingly few authors have investigated the effect of velocity on mussel filtration/feeding or growth rates. Walne (1972) claimed that growth rate in blue mussels *Mytilus edulis* was positively linked to increasing volumetric flow in the range tested ($50–700\,ml \cdot min^{-1}$ in 2.4-L-capacity boxes). On the other hand, Hildreth (1976) claimed that the pumping rate of blue mussels was unaffected by volumetric flows from 33 to $700\,ml \cdot min^{-1}$. These studies have led to some confusion (e.g. Vogel 1981), but, in both cases, the mixing characteristics within the experimental chambers, the degree of refiltration by the mussels, or velocities near the mussel inhalant were not determined, although the latter are likely to have been $<1\,cm \cdot s^{-1}$, and the results remain apparatus specific and not referable to the much wider range of velocities experienced in the field. Walne's (1972) results could be explained by small velocity increases helping to remove seston-depleted water near the mussel.

Preliminary measurements of blue mussel filtration rates as a function of velocity in the range $6–38\,cm \cdot s^{-1}$ have been made by Wildish and Miyares (1990) in a recirculating Vogel–LaBarbera flume of 90-L capacity. The velocity measurements were made with the aid of a small rotor of 11 mm diameter (see Chap. 2) placed in front of the mussel inhalant, with the mussel firmly attached by the valve to the flume floor so that the siphons opposed the flow. At velocities in the range $6–25\,cm \cdot s^{-1}$ and at a constant seston concentration of 10^4 cells of *Chroomonas salinus* $\cdot\, ml^{-1}$, filtration rates were inversely related to velocity. At velocities $>25\,cm \cdot s^{-1}$ and at the same seston concentration, the filtration rate was zero or a small percentage of the maximum possible at $6\,cm \cdot s^{-1}$. Although these results could be interpreted as part of a unimodal function of velocity where only the c and d stages were observed, the velocity range tested needs to be widened and seston concentrations tested need to be varied before a conclusion can be reached.

Growth studies in which velocity is varied while keeping seston concentration constant are not available for blue mussels. The flow range examined by Wildish and Kristmanson (1985) in growth experiments with blue mussels of $0.1–3.89\,cm \cdot s^{-1}$ is not considered wide enough to detect the unimodal response of feeding/growth to velocity. Bayne, Thompson, and Widdows (1976) suggested that the relationships among velocity, filtration rate, seston concentration, and respiration rate in mussels need to be elucidated, but as far as we are aware, this has not been done.

Seston concentration also affects the filtration rate of *Mytilus edulis*. An optimum rate at $75–200 \times 10^3$ cells of *Phaeodactylum tricornutum* $\cdot ml^{-1}$ was obtained in static seawater test conditions (Foster-Smith 1975). An indirect method was used to determine this with animals of 45- to 52-mm shell height and 2.56- to 4.00-g wet weight. The mean filtration rate was $1.47 \pm 0.20 L \cdot h^{-1}$ per individual or $0.39 \pm 0.04 L \cdot h^{-1} \cdot g^{-1}$ wet weight. Widdows, Fieth, and Worrall (1979), working with the same species and a similar indirect technique, showed that filtration rate was a unimodal function of the natural seston concentration (expressed as dry weight per litre, inclusive of sedimentary particles, and therefore not relatable to Foster-Smith's results). The optimum seston concentration increased as animal size became larger. Riisgård and Randlov (1981) also studied the filtration rate of *Mytilus edulis* over a wide range of concentrations of *Phaeodactylum tricornutum*. They found that filtration was independent of seston concentration between 1.5×10^3 and 30×10^3 cells $\cdot ml^{-1}$, but at higher concentrations there was a decrease in filtration. At seston concentrations lower than 1.5×10^3 cells of *P. tricornutum*, a decrease in filtration rate resulted from inhalant and valve closing.

Seston quality is a factor which has been extensively investigated in relation to filtration and feeding in the blue mussel. *Mytilus edulis* is considered to be an indiscriminate active suspension feeder whose filtration rate is not usually stimulated by ectocrines (Ward and Targett 1989), although seston quality, e.g. exudates from toxic dinoflagellates, may directly inhibit filtration or cause valve closures (see review by Shumway and Gainey 1992; Ward and Targett 1989). Some bacteria have also been reported to be inhibitory to filtration in the blue mussel (Birbeck and McHenery 1982; McHenery and Birbeck 1986; Birbeck, McHenery, and Nottage 1987). Ward and Targett (1989) have suggested that the response mechanism involves epicellular ectocrines which are cues for particle selection after capture but before ingestion. This response would be integrated by the nervous system, which then reduced, or shut, the valves, thus limiting or stopping filtration. Such a response would be consistent with the model shown in Fig. 4.6. One criticism of the work done on the effect of seston quality on bivalve filtration is that ambient flows around the feeding bivalve have not been simulated, and

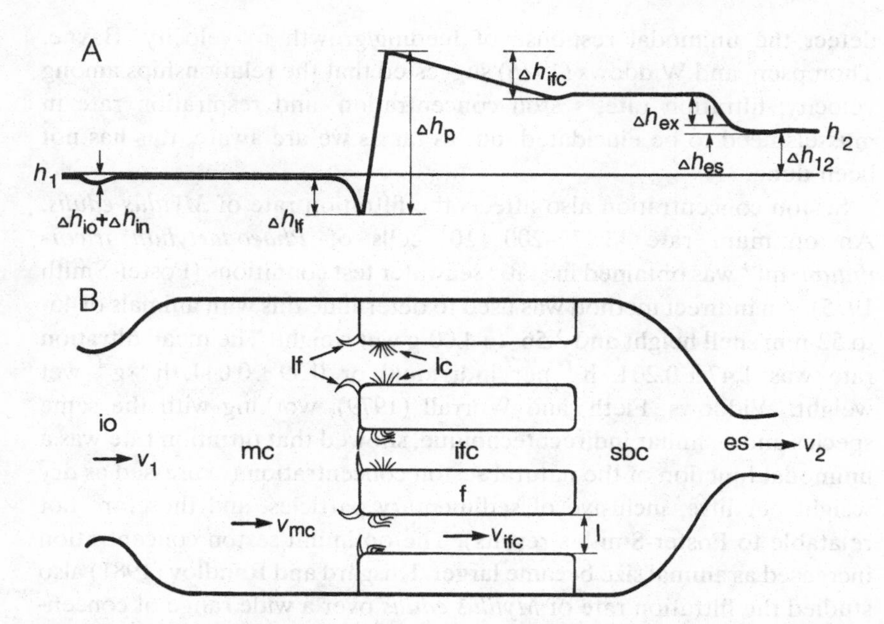

Figure 4.7 Diagrammatic representation of the blue mussel pump with typical static pressures (Jørgensen et al. 1986a) at points along the trophic fluid transport system: A, relative pressure heads; B, diagram of pump; h_1 and h_2 hydrostatic pressures at inlet and outlet; Δh_{12}, difference between h_1 and h_2 or back pressure; in, and ex, inlet and exit losses of pressure; io, inhalant opening; mc, mantle cavity; lf, laterofrontal cirri; p, pump; lc, lateral cilia; ifc, interfilament canal; f, filament; sbc, suprabranchial cavity; es, exhalant siphon; l, interfilament canal width; v_1 and v_2, velocities at inhalant and exhalant.

it remains to be seen whether this would change the conclusions already reached.

Jørgensen et al. (1986a) have provided a hydrodynamic model of the blue mussel pump useful in estimating inhalant and exhalant pressures. Pumping power is created by the metachronal beating of lateral cilia which are located at the upstream ends of the interfilament canals. There are resistances associated with various parts of the trophic fluid transport system, and these were estimated from their dimensions by Jørgensen et al. (1986a) to be of the order shown in Fig. 4.7. The pump characteristic was calculated from estimated resistances, pressure changes of the whole system, and volume flow rates to give the relationships among pump power, pressure heads, and volumetric flow rate (see Jørgensen et al. 1986a).

The measurement of bivalve pump characteristics, e.g. velocities at inhalant and exhalant, is rendered difficult by the rapidity of opening/closing of the mantle walls. According to Jørgensen and Ockelmann (1991), when this occurs, ciliary beating (in young, transparent-valved individuals) abruptly stops when the exhalant closes and immediately begins beating once the exhalant opens again. Apart from this, velocity distribution in the exhalant is that of laminar flow in a tube with mid-point velocities twice that of the bulk velocity. The difficulties involved in direct measurements of velocity or pressure at exhalant openings were emphasized in Chapter 2.

Pumping rate varies with the degree of valve gaping or, more precisely, with the degree of openness of the inhalant and exhalant siphons (Foster-Smith 1976). Maximally open siphons produce the maximum pumping rate. Retraction of the siphons causes a shortening of the gill axes and thus of the demibranchs, which results in a reduced lumen diameter of the interfilament canals (Jørgensen 1990). Complete siphon retraction and closing reduce the distance between the extended tips of opposing lateral cilia from ~10 μm to zero; this is suggested to be a general property of suspension feeding bivalve gills (Jørgensen and Riisgård 1988). Jørgensen (1990) concludes that pump power and volumetric capacity are probably direct functions of interfilament canal width, which is ultimately controlled by valve gaping and the linked degree of siphon retraction.

The relationship between seawater viscosity and the pumping rate of blue mussels (Jørgensen, Larsen, and Riisgård 1990) clearly shows the dependence of pumping rate on kinematic viscosity at different ambient temperatures. This effect was considered to be due to the resistance to water flow in the interfilament canals of the trophic fluid transport system. Jørgensen and Ockelmann (1991) showed that in young mussels lateral cilia beat with a mean frequency of ~10 Hz at 14°C to ~15 Hz at 20°–21°C, corresponding to a $Q_{10} \simeq 2$. Rapid changes of temperature in intact mussels are reported by Jørgensen (1990) not to lead to obvious effects on the pumping rate, despite the increased ciliary beating.

Physiologists interested in the energetics of filter feeding in bivalves such as mussels (e.g. Bayne and Newell 1983) have hypothesized that filtration is internally controlled, e.g. by gut fullness, to maintain a constant feeding rate over a wide range of environmental conditions. Although not supported by direct evidence, it is commonly held (Winter 1978) that ciliary pumping by bivalves is energetically expensive. Internal physiological control of filtration would thus be homeorheostatic by allowing the conservation of energy required for ciliary beating. An alternative view formulated in Jørgensen (1990) is that ciliary pumping (see also Jørgensen et al. 1986b) is inexpensive and, anyway, maintained at a constant rate irrespective of the pump output (see also Chap. 9, the section, Ultimate questions). Differences in pump output, and hence filtration rate, were a simple function of the degree of openness of the valves and siphons (Jørgensen 1990). Some additional evidence is available for Jørgensen's view; for example, Wildish and Saulnier (1993) found that the scallop exhalant opening area was a function of ambient velocity consistent with regulation of valve closure to control seston uptake and growth rates by velocity. Stenton-Dozey and Brown (1994) also found that in the South African clam *Venerupis corrugatus* the degree of siphon openness correlated positively with clearance rates and that increasing clearances were achieved without an increase in respiration rate. Both competing hypotheses outlined are compatible with the environmental–physiological model of feeding proposed by Wildish and Saulnier (1993).

Bryozoans

The phylum Bryozoa consists of colonial forms, usually attached to the substratum as an encrusting, stolonate, or bushy growth. The individual zooid of the colony is protected by a non-living chitinous skeleton, sometimes strengthened by calcareous secretions. The characteristic tentacle-bearing lophophore is the structure concerned with suspension feeding.

Observation of feeding Bryozoa in a static experimental system, where the seawater was slowly mixed by a peristaltic pump, was undertaken by Best and Thorpe (1983) in *Flustrellidra hispida*. This species is an epiphyte of the brown macroalga *Fucus serratus*, which is found in the lower intertidal zone of the Isle of Man. The experimental method involved viewing the lophophore at 1000× magnification with a stereomicroscope to observe the rate of movement of the seston particles (= *Tetraselmis suecica* of ~10-μm diameter at all densities up to 2×10^5 cells·ml^{-1}). Because the particles were moving quickly across the field of view, it was necessary to tune an oscilloscope trace to the same speed so that from the distance travelled (= field of view diameter), a particle speed could be calculated. Results showed that starved *Flustrellidra* could increase the seston particle speed if increased concentrations were supplied by up to 130%. Best and Thorpe (1983) noted that speeds were variable within the lophophore and further investigation revealed that particles located at the centre of the lophophore had the greatest speed. Rates of particle movement depended on ciliary currents generated by cilia present on the lophophore tentacles and ciliary speed varied in time, related to satiation. Best and Thorpe (1986) extended their investigations to five other species locally available around the Isle of Man. They found that starved zooids also responded to increased seston concentration by increasing the ciliary flow and that this resulted in a higher capture rate of a preferred food particle. They also found that feeding current velocities were positively linked to the lophophore height because increased tentacle length increases the number of cilia available to drive seawater through the lophophore.

Okamura (1984, 1985) has studied the effects of ambient flows on feeding of *Bugula stolonifera*, an arborescent species, and *Conopeum reticulum*, an encrusting species. *Bugula* and *Conopeum* were present on floating docks in the Pacific Ocean off California. Feeding success was determined by using latex beads (mean diameter of $11.9\,\mu m \pm 1.9\,\mu m$ standard deviation [SD]) offered at 1120–1190 $beads \cdot ml^{-1}$. Feeding experiments were of 20-min duration during which there was no settling out of the beads in the flume. The beads captured by individual zooids were counted after treatment with sodium hypochlorite, which left the exoskeleton and beads intact. Small *Bugula* colonies showed inhibition of feeding success for individual zooids at the higher flow velocity tested (10–$12\,cm \cdot s^{-1}$). Okamura (1984) suggests that this is because the sestonic particles are moving at a greater speed past the lophophore, making it more difficult to divert the streamlines into the lophophore. In addition, it is suggested that once they are in the lophophore, it may be more difficult to retain the particles for a sufficient time to effect transfer to the mouth as a result of the drag force caused by the current. Larger, more bushy *Bugula* did not show this effect of velocity on feeding because flows are decelerated as they pass through the branches, the colony changes its shape, and central or downstream, flow-protected zooids are the ones feeding at the greatest rate. In addition to flow protection, Okamura (1984) thought that the lack of flow inhibition in bushy forms could be due to a greater availability of seston because of the poor feeding effectiveness of upstream zooids and perhaps greater ciliary pumping by the flow-protected zooids.

In the encrusting bryozoan *Conopeum reticulum*, which were grown on glass microscope slides in the field, feeding experiments were conducted in the laboratory by placing slides normal to the flow in a small recirculating flow tank. Feeding success was determined as for *Bugula* and tests were made at three velocities: slow (1–$2\,cm \cdot s^{-1}$), intermediate

(4–6 cm · s^{-1}) and fast (10–12 cm · s^{-1}). The results show an inverse relation of feeding with velocity, and both large and small colonies were unable to accumulate beads at higher rates. At low and moderate velocities, zooids from larger colonies were more effective in feeding than small colonies, perhaps because the larger number of zooids could better divert seawater to the colony.

Subsequently, Okamura (1987, 1990) tested the interactive effects of particle size and velocity on the feeding success of Californian Bryozoa. She was able to show in *Bugula stolonifera* that medium and large particles (14.6 ± 1.0 and 19.1 ± 1.1 μm) were ingested at rates greater than small particles (9.6 ± 0.5 μm). Flow influenced the particle size ingested by *B. stolenifera* so that this bryozoan fed preferentially on larger particles at the highest flow tested (10–12 cm · s^{-1}). Particle size alone affected the feeding success of *Bugula neritana*; the medium and larger particles were preferred. The mechanism of selective feeding may involve size-dependent particle behavior, active selection, or rejection involving alternative feeding mechanisms such as tentacle flicking.

Sieving suspension feeders

Mucous filter nets have been found in a wide range of taxa, including ascidians, appendicularians (Flood and Fiala-Médioni 1982), some vermetid gastropods (Hughes 1978), and polychaetes such as *Nereis diversicolor* and *Chaetopterus variopedatus* (Wells and Dales 1951; Riisgård 1989, 1991).

Ascidians, or sea squirts, are attached benthic animals which may be solitary or colonial and are characteristically suspension feeders. A single feeding zooid of the colonial ascidian *Clavelina* species (Fig. 4.8) has two openings – the buccal siphon for inhalant flow and the atrial siphon for exhalant flow. The pharyngeal side walls are punctuated by gill slits or stigmata. The inhalant flow is generated by cilia lining the stigmata which cause inhaled seawater to pass around the pharynx, through the stigmata to the atrium, then out via the atrial siphon. Sestonic particles are captured on a mucous net which is secreted by the endostyle and the whole thing periodically ingested. When feeding conditions are suitable, the endostyle continuously produces the mucous net.

Randløv and Riisgård (1979) have studied the effect of sestonic particle size (in the range 1- to 7-μm equivalent spherical diameter) on the fractions retained by *Ascidiella aspersa*, *Molgula manhattensis*, *Clavelina lepadiformis*, and *Ciona intestinalis*. The small particles used originated

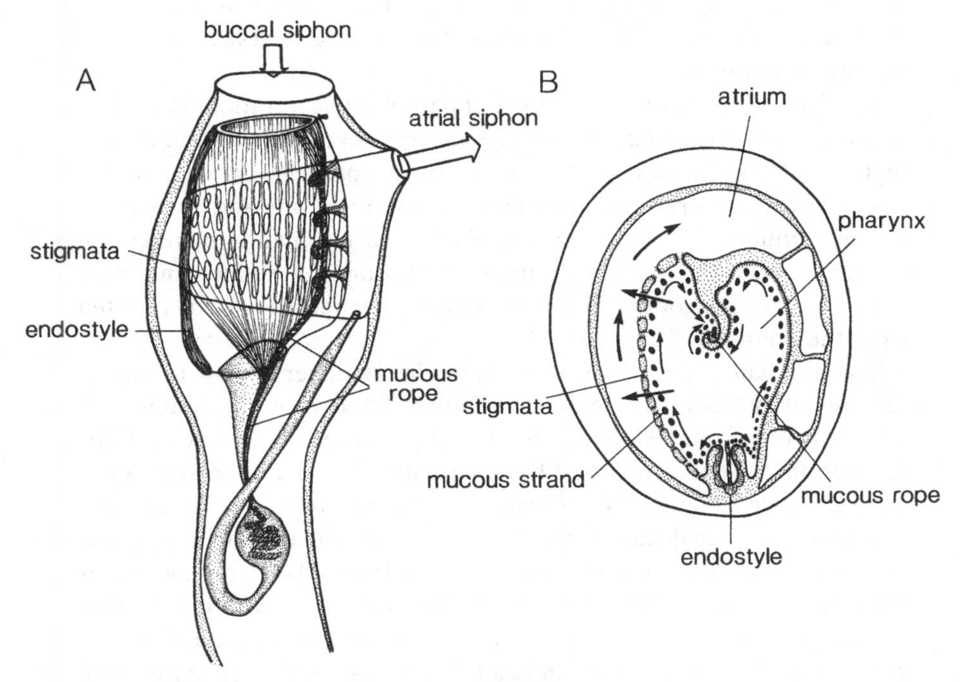

Figure 4.8 A, Single feeding zooid of *Clavelina*; B, transverse section of *Clavelina* (Carlisle 1979).

from those already present in the seawater system. Cultures of *Monochriysis lutheri* and *Dunaliella marina* were added at a concentration of 1×10^4 cells·ml^{-1} to increase the seston size range. Samples were collected from the inhalant and exhalant areas as in Møhlenberg and Riisgård (1978) and the concentration determined in an electronic particle counter/sizer. Results showed that seston of equivalent diameter in the range 2–3 to 7 µm were completely retained, while particles of <1 µm diameter had a particle retention efficiency of ~70%. Within the experimental error likely in measuring mesh size and retention efficiency, the results provide support for the hypothesis that food capture in these ascidians is by sieving. Fiala-Médioni (1978), working with at least one of the same species, *C. intestinalis*, suggested that up to one-third of *Monochrysis lutheri* (6–10 × 2–3 µm in size), offered as food at 20×10^6 cells·l^{-1} and 15 ± 1°C, passed through the filter. As Randløv and Riisgård (1979) point out, this result was not obtained by direct observation of retention efficiency but by calculation from direct observation of pumping rate and indirect observations based on particle clearance rates, and so may be inaccurate.

Robbins (1984) studied the effect of particle concentration on feeding of *Ciona* and *Ascidiella* spp. and found that ingestion was reduced at higher concentrations of fuller's earth (mean particle diameter of 25–30 μm). This was a result of a reduction in pumping rate and an increase in the frequency of squirting, but there was no change in retention efficiency, which remained at 100%. Squirting is a sudden muscular contraction of the zooid which causes excess sestonic food to be rejected from the pharynx.

The ascidian pump has been analyzed in a manner similar to that of the mussel by Riisgård (1988). The pump in *Styela clava*, by comparison with that in the mussel, is a much lower pressure/energy system. Flow through the filter is calculated to be $0.03\,mm \cdot s^{-1}$ and the pressure drop across it to be 0.069 mm H_2O (using a net dimension of $1.4 \times 5.4\,\mu m$, this value has been recalculated; see Riisgård 1989). The maximum pressure rise in the trophic canal system is 1.2 mm H_2O. The pumping rate is reduced by closing the atrial siphon, but this does not automatically reduce pressure, as it does in bivalves, because the stigmata cilia may continue to beat. Factors which result in the cessation of ciliary beating include a lack of sestonic particles or direct physical disturbance of the ascidian.

Most of the ascidians which have been studied are found in the nearshore continental shelf area. Carlisle (1979) reports that some continental slope stalked ascidians such as *Bolteniopsis* are specialized for opposing a steady external current and passively receive seawater and seston in a much enlarged buccal siphon. Concomitantly, there is a great reduction in the branchial basket, which may no longer be needed as a pump. Another example is provided by a stalked ascidian, *Styela monteryensis*, which is adapted to oscillating wave surge channels of exposed shores of the Pacific Ocean on the U.S. West Coast (Young and Braithwaite 1980). Characteristic of this species of *Styela* is a spar buoy system (see Chap. 6) in which wave forces passively reorient the stalked ascidian (Fig. 4.9) which is articulated near its base so that the buccal siphon faces the flow either on the onshore wave or on its back surge. It was shown by field experiments that the time required for seawater to pass through the trophic fluid transport system depends on ambient velocity conditions. It was also found that seawater was pushed through by dynamic force from waves, perhaps assisted by viscous entrainment. It

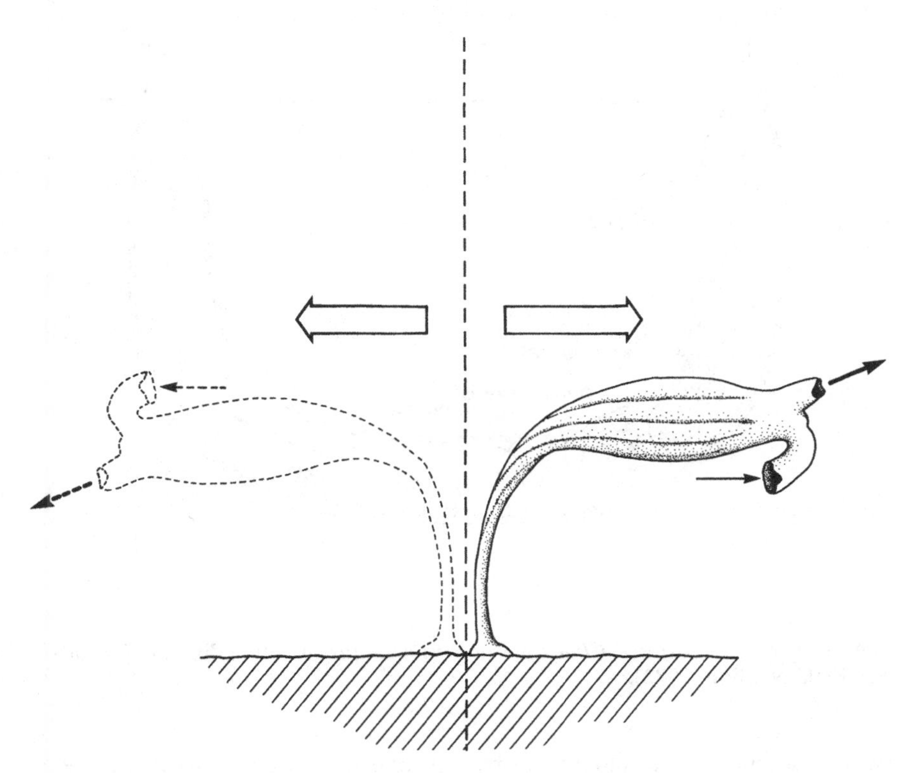

Figure 4.9 *Styela monteryensis*: responses to oscillating flows (Young and Braithwaite 1980).

is thus possible that these specialized ascidians are combined passive–active suspension feeders. Manahan et al. (1982) have shown that this species can remove free amino acids from seawater, although the importance of dissolved organic matter in its nutrition remains unclear.

Other species of considerable interest are polychaetes such as *Chaetopterus variopedatus* and *Nereis diversicolor* which are both capable of producing mucous nets and capturing microscopic food, probably by sieving. *Chaetopterus* lives in a parchmentlike tube secreted by the epidermis and attached to stones, or in soft sediment in a U-shaped burrow (Fig. 4.10). The parapodia of segments 14, 15, and 16 are disc shaped and beat to produce a flow of water through the tube. Flow direction is always from the head to the tail. The mucous net used in sieving food particles is produced by the modified parapodia of the

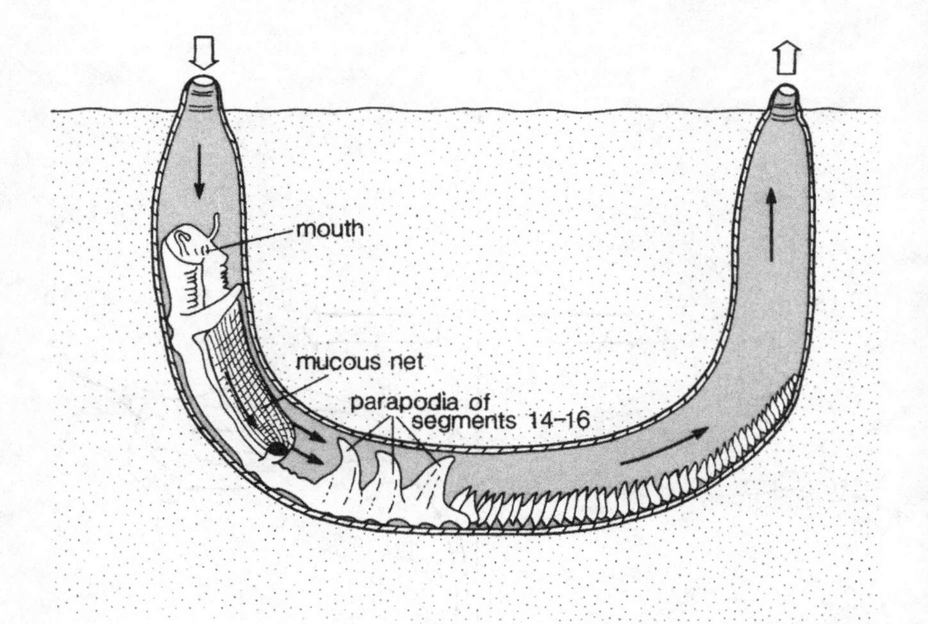

Figure 4.10 Life form of *Chaetopterus variopedatus* in soft sediments (Flood and Fiala-Médioni 1982).

twelfth segment. The ciliated dorsal groove and cuplike organ of the thirteenth segment gather the mucus and roll it into a food ball. When the ball is sufficiently large, the parapodia stops pumping and the food, plus mucus, is transported anteriorly by cilia in the dorsal groove to the mouth for ingestion.

The *Chaetopterus variopedatus* pump is a muscular one. Riisgård (1989) has used his constant-level chamber to investigate these worms from the Swedish west coast. The methods employed were similar to those used with mussels (Jørgensen et al. 1986a) and ascidians (Riisgård 1988). Video recordings of *Chaetopterus* confirmed that the mechanism involved a positive displacement pump and that parapod beating decreased with increasing back pressures. Imposed back pressures set up in the constant-level chamber caused the worm to reverse itself in the tube. The maximum pressure rise in the tube and the total head loss are approximately twice those in bivalves and five times those in ascidians, consistent with the filter area's being relatively smaller than in those species. The pressure drop across the mucous net as estimated by the Tamada–Fujikawa equation requires knowledge of mesh opening size and diameters of the mesh fibres. Estimates made by Jørgensen et al. (1984) of mesh size based on observations of particle retention efficiency at different particle sizes suggest that the *C. variopedatus* mucous net meshes are three times larger than suggested by Flood and Fiala-Médioni (1982). This results in a pressure drop estimate of 0.7 mm (Jørgensen's estimate) versus 3.0–4.5 mm of water for the mesh size suggested by Flood and Fiala-Médioni (1982). Even with the lower estimate, the pressure drop across the net in the tubicolus worm is several times greater than in the ascidian.

A field study of the pumping activity of the facultative sieving suspension feeder *Nereis diversicolor* was reported by Videl et al. (1994), who used a phototransducer system to record long-term seasonal patterns of pumping activity indicating filtration activity. Observations showed that the worms also go to the sediment surface to deposit-feed on benthic diatoms just after low tide at one, but not the other, Danish location studied. During the summer, *Nereis* was suspension feeding for 50%–100% of the time, and this behavior seemed to be triggered by a threshold chlorophyll a concentration of 1–3 $\mu g \cdot L^{-1}$. During the spring and autumn when chlorophyll a was scarce, the filtration activity occurred only 5%–20% of the time.

Facultative active suspension feeders

In some benthic suspension feeders, the filtration-feeding behavior changes with velocity so that at low velocities active filtration occurs, while above a critical higher velocity passive suspension feeding occurs. Although LaBarbera (1984) lists some ascidians – e.g. *Styela monteryensis* (Fig. 4.9), some brachiopods, and some sponges – in this category, their filtration/feeding responses to velocity have not been adequately examined. The status of these forms is considered to be unclear, although tentatively they are placed in a separate combined passive–active category (discussed later, in the section, Combined passive–active suspension feeders). The taxa best known as facultative active suspension feeders are the cirripede barnacles, but some decapod Crustacea also have representatives in this category (Trager et al. 1992).

Barnacles

It was determined by Crisp and Southward (1961) that changes in velocity can alter the type of feeding behavior of a number of adult cirripede

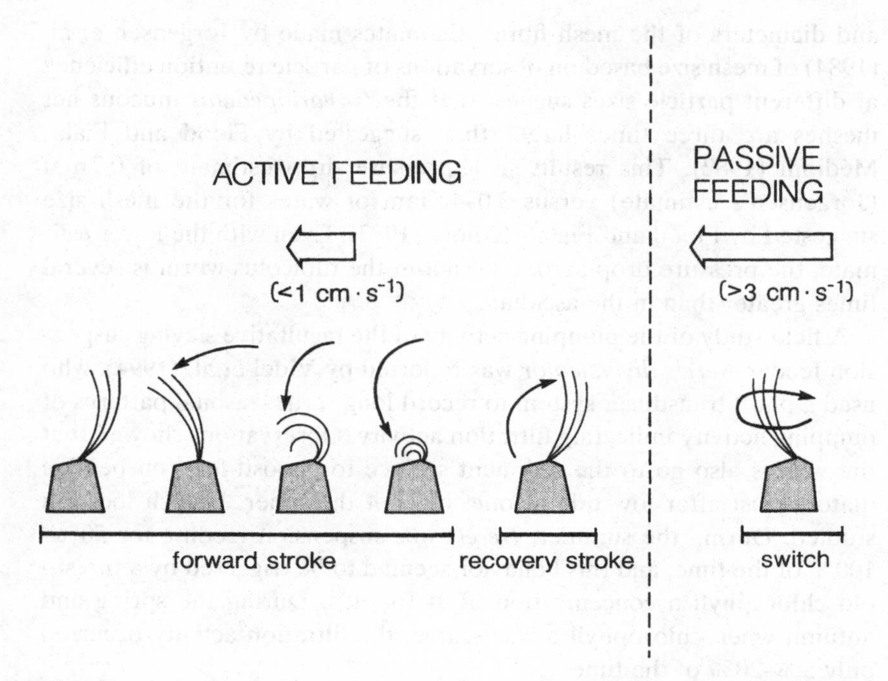

Figure 4.11 Active and passive feeding in *Semibalanus balanoides* (Trager et al. 1990).

barnacles. They described a fast-beat rhythmic sweeping of the cirri through slowly moving seawater and a non-rhythmic full extension of the cirri held rigidly opposed to the direction of flow when the ambient flow was more energetic. These different means of suspension feeding correspond to active suspension feeding at low velocity and passive suspension feeding at higher velocities. Barnacles, such as *Semibalanus balanoides*, are found on those rocky intertidal Atlantic Ocean shores where tidal currents and wind–wave activity are usually energetic and are constantly changing. In a flume study of this species, Trager et al. (1990) investigated the effect of velocity near the barnacle of up to $9\,cm \cdot s^{-1}$ on feeding behavior. The recirculating flume (capacity 15 L) had a sophisticated electronic controller which could reverse the propeller, thereby creating oscillating flows with sine wave acceleration or deceleration. Velocity was measured from dark field, laser optical pathway video frames of sestonic particle movement taken at known intervals. The mean velocity at which adult cirripedes changed from active to passive feeding was determined (Fig. 4.11) to be $3.10 \pm 0.73\,cm \cdot s^{-1}$ (±SD). The actual range of switch velocities from 22 experiments was 1.84–4.83 $cm \cdot s^{-1}$. From Fig. 4.11, it can be seen that the active cirral beating involved rhythmic movements in the same direction as the slow flows (individuals positioned carinorostrally to the flow). In passive feeding, the cirral fan is held rigidly into the flow so that the concave surface is opposed to it. If the flow direction and position of *S. balanoides* were as described previously, passive feeding involves reversal of the cirral fan through 180°. Trager et al. (1990) could make the cirripedes "dance" (change the orientation of their cirral fans) by rapidly changing the flow direction in the flume. This suggests that the response is controlled by the nervous system with accelerating or decelerating flows acting as the stimulus to cause the reflex responses observed. Further work by Trager, Hwang, and Strickler (1992) using a recirculating computer-controlled oscillating flow tank confirmed that in two species of barnacles, *S. balanoides* and *Sarignium milleporum* (from the Red Sea), flow oscillation frequencies in the range 0.16–0.65 Hz, with flow velocities ranging from 6.2 to 12.1 $cm \cdot s^{-1}$, stimulated reorientation of the cirral fan. Reorientation occurred during passive feeding so that the concave surface of the fan always faced upstream, where it is known to be most efficient at capturing suspended particles. Behavioral anticipation of flow change appeared to be based on water deceleration that occurred during the oscillation cycle. Trager et al. (1992) also showed that barnacles subjected to increasing oscillation frequencies spent less time in passively filtering and relatively more in removing particles filtered and in re-orienting the fan. Thus, the ratio between the time spent feeding to non-feeding activity depended on the wave oscillation frequency so that at high oscillation frequencies less time was available for filtration.

Hunt and Alexander (1991) have used an endoscope developed for medical research to observe movements of the mouth parts and cirral fan of a tropical cirripede, *Tetraclita squamosa*, collected from the east coast of Australia. They were able to show that a wide range of particle sizes, including zooplankters and microplankton, were filtered and utilized as food. They include a description of the ultrastructure of the thoracic appendages and how they are used in capturing the various prey sizes. A more detailed description is provided of six different cirral activity patterns, including that of the passive and active cirral fan behavior described by previous authors.

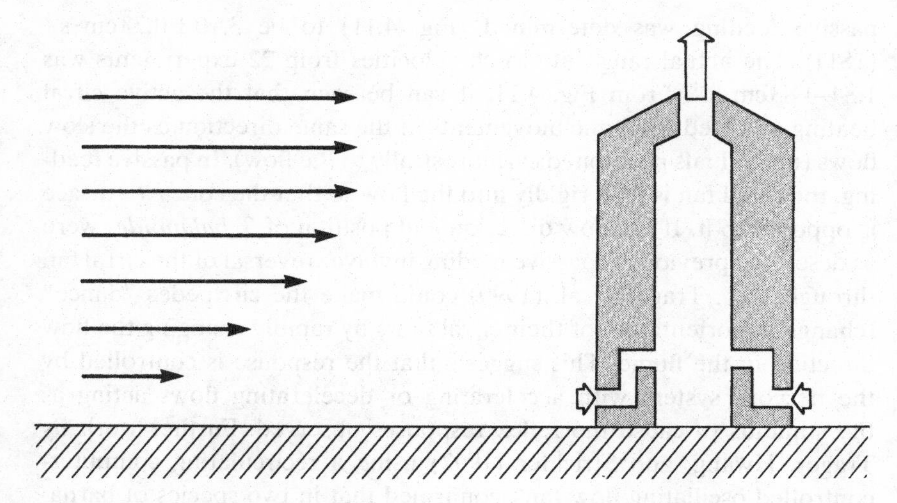

Figure 4.12 Diagrammatic view of a model sponge in a flow where a benthic boundary layer is present (Vogel and Bretz 1972).

Combined passive–active suspension feeders

Unlike in facultatively active suspension feeding, in which ambient velocity conditions provide the environmental switch to signal change from, or to, passive feeding, in combined passive–active suspension feeding the passive and active mechanisms appear to continue simultaneously.

Sponges

Model studies were undertaken by Vogel and Bretz (1972) in sponges, which have many small incurrent openings through the body wall. These openings connect to the spongocoel and an excurrent opening or osculum which opens dorsally in the animal so that it is high in the benthic boundary layer (Fig. 4.12). Besides an active flagellar pumping, passive flow in the sponge is induced by the Bernoulli effect resulting from the velocity and consequent pressure differences at the incurrent and excurrent openings caused by their different positions within the benthic boundary layer. Experiments with live erect sponges *Halichondria bowerbanki* obtained from the Massachussets coast, were made at ambient flows from 0 to 10 $cm \cdot s^{-1}$ (Vogel 1974). Flow measurements at 4 mm into the osculum and 3 cm in front of the sponge were made to record the osculum and ambient velocities, respectively. By immersing live sponges in freshwater, it was possible to inhibit choanocyte pumping. At zero velocity, live *Halichondria* of 3.1 cm height × 1.2 cm diameter produced an internal flow measured at the osculum of $2.5\,cm \cdot s^{-1}$. As ambient flows were increased, both normal live and choanocyte-inhibited live *Halichondria* increased internal flows at the same rate proportional to ambient velocity. At ambient flows $>6\,cm \cdot s^{-1}$, most of the total flow through the spongocoel was due to passive flow induction. Vogel (1974) thought that active pumping in *Halichondria* was unaffected by ambient flow and merely added on to the passive flow, which was directly related to the Bernoulli effect and ambient velocities. Further results using similar methods (Vogel 1977), but on eight species of live sponges obtained from Bermuda, showed similar results. In this case, the velocity measurements were made in the field at depths of 1–2.5 m, without removing the sponges. The osculum flow rates (when ambient flow = 0) obtained were much higher than obtained in the lab studies with *Halichondria*. For example, in *Verogia fistularis* and *Haliclona viridis*, the osculum flow rate was $21\,cm \cdot s^{-1}$, whereas in *Ircinia fasciculata* it was $11\,cm \cdot s^{-1}$. These results may have differed from the laboratory ones because of difficulties in handling the sponge and in simulating suitable environmental conditions.

Riisgård et al. (1993) examined the effects of temperature on filtration rates in the sponges *Halichondria panicea* and *Haliclona urceolus* by using a direct and an indirect method. The sponge pump characteristics were also estimated by a model and methods previously employed (Famme, Kristensen, Larsen, Møhlenberg, and Riisgård 1986a). The maximal operating point of the pump was 0.673 mm H_2O. The pump work was less than 1% of the respiratory output, suggesting a very low energy cost for filtration in sponges. Because of this low cost, Riisgård et al. (1993) considered it unlikely that passive increments to flow as postulated by Vogel and Bretz (1972) would be relatively important.

Vogel (1978) has postulated that a now extinct fossil group present in the Cambrian period – the Archaeocyanthids – may have been purely passive feeders relying on strong local currents to supply seawater for passive suspension feeding in the manner described for modern sponges.

Lampshells

LaBarbera (1977) proposed that active suspension feeders such as lampshells or brachiopods may use ambient flows to "augment or even supplant their active pump," as has been suggested for sponges. Model studies showed that induction of water movement through the dead valves of *Terebratulina unquicula*, *Terebratalia transversa*, and *Laqueus californianus* did passively occur at ambient flows of 2–4 $cm \cdot s^{-1}$. It is also known that brachiopods may actively pump at zero flows. However, as pointed out by LaBarbera (1977), there is no evidence available to show that the filtration rate increases as ambient velocities are increased.

The bilateral symmetry of brachiopods (Chap. 6) with two incurrent openings 180° apart allows them to utilize oscillating flows if the anterior–posterior axis of the valves is aligned parallel to the current direction. There is also evidence that some lampshells can behaviorally reorient to flow direction by means of pedicel musculature (see Chap. 6). LaBarbera (1977) proposed that passive flow augmentation through the brachiopod lophophore resulted in increased feeding rates and/or in increased availability of dissolved oxygen and removal of wastes. Vogel and Bretz (1972) and LaBarbera (1977) suggested that their *ambient velocity passive enhancement theory* could apply to other species, e.g. scallops. An independently derived theory, the *inhalant/exhalant pressure field theory*, was proposed by Wildish et al. (1987) to explain behavioral orientation to flow direction by scallops, which might also apply to brachiopods. Certainly, in future studies, both theories should be considered in designing experiments and testing specific hypotheses. According to the inhalant/exhalant pressure field theory applied to brachiopods, the lampshells would be considered as active suspension feeders, and the low exhalant flows measured by LaBarbera (1981) in still water indicate that the ciliary pump is much less effective than in bivalves. Exhalant velocities of *Terebratulina unquicula, Terebratalia transversa*, and *Laqueus californianus* ranged from 0.2 to 1.2 $cm \cdot s^{-1}$ and flow within the trophic fluid transport system was laminar. Referring to Fig. 6.5, the following orientation mechanisms are possible:

A. When the exhalant flow directly opposes the ambient flow direction and when the ambient flow pressure exceeds the ciliary pump pressure either extra work must be done by the pump to overcome it or the valves partially close (as in scallops) to reduce the differential.
B. With the exhalant downstream with respect to ambient flow direction and as the flow builds up so that the external pressure tends to force flow across the ciliary pump at a rate faster than that maintained by the cilia alone, then either the valves partially close to reduce it or filtration retention efficiency declines.
C. In this orientation the valves present the maximum amount of drag opposing the flow.

D. The valves present minimal drag to the flow and, as in C, the inhalant and exhalant can be oriented optimally for either uni- or bidirectional flows.

These represent suggestions (see Chap. 6, the section, Sessile epifauna, for an alternative view) and should be worked up further as null and alternative hypotheses for experimental testing.

Deposit–suspension feeders

Some species of benthic infauna have the ability to switch between deposit and suspension feeding. Ambient flow velocity is the environmental cue thought to trigger the change of behavior. Thus at low velocities, these animals feed on particles present on the sedimentary surface, while at higher velocities, they switch to suspension feeding.

Spionids

The tube-building spionid polychaetes are characteristically able to switch from deposit to suspension feeding. The switch seems to encompass a range of velocities from <2 to $5\,cm \cdot s^{-1}$ when a worm population is sampled (Taghon, Nowell, and Jumars 1980; Taghon and Greene 1992). For *Pseudopolydora kempi japonica* and *Boccardia pugettensis*, increasing velocity causes an asymptotic rise in the number of worms which are suspension feeding. Thus 50% of worms are suspension feeding at ~1.7 for the former species and $4.5\,cm \cdot s^{-1}$ for the latter species (measured 0.3 cm above the bottom). Partly because a wide enough range of velocities was not used in the flume experiments of Taghon and Greene (1992), their results do not provide sufficient detail to determine whether a unimodal feeding response to velocity exists.

Miller, Bock, and Turner (1992) have extended observations to *Spio setosa* in a flume oscillating flow. The sinusoidal flows produced in the working section were characterized by the maximum free-stream velocity reached in each excursion. The percentage of worms suspension feeding (that is palp coiling) in still water is zero, rising to ~90% at $6.5\,cm \cdot s^{-1}$ and thereafter dropping off (maximum velocity tested of $\sim 10\,cm \cdot s^{-1}$). The feeding response shown by populations of *S. setosa* appears to be unimodal with respect to maximum oscillation velocity, but the absence of higher velocity tests prevents drawing conclusions. In another experiment, the percentage of *Spio* showing palp coiling was observed at five maximum free stream velocities ($6.8–30\,cm \cdot s^{-1}$). In this experiment, the flows were increased every 10 min and the palp coiling of 19 individual *Spio* observed. Again, a unimodal response is suggested with a lower percentage of worms coiling at 6.8 and the two highest velocities tested: 24 and $30\,cm \cdot s^{-1}$. To establish firmly the oscillating velocity threshold from deposit to suspension feeding and the worms' responses to higher flows, a wider range of velocities should be tested. When one, or both, tentacles of spionids are lost to sublethal predation, Lindsay and Woodin (1995) have shown experimentally that such worms can still deposit feed directly from sediment surfaces.

Bivalves

Some species of the bivalve molluscan family Tellinacea, although mainly sediment interface deposit feeders, may, under some environmental conditions, subsurface deposit feed or suspension feed. Levinton (1991) suggests the following feeding classification for tellinaceans:

- *Swash rider suspension feeders*, intertidal species such as *Donax* which migrate on sandy beaches to maintain moist and relatively unexposed sediments
- *Sandy bottom suspension feeders*, intertidal and nearshore subtidal species with relatively large ctenidia specialized for suspension feeding
- *Sandy–mud bottom switching feeders*, intertidal and nearshore subtidal sandy-mud species such as *Macoma* and *Scrobicularia*, some species of which can switch between deposit and suspension feeding.

Of the latter group, Hughes (1969) studied the suspension feeding of *Scrobicularia plana* by using the direct method of measuring pumping rate in apparatus designed by Drinnan (1964). Filtration rates were estimated independently by indirect measurement of seston depletion in 100-ml dishes containing one *Scrobicularia*. In these studies, no attempt was made to investigate the effect of flow and the environmental factors causing deposit to change to suspension feeding. In the study by Levinton (1991), the feeding behavior of three Pacific species of *Macoma* from the San Juan Archipelago – *M. nasuta*, *M. secta*, and *M. unquinata* – was studied with respect to water flow and sediment transport. These studies were done in an annular flume and showed that deposit feeding was inhibited by increasing velocities (only three free-stream velocities, ~10, ~25, and $33–36\,cm \cdot s^{-1}$ were tested). This involved a decrease in deposit feeding radius from the siphon hole because the inhalant was withdrawn as velocity was increased. Levinton (1991) observed a loss of control of the inhalant siphon (quivering/flopping over) at a free-stream velocity between 10.5 and $12.0\,cm \cdot s^{-1}$, suggesting that drag on the siphon is sufficient to overcome its normal functioning.

Summary

The following criteria can be used for assessing the completeness, or lack thereof, of our knowledge of the physiological mechanisms of filtration in benthic suspension feeders:

- That velocity has been accurately measured near the appropriate filter or inhalant opening of the trophic fluid transport system over a sufficiently wide range of ambient velocities and directions
- That the volumetric flow which enters the filter or inhalant is known
- That the retention efficiency of the filter is known.

Most suspension feeders have not been studied over a sufficiently wide range of velocities to determine effects on filtration, feeding, or growth and, consequently, conclusions about the general applicability of the unimodal function model may be premature. Some hypotheses concerning filtration of benthic macrofauna which require further experimental testing are summarized in Table 4.5.

Of the few species that have been studied adequately with respect to velocity, inclusive of passive (the sea pen *Ptilosarcus querneyi*, gorgonians of the genus *Pseudopterogorgia*, the crinoids *Oligometra serripinnia* and *Antedon mediterrana*) and active suspension feeders (scallops, *Argopecten irradians*, *Placopecten magellanicus*, plus a few species of Bryozoa), all appear to fit the unimodal model. The cause–effect mechanisms at each response stage of the model are poorly understood and require further experimental investigations. These studies need access to controlled flows in flumes (see Chap. 2) with well-defined boundary layer flows which simulate field flows as closely as possible. In general, in stage a transport limitation of filtration can be expected, in stage b this effect is absent and ambient velocity has no influence on filtration, whereas in stage c filtration inhibition related to velocity can be expected. In active suspension feeders, the filtration responses may also be interactively linked to seston concentration and quality, making them complex and often difficult to interpret.

As far as we are aware, there have been no studies which investigate the effect of a wide range of flume-controlled ambient velocities on filtration or feeding of sieving suspension feeders (Table 4.5, number 1), and a unimodal response hypothesis probably applies to most groups considered. The exact nature of the compound passive–active suspension feeding group is still in doubt because, in order to establish firmly that passive supply of seawater to the organism is of physiological consequence, it must be demonstrated in live animals that enhanced filtration associated with enhanced ambient velocity results in increased growth

Table 4.5. *Some hypotheses related to filtration of benthic suspension feeders which need further experimental testing.*

No.	H_0/H_1	Description	Reference
1	H_0	The filtration rate of sieving suspension feeders is unaffected by velocity	Vogel (1981) LaBarbera (1984)
	H_1	The filtration rate of sieving suspension feeders is a unimodal response of velocity	
2	H_0	Combined passive–active pumping does not enhance filtration/growth rates	This analysis
	H_1	Combined passive–active pumping enhances filtration/growth in some sponges and brachiopods	
3	H_0	The energetic cost of ciliary pumping in bivalves is directly proportional to pump power output	Winter (1978) Jørgensen (1990)
	H_1	The energetic cost of ciliary pumping in bivalves is small and does not vary with pump power output	
4	H_0	Filtration or growth rate in brachiopods and scallops is not affected by their orientation to flow	Vogel and Bretz (1972), LaBarbera (1977) Wildish et al. (1987), Wildish and Saulnier (1993)
	H_1	Filtration or growth in brachiopods and scallops is controlled by their orientation with respect to flow direction	

(Table 4.5, number 2). The critical experiments required to establish this have not yet been carried out. The question of the energetic costs associated with ciliary pumping in bivalves (number 3, Table 4.5) is peripheral to the arguments presented in this chapter, although we return to this subject in relation to the evolution of the ciliary pump in Chapter 9 (the section, Ultimate questions).

Another interesting unanswered question concerns the ambient velocity passive enhancement and inhalant/exhalant pressure field theories and their possible applicability to both brachiopods and scallops. As proposed, the hypothesis (Table 4.5, number 4) does not appear to be incisive in its ability to provide support for one or the other of the two competing theories. Published data of seston uptake in scallops (Wildish et al. 1987) do support H_1 in hypothesis number 4 (Table 4.5). Utilizing the principle of multiple hypotheses – if the unimodal response for scallops outlined earlier is valid – the most likely applicability of the passive enhancement theory is in the "a part" of the unimodal response, i.e. from 0 to ~3 $cm \cdot s^{-1}$. Thus the goal of an experimental test would be to determine whether small increases in flow serve only to remove localized seston depletion due to dilution with pre-filtered seawater around the animal, or whether the flow provides a direct subsidy to the scallop's active pumping, as is required by the passive enhancement theory.

References

Bayne, B. L., and R. C. Newell. 1983. Physiological energetics of marine molluscs, pp. 407–515. *In* K. M. Wilbur (ed.) The mollusca. Vol. 4. Academic Press, New York.

Bayne, B. L., R. J. Thompson, and J. Widdows. 1976. Physiology I, pp. 121–206. *In* B. L. Bayne (ed.) Marine mussels, their ecology and physiology. Cambridge University Press, London.

Beninger, P. G. 1991. Structures and mechanisms of feeding in scallops: paradigms and paradoxes, pp. 331–340. *In* S. E. Shumway and P. A. Sandifer (ed.) World Aquaculture Society, Baton Rouge, La.

Best, B. A. 1988. Passive suspension feeding in a sea pen: effects of ambient flow on volume flow rate and filtering efficiency. Biol. Bull. 175: 332–342.

Best, M. A., and J. P. Thorpe. 1983. Effects of particle concentration on clearance rate and feeding current velocity in the marine bryozoan, *Flustrellidra hispida*. Mar. Biol. 77: 85–92.

1986. Effects of food particle concentration on feeding current velocity in six species of marine Bryozoa. Mar. Biol. 93: 255–262.

Birbeck, T. H., J. G. McHenery, and A. S. Nottage. 1987. Inhibition of filtration in bivalves by marine vibrios. Aquaculture 67: 247–248.

Birkbeck, T. H., and J. G. McHenery. 1982. Degradation of bacteria by *Mytilus edulis*. Mar. Biol. 72: 7–15.

Bricelj, V. M., A. E. Bass, and G. R. Lopez. 1984. Absorption and gut passage time of microalgae in a suspension feeder: an evaluation of the $^{51}Cr:^{14}C$ twin tracer technique. Mar. Ecol. Prog. Ser. 17: 57–63.

Cahalan, J. A., S. E. Siddall, and M. W. Luckenbach. 1989. Effects of flow velocity, food concentration and particle flux on growth rates of juvenile bay scallops *Argopecten irradians*. J. Exp. Mar. Biol. Ecol. 129: 45–60.

Carlisle, D. B. 1979. Feeding mechanisms in tunicates. Sci. Ser. 103, Canada Inland Waters Directorate.

Crisp, D. J., and A. J. Southward. 1961. Different types of cirral activity of barnacles. Phil. Trans. R. Soc. (Ser. B) 243: 271–308.

Dai, C.-F., and M.-C. Lin. 1993. The effects of three gorgonians from southern Taiwan. J. Exp. Mar. Biol. Ecol. 173: 57–69.

Drinnan, R. E. 1964. An apparatus for recording the water-pumping behaviour of lamellebranchs. Neth. J. Sea Res. 2: 223–232.

Eckman, J. E., C. H. Peterson, and J. A. Cahalan. 1989. Effects of flow speed,

turbulence and orientation on growth of juvenile bay scallops, *Argopecten irradians concentricus* (Say). J. Exp. Mar. Biol. Ecol. 132: 123–140.
Fenchel, T. 1980. Suspension feeding in ciliated Protozoa: feeding rates and their ecological significance. Microb. Ecol. 6: 13–25.
Fiala-Médioni, A. 1978. Filter feeding ethology of benthic invertebrates (Ascidians). V. Influence of temperature on pumping, filtration and digestion rates and rhythms in *Phallusia mamillata*. Mar. Biol. 48: 251–259.
Flood, P. R., and A. Fiala-Médioni. 1981. Ultrastructure and histochemistry of the food trapping mucous film in benthic filter feeders (Ascidians). Acta. Zool. Stockh. 62: 53–65.
Flood, P. R., and A. Fiala-Médioni. 1982. Structure of the mucous filter of *Chaetopterus variopedatus* (Polychaeta). Mar. Biol. 72: 27–33.
Foster-Smith, R. L. 1975. The effect of concentration of suspension on the filtration rates and pseudofaecal production for *Mytilus edulis* L., *Cerastoderma edule* (L.) and *Venerupis pullastra*. J. Exp. Mar. Biol. Ecol. 17: 1–22.
1976. Pressures generated by the pumping mechanism of some ciliary filter-feeders. J. Exp. Mar. Biol. Ecol. 25: 199–206.
Grizzle, R. E., and P. J. Morin. 1989. Effect of tidal currents, seston, and bottom sediments on growth of *Mercenaria mercenaria*: results of a field experiment. Mar. Biol. 102: 85–94.
Hildreth, D. I. 1976. The influence of water flow rate on pumping rate in *Mytilus edulis* using a refined direct measurement apparatus. J. Mar. Biol. Assoc. U.K. 56: 311–319.
Hughes, R. N. 1969. A study of feeding in *Scrobicularia plana*. J. Mar. Biol. Assoc. U.K. 49: 805–823.
1978. The biology of *Dendropoma corallinaceum* and *Serpulorbis natalensis*, two South African vermetid gastropods. Zool. J. Linn. Soc. 64: 111–127.
Hunt, M. J., and C. G. Alexander. 1991. Feeding mechanisms in the barnacle *Tetraclita squamosa* (Brugière). J. Exp. Mar. Biol. Ecol. 154: 1–28.
Jørgensen, C. B. 1966. Biology of suspension feeding. Pergamon Press, Oxford.
1990. Bivalve filter feeding: hydrodynamics, bioenergetics, physiology and ecology. Olsen and Olsen, Fredensborg, Denmark.
Jørgensen, C. B., P. Famme, H. S. Kristensen, P. S. Larsen, F. Møhlenberg, and H. U. Riisgård. 1986a. The bivalve pump. Mar. Ecol. Prog. Ser. 34: 69–77.
Jørgensen, C. B., T. Kiørboe, F. Møhlenberg, and H. U. Riisgård. 1984. Ciliary and mucus-net filter feeding, with special reference to fluid mechanical characteristics. Mar. Ecol. Prog. Ser. 15: 283–292.
Jørgensen, C. B., P. S. Larsen, and H. U. Riisgård. 1990. Effects of temperature on the mussel pump. Mar. Ecol. Prog. Ser. 64: 89–97.
Jørgensen, C. B., F. Møhlenberg, and O. Sten-Knudsen. 1986b. Nature of relation between ventilation and oxygen consumption in filter feeders. Mar. Ecol. Prog. Ser. 29: 73–88.
Jørgensen, C. B., and K. Ockelmann. 1991. Beat frequency of lateral cilia in intact filter feeding bivalves: effect of temperature. Ophelia 33: 67–70.
Jørgensen, C. B., and H. U. Riisgård. 1988. Gill pump characteristics of the soft clam *Mya arenaria*. Mar. Biol. 99: 107–109.
Kerswill, C. J. 1949. Effects of water circulation on the growth of quahaugs and oysters. J. Fish. Res. Board Can. 7: 545–551.

Kirby-Smith, W. W. 1972. Growth of the bay scallop: the influence of experimental water currents. J. Exp. Mar. Biol. Ecol. 8: 7–18.
LaBarbera, M. 1977. Brachiopod orientation to water movement. I. Theory, laboratory behaviour and field orientations. Paleobiology 3: 270–287.
1981. Water flow patterns in and around three species of articulate brachiopods. J. Exp. Mar. Biol. Ecol. 55: 185–206.
1984. Feeding currents and particle capture mechanisms in suspension feeding animals. Am. Zool. 24: 71–84.
Leonard, A. B. 1989. Functional response in *Antedon mediterranea* (Lamarck) (Echinodermata: Crinoidea): the interaction of prey concentration and current velocity on a passive suspension-feeder. J. Exp. Mar. Biol. Ecol. 127: 81–103.
Leonard, A. B., J. R. Strickler, and N. B. Holland. 1988. Effects of current speed on filtration during suspension feeding in *Oligometra serripinna* (Echinodermata: Crinoidea). Mar. Biol. 97: 111–125.
Leversee, G. J. 1976. Flow and feeding in fan-shaped colonies of the gorgonian coral, *Leptogorgia*. Biol. Bull. 151: 344–356.
Levinton, J. S. 1991. Variable feeding behaviour in three species of *Macoma* (Bivalvia: Tellinacea) as a response to water flow and sediment transport. Mar. Biol. 110: 375–383.
Lindsay, S. M., and S. A. Woodin. 1995. Tissue loss induces switching of feeding mode in spionid polychaetes. Mar. Ecol. Prog. Ser. 125: 159–169.
Manahan, D. T., S. H. Wright, G. C. Stephens, and M. A. Rice. 1982. Transport of dissolved amino acids by the mussel *Mytilus edulis*: demonstration of net uptake from natural seawater. Science 215: 1253–1254.
McHenery, J. G., and T. H. Birkbeck. 1986. Inhibition of filtration in *Mytilus edulis* L. by marine vibrios. J. Fish. Dis. 9: 256–261.
Miller, D. C., M. J. Bock, and E. J. Turner. 1992. Deposit and suspension feeding in oscillatory flows and sediment fluxes. J. Mar. Res. 50: 489–520.
Mohlenberg, F., and H. U. Riisgård. 1978. Efficiency of particle retention in 13 species of suspension feeding bivalves. Ophelia 17: 239–246.
Okamura, B. 1984. The effects of ambient flow velocity, colony size, and upstream colonies on the feeding success of Bryozoa. I. *Bugula stolonifera* Ryland, an arborescent species. J. Exp. Mar. Biol. Ecol. 83: 179–193.
1985. The effects of ambient flow velocity, colony size, and upstream colonies on the feeding success of Bryozoa. II. *Conopeum reticulum* (Linnaeus), an encrusting species. J. Exp. Mar. Biol. Ecol. 89: 69–80.
1987. Particle size and flow velocity induce an inferred switch in bryozoan suspension-feeding behaviour. Biol. Bull. 173: 222–229.
1990. Particle size, flow velocity, and suspension-feeding by the erect bryozoans *Bugula neritina* and *B. stolonifera*. Mar. Biol. 105: 33–39.
Patterson, M. R. 1991. The effects of flow on polyp-level prey capture in an octocoral, *Alcyonium siderium*. Biol. Bull. 180: 93–102.
Randløv, A., and H. U. Riisgård. 1979. Efficiency of particle retention and filtration rate in four species of Ascidians. Mar. Ecol. Prog. Ser. 1: 55–59.
Riisgård, H. U. 1988. The ascidian pump: properties and energy cost. Mar. Ecol. Prog. Ser. 47: 129–134.
1989. Properties and energy cost of the muscular piston pump in the suspension feeding polychaete *Chaetopterus variopedatus*. Mar. Ecol. Prog. Ser. 56: 157–168.

1991. Suspension feeding in the polychaete *Nereis diversicolor*. Mar. Ecol. Prog. Ser. 70: 29–37.
Riisgård, H. U., and A. Randløv. 1981. Energy budgets, growth and filtration rates in *Mytilus edulis* at different algal concentrations. Mar. Biol. 61: 227–234.
Riisgård, H. U., S. Thomassen, H. Jakobsen, J. M. Wicks, and P. S. Larsen. 1993. Suspension feeding in marine sponges *Halichondria panicea* and *Haliclona urceolus*: effects of temperature on filtration rate and energy cost of pumping. Mar. Ecol. Prog. Ser. 96: 177–188.
Robbins, I. J. 1984. The regulation of ingestion rate, at high suspended particulate concentrations, by some phleobranchiate ascidians. J. Exp. Mar. Biol. Ecol. 82: 1–10.
Shumway, S. E., T. L. Cucci, L. Gainey, and C. M. Yentsch. 1985a. A preliminary study of the behavioural and physiological effects of *Gonyaulax tamarensis* on bivalve molluscs, p. 389–394. *In* D. M. Anderson, A. W. White, and D. G. Baden (eds.) Toxic dinoflagellates. Elsevier, Amsterdam.
Shumway, S. E., T. L. Cucci, R. C. Newell, and C. M. Yentsch. 1985b. Particle selection, ingestion, and absorption in filter-feeding bivalves. J. Exp. Mar. Biol. Ecol. 91: 77–92.
Shumway, S. E., and L. F. Gainey. 1992. A review of physiological effects of toxic dinoflagellates on bivalve molluscs. Proc. Ninth Int. Malac. Congress: 357–362.
Sponaugle, S. 1991. Flow patterns and velocities around a suspension-feeding gorgonian polyp: evidence from physical models. J. Exp. Mar. Biol. Ecol. 148: 135–145.
Sponaugle, S., and M. LaBarbera. 1991. Drag-induced deformation: a functional feeding strategy in two species of gorgonians. J. Exp. Mar. Biol. Ecol. 148: 121–134.
Stenton-Dozey, J. M. E., and A. C. Brown. 1994. Short-term changes in the energy balances of *Venerupis corrugatus* (Bivalvia) in relation to tidal availability of natural suspended particles. Mar. Ecol. Prog. Ser. 103: 57–64.
Taghon, G. L., and R. R. Greene. 1992. Utilization of deposited and suspended particulate matter by benthic "interface" feeders. Limnol. Oceanogr. 37: 1370–1391.
Taghon, G. L., A. R. M. Nowell, and P. A. Jumars. 1980. Induction of suspension feeding in spionid polychaetes by high particulate fluxes. Science 210: 562–564.
Timko, P. L. 1976. Sand dollars as suspension feeders: a new description of feeding in *Dendraster excentricus*. Biol. Bull. 151: 247–259.
Trager, G. C., D. Coughlin, A. Genin, Y. Achituv, and A. Gangopadyay. 1992. Foraging to a rhythm of ocean waves: porcelain crabs and barnacles synchronize feeding motions with flow oscillations. J. Exp. Mar. Biol. Ecol. 164: 73–86.
Trager, G. C., J.-S. Hwang, and J. R. Strickler. 1990. Barnacle suspension-feeding in variable flow. Mar. Biol. 105: 117–129.
Videl, A., B. B. Andersen, and H. U. Riisgård. 1994. Field investigations of pumping activity of facultatively filter feeding polychaete *Nereis diversicolor* using an improved infrared phototransducer system. Mar. Ecol. Prog. Ser. 103: 91–101.
Vogel, S. 1974. Current-induced flow through the sponge, *Halichondria*. Biol. Bull. 147: 443–456.

1977. Current-induced flow through sponges *in situ*. Proc. Nat. Acad. Sci., U.S.A. 74: 2069–2071.
1978. Organisms that capture currents. Sci. Am. 239: 128–139.
1981. Life in moving fluids: the physical biology of flow. Willard Grant Press, Boston.
Vogel, S., and W. L. Bretz. 1972. Interfacial organisms: passive ventilation in the velocity gradients near surfaces. Science 175: 210–211.
Walne, P. R. 1972. The influence of current speed, body size and water temperature on the filtration rate of five species of bivalves. J. Mar. Biol. Assoc. U.K. 52: 345–374.
Ward, J. E., H. K. Cassel, and B. A. MacDonald. 1992. Chemoreception in the sea scallop *Placopecten magellanicus* (Gmelin). I. Stimulatory effects of phytoplankton metabolites on clearance and ingestion rates. J. Exp. Mar. Biol. Ecol. 163: 235–250.
Ward, J. E., and N. M. Targett. 1989. Influence of marine microalgal metabolites on the feeding behaviour of the blue mussel *Mytilus edulis*. Mar. Biol. 101: 313–321.
Warner, G. F. 1977. On the shapes of passive suspension feeders, p. 567–576. *In* B. F. Keegan, P. Ó. Céidigh, and P. J. S. Boaden (eds.) Biology of benthic organisms. Pergamon Press, Oxford.
Warner, G. F., and J. D. Woodley. 1975. Suspension-feeding in the brittle star, *Ophiothrix fragilis*. J. Mar. Biol. Assoc. U.K. 55: 199–210.
Wells, G. P., and R. P. Dales. 1951. Spontaneous activity patterns in animal behaviour: the irrigation of the burrow in the polychaetes *Chaetopterus variopedatus* Renier and *Nereis diversicolor*. J. Mar. Biol. Assoc. U.K. 29: 661–680.
Widdows, J., P. Fieth, and C. M. Worrall. 1979. Relationships between seston, available food and feeding activity in the common mussel *Mytilus edulis*. Mar. Biol. 50: 195–207.
Wildish, D. J., and D. D. Kristmanson. 1985. Control of suspension-feeding bivalve production by current speed. Helgol. Wiss. Meeresunters. 39: 237–243.
Wildish, D. J., D. D. Kristmanson, R. L. Hoar, A. M. DeCoste, S. D. McCormick, and A. W. White. 1987. Giant scallop feeding and growth responses to flow. J. Exp. Mar. Biol. Ecol. 113: 207–220.
Wildish, D. J., D. D. Kristmanson, and A. M. Saulnier. 1992. Interactive effect of velocity and seston concentration on giant scallop feeding inhibition. J. Exp. Mar. Biol. Ecol. 155: 161–168.
Wildish, D. J., and M. P. Miyares. 1990. Filtration rate of blue mussels as a function of flow velocity: preliminary experiments. J. Exp. Mar. Biol. Ecol. 142: 213–220.
Wildish, D. J., and A. M. Saulnier. 1992. The effect of velocity and flow direction on the growth rate of juvenile and adult giant scallops. J. Exp. Mar. Biol. Ecol. 155: 133–143.
1993. Hydrodynamic control of filtration in the giant scallop. J. Exp. Mar. Biol. Ecol. 174: 65–82.
Winter, J. E. 1978. A review on the knowledge of suspension-feeding in lamellibranchiate bivalves with special reference to artificial aquaculture systems. Aquaculture 13: 1–33.
Young, C. M., and L. F. Braithwaite. 1980. Orientation and current-induced flow in the stalked ascidian, *Styela montereyensis*. Biol. Bull. 159: 428–440.

5

Mechanisms of seston capture

It is the purpose of this chapter to address the second stage of suspension feeding, as defined in Chapter 4: specific particle capture mechanisms at the appropriate suspension-feeding collecting surface. The theory for our approach to this is borrowed from physics and engineering literature, inclusive of sieving and the aerosol theory mechanisms. We are especially indebted to reviews by Rubenstein and Koehl (1977), LaBarbera (1984), and Shimeta and Jumars (1991), which help interpret this in a biological context. We also include a novel mechanism involving hydromechanical shear forces at the bivalve gill proposed by Jørgensen (1981a, 1983, 1990), which is not part of the classical particle capture theory. Other topics discussed here include the nature of seston, how sestonic particles reach the collector, and a brief consideration of those species where calculations of seston encounter efficiencies according to aerosol theory have been made.

Nature of seston

The type of particles which may be collected by suspension feeders are highly variable and dependent on the local conditions. Non-viable particles of silt, clay, sand, and detritus may be processed for the attached microbiota. Viable particles include bacteria, phytoplankton, invertebrate larvae, and eggs, many of which are weakly motile. Collectively, this material is referred to as seston (Fig. 5.1); it ranges in size from less than 1 μm up to 1000 μm or more.

In order to understand the mechanics of collection of sestonic particles properly, their size, shape, density, and settling velocities should be known. Shapes range from spherical to very complex. Inorganic particles are of densities often well known from their mineral content, and settling

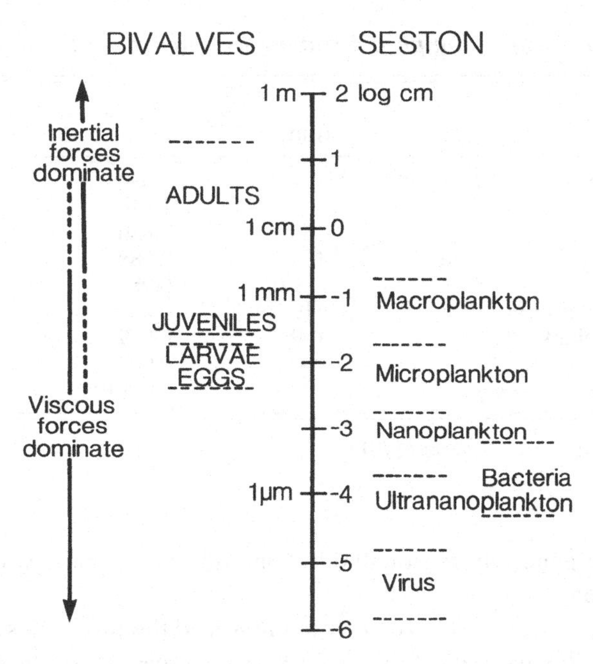

Figure 5.1 Comparative sizes of suspension-feeding bivalves and their potential food (= seston) (Wildish and Kristmanson 1993).

velocities can often be estimated from Stokess law. Typically, particles of biological origin have density close to that of seawater. It is particularly difficult to generalize about their settling velocities. Even for a particular species, density will vary between individuals and may be modified by changes in lipid content or ionic concentrations of the cell. A compilation of literature on the sinking rates for phytoplankton (Parsons, Takahashi, and Hargrave 1977) includes living phytoplankton (25 species), 0–30 $m \cdot d^{-1}$, and phytoplankton, dead, intact (10 species): <1–510 $m \cdot d^{-1}$. This range gives an indication of the variability of this parameter.

Swimming speeds of motile species present in the plankton are also variable and resistant to generalisations. Some small flagellates swim at a speed of the order of a body length per second, large dinoflagellates at 25–50 times body length per second (Parsons et al. 1977). In most published accounts of the mechanics of particle collection discussed here, it has been assumed that the particles are of low motility and move with the fluid streamlines without significant settling. This assumption

Table 5.1. *Typical seston particles and encounter rate estimates.*

Particle	d_p (μm)	$\rho_p - \rho$ $g \cdot cm^{-3}$	w_s $cm \cdot s^{-1}$
Medium clay	1.0	1.630	8.9×10^{-5}
Medium silt	20	1.630	3.5×10^{-2}
Fine sand	150	1.630	2.0
Organic–mineral aggregate	100	0.088	5.0×10^{-2}
Bacterium	1.0	0.050	2.7×10^{-6}
Small phytoplankton	10	0.069	3.4×10^{-4}
Large phytoplankton	100	0.009	5.0×10^{-3}
Invertebrate larvae	500	0.0073	0.1
Marine snow	5000	6.5×10^{-5}	8.8×10^{-2}

Source: Shimeta and Jumars (1991).

seems to be generally reasonable but should not be made without due consideration.

Shimeta and Jumars (1991) have summarized the properties of typical seston particles used in their analysis of examples of particle capture which will be discussed later (Table 5.1).

Fluxes, encounter rates, and collection efficiencies

If we consider an object fixed in a steady flow of water containing seston, then the undisturbed flow upstream of the object can be characterized by a mean velocity, U (convenient units of centimetres per second [$cm \cdot s^{-1}$]). For the purposes of this presentation, the effects of shear and turbulence, though typical of the benthic boundary layer, will not be considered. Our purpose is to define the relationships among flux, encounter rates, and collection efficiency.

The concentration of seston in the approaching stream, c (the number of particles per unit volume), is assumed to be constant in time and space. Then, the flux of seston at a point upstream of the organism can be defined as the product of velocity and seston concentration, cU. The frontal area of the object, A, is the projection of its outline on a plane normal to the flow vector. The engineering definition of the collection efficiency (ε) of a suspension feeder would be the ratio of the number of particles striking it to the number which would strike it if the stream were not diverted (Dorman 1966). This definition is based on the assumption that all particles which strike the collector are retained. In engineering applications, an adhesion coefficient (ε_A) is defined to allow for any resuspension of particles so that the overall collection efficiency is the product of the adhesion coefficient and the engineering collection efficiency.

The rate at which particles are collected is, therefore, given by

$$\varepsilon c U A \varepsilon_A$$

Shimeta and Jumars (1991) argue that the engineering efficiency of biological filters is a less useful term than the *encounter rate*, defined as the rate at which particles reach the surface of the collector:

$$F = \varepsilon c U A \tag{5.1}$$

The advantage of using encounter rates is that this quantity is a direct measure of the potential food available to the organism and takes into account the seston concentration. The rate at which food is collected is obtained by multiplying the encounter rate by the retention coefficient (Shimeta and Jumars 1991), equivalent to the adhesion coefficient used in engineering.

The clearance rate, R, as used in Chapter 2, is generally defined as the volume of fluid cleared of particles per unit time. This term is equivalent to the encounter rate times the retention coefficient divided by the upstream concentration, or

$$R = \varepsilon \varepsilon_A U A$$

Following Shimeta (1993) we will also use a clearance rate for a circular cylinder defined in units of volume cleared per unit time per unit length of cylinder.

The passive suspension feeder is often configured as a large number of individual collecting elements in a supporting structure. A typical example is the sea pen (see the section, Passive suspension feeders, Chap. 4). Particle collection theory is mainly based on the analysis of capture by isolated elements of simple geometry or by filter beds composed of layers of cylinders or spheres. The preceding definitions were written with a single solid obstacle such as a cylinder in mind, but with suitable reservations they can be applied to the whole structure, in which case the collection efficiency represents the sum of the contributions of each of the collecting elements in the entire organism. The analysis can also be applied to a single collecting element. Each collecting element is characterised by its own approach velocity and upstream concentration, and an estimate of its collection efficiency may be made from theory, as will be detailed later. It is important to note, however, that the collection efficiency of the organism as defined is in general different from that of the individual collecting elements and is not necessarily easily estimated from the former.

Mechanics of capture

Suspension feeders must process large volumes of water to extract sufficient seston as food. The food is in the form of particulate matter, usually in low concentrations, often less than a few milligrams per liter ($mg \cdot L^{-1}$), and must be removed from the water being processed. The means by which this is done by benthic suspension feeders are many and varied. We have discussed the classification of the major methods in Chapter 4. Classification may also be based on the mechanisms by which the particles are removed (LaBarbera 1984):

- Biological, involving active response of the animal
- Physical

The latter can be further classified according to the dominant mechanisms involved. Another mechanism of particle segregation which does not depend on a biological response or on physical contact with a biological structure has been proposed by Jørgensen (1983). It has been argued that in the bivalve gill, particles migrate across streamlines in pulsatile creeping flows close to the frontal surface of the gill. It has been suggested by Nielsen et al. (1993) that this is a third category, fluid/mechanical, which should be added to LaBarbera's scheme.

Two conflicting demands must be satisfied to handle this fluid processing problem: pressure losses must be small and, at the same time, a large proportion of the particles present in the stream to be processed must be captured. Particulate matter should be removed from the water as efficiently as possible in order to minimize the amount of water which has to be handled. The most important component of seston, in terms of its contribution to the ecology of coastal waters, is the primary producers, the phytoplankton. Seston particles, inclusive of phytoplankton, are generally small, and the need for high overall particle removal efficiency would suggest that the filtration structures would be of small dimensions and, as a consequence, of high flow resistance. In the case of many passive filter feeders, for instance, an array of small collecting elements is presented to the flow. The array must be close enough together to be an efficient collector without unduly restricting the flow through it. A balance must be struck between the two competing effects.

The same optimization problem arises in the design of filters by engineers for purposes such as air cleaning. For example, filters used for cleaning air in domestic heating systems are made of beds of loosely packed fibres. The designer must decide on the diameter of the fibres used and the density at which they are packed in order to achieve a desired level of collection efficiency. The criterion to be satisfied is that the design gives a minimal total cost for the performance specified. The costs expressed on an annual basis arise from an amortized charge to pay for the construction of the filter and the cost of energy required to force the air across the filter. Broadly speaking the designer looks for a balance between a large and loosely packed filter and a small, more tightly packed version. Fibre diameter must also be specified. Small-diameter fibres will give good collection efficiency but will require more energy for gas transport, so again a compromise is necessary in the setting of this parameter of the design.

In the same way, the structures which suspension feeders use to collect seston can only be effective if a balance is reached between the conflicting demands of high removal efficiency and the costs of building and maintaining collecting structures. The cost associated with pumping seawater to the filtration surfaces will play a part in the metabolism of an active suspension feeder, but its relative importance is disputed (see Chap. 9, the section, Ultimate questions).

The theoretical analyses which guide the industrial designer of filtering devices were largely developed to solve problems of gas/solid separations including gas cleaning, industrial hygiene, and medicine. In the biological literature, these developments are generally called the *aerosol theory*; we will continue to use this term. Many of these findings can help us begin to understand the mechanisms which might be important in suspension feeding, as was first pointed out by Rubenstein and Koehl (1977). Although the medium of interest for us is water, not air, much of the aerosol theory can be applied with little or no modification.

The student of suspension feeding, however, faces a more difficult problem than the physical scientist or engineer, who generally deals with steady flows, rigid structures, well-defined geometry, relatively simple particulate properties, and no feedback from the filtering system to changes in external variables such as velocity and concentration. Biological suspension feeding systems are more complicated in structure and often surprising in their subtlety of operation. Filter elements may be much more complex in their geometry than the simple shapes, e.g. circular cylinders and spheres, commonly addressed by aerosol theory. Biological filtering elements may assume different orientations with changes in velocity (Harvell and LaBarbera 1985) and may systematically sweep the fluid around them (Trager, Hwang, and Strickler 1990). Secondary flows may be exploited to divert food particles to collecting surfaces as, for instance, in the bivalve filtering system (Jørgensen 1981a). In addition, as pointed out by Leonard (1989), seston capture rates not only are determined by fluid mechanics, but also may depend on "complex behavioral modulation of the capture response." Notwithstanding these difficulties in the extension of aerosol theory to biological problems, insights into the limitations to feeding rates which may control the growth of filter feeders may sometimes be obtained by the use of these theoretical principles.

Sieving

Of all the physical mechanisms, sieving is the most obvious and simple to understand. Water containing seston is passed through a porous obstacle and particles larger than the passages for flow are retained. Although this is a simple concept, the reality is more complicated. Most suspension feeders utilize seston of a wide range of sizes and while large particles may be sieved, smaller particles will also be separated simultaneously at lesser collection efficiencies by other mechanisms. There are many examples, however, of suspension feeders in which sieving is the dominant mechanism. The use of mucous nets by ascidians and polychaetes has been discussed in the previous chapter (see the section, Sieving suspension feeders). These nets are used within an animal, and the pressure differences required to pump water through the mesh have been calculated and in some cases reconciled with the mesh sizes inferred from particle retention efficiency measurements.

In some freshwater aquatic insect larvae, seston is collected by presenting a filter net to a passing flow. Loudon (1990) and Loudon and Alstad (1990) have discussed the mechanics of particle collection by freshwater caddisfly larvae. The filter is external to the animal, and the seston particles after capture are collected by the forelimbs for passing to the mouth. These filtering nets are typically arrays of cylindrical obstacles and separate particles both larger and smaller than the openings between the elements of the filter. Silvester (1983) and Cheer and Koehl

(1987) have analyzed the fluid mechanics of flows through biological filters and have described some applications in suspension feeding.

A pelagic tunicate, *Oikopleura vanhoefferi*, uses an unusual form of sieving to separate particles (Morris and Diebel 1993). Instead of collecting particles, a mesh system is used to remove water from the inlet stream to concentrate the particles before acceptance by the food tube. The stream is, so to speak, de-watered. This process is analogous to that of a man-made tangential, or cross-flow, filter. Concentration factors of 100–1000 were measured in these experiments.

The aerosol theory and benthic suspension feeders

Many suspension feeders do not remove seston from water by a simple sieving mechanism. When the structures which effect the separation are examined, it is seen that they are not formed as nets, and the gaps between collecting elements are often larger than particles which are observed to be collected efficiently. There are obviously other mechanisms at work.

In gas-cleaning technology, much the same situation is met. When particles are removed by filters which consist of beds of fibres, the passages in the filter are generally much larger than the smallest particles collected. Particles are collected within the bed, rather than preferentially at the surface, as would be the case if the passages were smaller than the particles. The fibres in the bed must collect particles by mechanisms other than sieving. These involve aerosol mechanisms occurring within the fibre bed filter, as pointed out by Patterson (1991). The basic principles of aerosol filtration have been reviewed by many authors, including Pich (1966), Licht (1980), and Fuchs (1986).

Aerosol mechanisms should be applicable to the similar problem of suspension feeding by attached benthic animals. The particles of interest are of similar size, although their densities are much closer to that of the fluid, and the collecting elements of suspension feeders are often of cylindrical shape and in the same size range as aerosol filters. The velocities are generally lower in biological systems, but water has a kinematic viscosity about $^1/_{15}$ that of air so that the Reynolds numbers of both types of applications are often in the same range. In fact, the creeping flow regime, which is usually encountered at the scale of the collecting elements of biological systems, is perhaps the best understood Reynolds number range of aerosol theory. The application of these ideas to suspension feeding in benthic animals was developed systematically in the important paper of Rubenstein and Koehl (1977). They showed that by considering in detail the mode of particle capture by an isolated fibre, it was possible to determine the dominant mechanisms and to "provide insights for those investigating the efficiency of various modes of filter feeding and the mechanisms of size-selective filter feeding."

Consider the problem of a fluid stream containing suspended particles which flows towards an isolated circular cylinder oriented with its axis normal to the flow vector. Some of the particles in the approaching stream will strike the obstacle, and if the adhesion or retention coefficient is unity, these particles will be collected. The engineering efficiency of collection must be predicted, and this is done according to equation (5.1).

In order to analyze seston encounter efficiencies, some simplifying assumptions are necessary. The particles are assumed to be much smaller than the obstacle and to be evenly distributed in the approaching stream. The surface of the obstacle is assumed to be smooth and the flow to be steady, uniform, and turbulence-free. The kinematics of the flow around the obstacle, i.e. the pattern of streamlines, is assumed to be known. The adhesion coefficient is assumed to be unity. Emphasis will be placed on the low Reynolds number case because of its practical importance in suspension feeding. Additional assumptions may be added in specific cases as the arguments are developed.

Each of the five mechanisms of capture of particles from aerosols (direct interception, gravitational deposition, inertial impaction, Brownian diffusional deposition, electrostatic attraction) will be examined in turn, and the functional dependencies of the collection mechanisms will be described. Equations for the engineering efficiency and the encounter rate will be developed. As well, indices will be defined which will provide a simple means of judging whether or not the conditions under investigation are favorable to the mechanism. The treatment follows that of Pich (1966), Rubenstein and Koehl (1977), Spielman (1977), and Shimeta and Jumars (1991).

Direct interception

Consider a particle of neutral buoyancy in a steady flow around a right circular cylinder. This particle will move with the fluid if there are no other forces acting on it. Its path will coincide with the fluid streamline it was on upstream of the cylinder. It will collide with the cylinder if this stream line comes within one particle radius of the surface of the cylinder. If the retention efficiency is unity, the particle will be captured. This mode of capture is called direct interception.

Direct interception by a right circular cylinder is shown schematically in Fig. 5.2A, with the flow assumed to be steady and two-dimensional. Well upstream of the cylinder, the flow is uniform, meaning that the velocity does not vary either in the flow direction or in the cross-flow direction. The motion of the fluid is visualized by the use of "streamlines," which represent the trajectories of infinitesimal fluid elements marked at the upstream edge of the diagram. Molecular diffusion is ignored. The streamlines must be everywhere tangent to the velocity vector, so that fluid cannot cross streamlines. As a consequence of the law of conservation of mass, the product of the average velocity between adjacent streamlines and their spacing is everywhere constant and equivalent to the volumetric flow per unit depth into the paper between streamlines. By convention, the streamlines are evenly spaced in the uniform flow upstream of the obstacle: i.e. the "stream tubes" defined by adjacent streamlines all carry equal volumetric flows. Streamlines which come closer indicate an accelerating flow, whereas any reduction in velocity will result in the streamlines' diverging.

The capture efficiency and the encounter rate can be deduced if the streamlines are defined and it is assumed that the particles follow them without modifying the flow. All particles (diameter d_p) carried in the stream tube defined by the streamlines which pass above and below the cylinder (diameter d_c) at a distance of one particle radius will be removed (see Fig. 5.2A). This defining streamline (the "limiting streamline") has a position far upstream with respect to the axis of flow which is defined by the cross-flow distance (δ). Since the width of the stream tube which would have flowed through the space occupied by the cylinder in its absence is d_c, the cylinder diameter, the engineering collection efficiency for direct interception alone, E_R, must be

$$E_R = \frac{\delta}{d_c} \tag{5.2}$$

The encounter rate for a cylinder of length l_c is given by

$$F_R = cUl_c d_c\left(\frac{\delta}{d_c}\right) = cUl_c\delta \tag{5.3}$$

and a convenient index to indicate the relative efficiency can be defined as

$$N_R = E_R \tag{5.4}$$

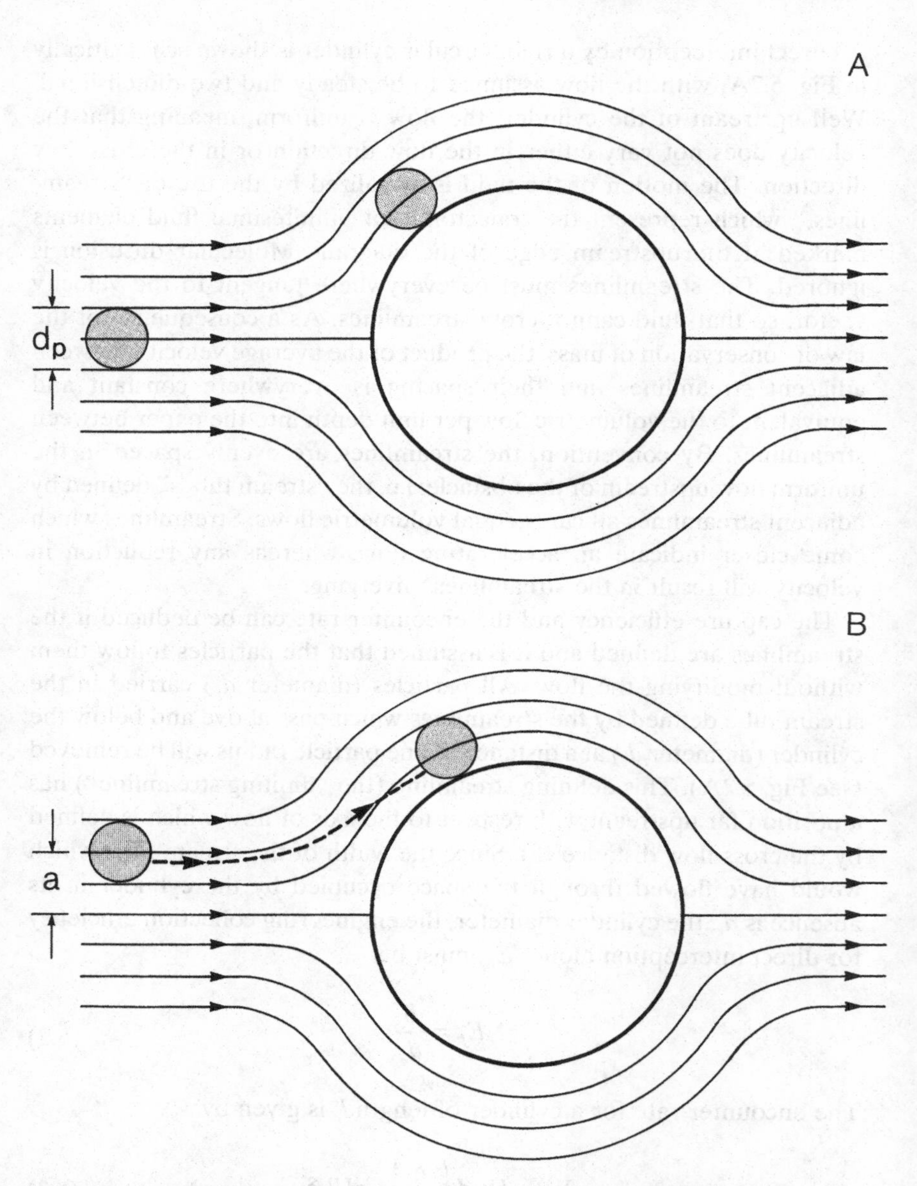

Figure 5.2 Schematics of capture of sestonic particles of diameter d_p by a cylinder of diameter d_c in a flow of $Re_c < 1$: A, direct interception; B, inertial impaction.

The distance, δ, will be dependent on the flow field around the cylinder. If the spacing of the streamlines is assumed to be unchanged as the flow passes around the cylinder, δ can be replaced by the particle radius, giving

$$E_R = \frac{d_p}{d_c} \tag{5.5}$$

This approach is used by Rubenstein and Koehl (1977) and Shimeta and Jumars (1991) as a first approximation to the solution of the problem. In creeping flow, however, the fluid velocity near the cylinder is less than the upstream velocity. By the principle of conservation of mass, the volumetric flow between any two streamlines must be constant everywhere in the flow field. It follows that the spacing between the stream line to the stagnation point and the limiting stream line must increase in the downstream direction since the average velocity in this stream tube decreases.

A comment on the mechanics of steady flow around submerged cylinders is needed at this point. The fluid flowing around the cylinder is subject to forces which arise from the inertia of the fluid elements and also as a result of the viscosity of the fluid, which resists any deformations. If the inertial forces are small with respect to the viscous forces, it may be possible to ignore them and find a solution for the streamlines dependent on the viscous properties of the fluid. The criterion for this condition is that the "Reynolds number" of the flow is small. The Reynolds number is defined as

$$\mathrm{Re} = DU\upsilon^{-1}$$

where D and U are, respectively, a characteristic size parameter, here conveniently the diameter of the cylinder or sphere and the upstream velocity, and υ is the kinematic viscosity of the test fluid used. For Reynolds numbers well below unity for spheres, the flow can be represented by a "creeping flow" solution. In the case of cylinders, the assumption that inertial forces are everywhere much less than viscous forces at low Reynolds numbers raises mathematical difficulties. An approximate solution accurate near the surface of the cylinder has been provided by Lamb (1962).

A more accurate formulation of the problem can be made if the Lamb solution for the velocity field around the cylinder at low Reynolds number is used. For the flow close to the cylinder, the following equations apply (Spielman 1977):

$$E_R = 2A_F\left(\frac{d_p}{d_c}\right)^2 \tag{5.6}$$

and

$$F_R = 2A_F cUl_c d_p^2 d_c^{-1} \tag{5.7a}$$

where

$$A_F = 0.5\left(2 - \ln\left(\text{Re}_c\right)\right)^{-1} \tag{5.7b}$$

$$\text{Re}_c = \frac{Ud_c}{\upsilon}$$

The term A_F is a dimensionless coefficient and is here evaluated for the specific case of an isolated right circular cylinder in a low Reynolds number flow.

Unfortunately, this solution of the problem is not the whole picture. As the particle approaches the surface of the collector, the implicit assumption in this treatment that the presence of the particle does not alter the flow around the cylinder becomes less justifiable. It can be expected that the particle's approach to the surface will be restrained by the outflow of fluid displaced by the particle. In addition, as the particle approaches the collector it is subject to attractive van der Waals forces, which tend to counteract the effect of fluid drainage around the particle. Contact with the surface is thus mediated by intersurface forces. These complications are described by Spielman (1977) and will be discussed in a later section.

Gravitational deposition

A particle which has a greater density than the surrounding fluid will sink at its terminal settling velocity, w_s, and its trajectory will cross the fluid streamlines, resulting in capture if the surface of the collector is reached. The encounter rate for a horizontal cylinder is given (Shimeta and Jumars 1991) by

$$F_G = cw_s l_c\left(d_c + d_p\right) \tag{5.8}$$

Most particles of interest will settle at a Reynolds number considerably smaller than 0.2, for which the settling velocity can be calculated from Stokess law by

$$w_s = \frac{d_p^2\left(\rho_p - \rho\right)g}{18\mu} \tag{5.9}$$

where

ρ_p = density of the particle
ρ = density of the fluid
μ = absolute viscosity of the fluid
g = gravitational constant

given that the particle is spherical and of known density. The encounter rate for particles small with respect to the obstacle is then approximately proportional to the square of the particle diameter and is independent of the ambient velocity. Equation (5.9) cannot be used if the density of the particle is not precisely known and inaccuracies may arise if the particle deviates from a spherical shape.

The engineering efficiency can be calculated as a fraction of the upstream flux of particles:

$$E_G = \frac{w_s\left(1 + d_p/d_c\right)}{U} \tag{5.10}$$

and the efficiency index can be taken to be

$$N_G = \frac{w_s}{U} \tag{5.11}$$

which can be considered as being equivalent to the efficiency for particles small with respect to the obstacle, or as a measure of the drift of particles across the fluid streamlines towards the upper surface of the obstacle.

Inertial impaction

When fluid containing particles approaches an obstacle, the streamlines curve and particles of densities greater than the fluid will tend to move in straight lines because of their inertia. These particles are more likely to be captured. This mechanism of capture is called inertial impaction and is shown schematically in Fig. 5.2B.

The importance of inertial impaction is determined by the stopping distance of the particle in the fluid, which is defined as the distance the particle, after release in the fluid at a certain velocity, will move before it stops. The stopping distance of interest here is that for the approach velocity of the fluid. It is given by

$$l_s = \frac{U d_p^2 \left(\rho_p - \rho\right) g}{18\mu} \tag{5.12}$$

An index of efficiency for inertial impaction may be formed by the ratio of the stopping distance to the diameter of the collecting cylinder. This ratio is called Stokes number.

$$N_I = \frac{l_s}{d_c} = \text{Stokes number} \tag{5.13}$$

For low values of Re_c, it can be argued (Shimeta and Jumars 1991) that if the stopping distance is less than the radius of the cylinder, the width of the stream tube cleaned of particles is increased by twice the stopping distance and so the encounter rate, in excess of that by direct interception, is given by

$$F_I = 2cUl_sl_c \tag{5.14}$$

If, on the other hand, the stopping distance is much greater than the radius of the cylinder, the particles will continue in a straight line as the streamlines curve around the obstacle and all the particles in a stream tube of the diameter of the cylinder will be collected, giving

$$F_I = cUd_cl_c \tag{5.15}$$

At other values of the stopping distance, prediction of the encounter rate is more difficult. It is determined by Stokes number and also by the Reynolds number of the flow around the cylinder. At Reynolds numbers less than about 0.2, the efficiency can be estimated by hydrodynamic arguments. Davies and Peetz (1955) have presented a plot of efficiency against Stokes number for an Re_c of 0.2 with the particle to cylinder diameter ratio as a parameter. They have also predicted efficiency at higher Reynolds numbers.

Inertial impaction is of considerable practical importance over a wide range of applications in aerosol science at higher Reynolds number flows. Langmuir and Blodgett (1946) analyzed this problem in a study of aircraft wing icing and provided predictions of collection efficiency as a function of Stokes number. Brun et al. (1955) refined and verified these calculations. Experimental measurements of collection efficiencies of cylinders, ribbons, and discs which were made by May and Clifford (1967) at Re_c of 165–8500 generally verify the theoretical predictions for cylinders. These results are important in aerosol mechanics but are of limited use in the understanding of suspension feeding by benthic animals, which are generally characterized by smaller values of N_I and Re_c.

Collection by Brownian diffusion

Particles of interest in suspension feeding are often small enough to be affected by Brownian diffusion. These particles will move off the streamlines and some will strike the collecting surface. A diffusion coefficient for non-motile particles can be estimated from

$$D = \frac{KT}{3\pi\mu d_p} \tag{5.16}$$

where K is Boltzmann's constant, T is temperature (Kelvin), and μ is the absolute viscosity of the fluid.

One way of evaluating the relative importance of the diffusion mechanism in particle collection is to determine the Peclet number

$$\mathrm{Pe} = \frac{Ud_c}{D} \tag{5.17}$$

where U is the approach velocity and d_c is the diameter of the cylinder, here taken to be the collector of interest. The Peclet number is an indicator of the ratio of the advective flux of particles from upstream and the diffusive flux to the cylinder. Pich (1966), Rubenstein and Kohl (1977), and others have used the reciprocal of the Peclet number as an index for the intensity of collection by diffusion: $N_D = 1/\mathrm{Pe}$.

The analysis of the mechanics of diffusion of particles to a collecting surface has been done in an approximate way by assuming that the particles can be represented by points and that they are present in sufficient numbers that their concentration can be assumed to approximate a continuum. Given these assumptions the problem can be analyzed in a similar manner to that for problems in the transfer of dissolved solutes or heat. The simplest case is of diffusion to a cylinder or sphere in still water. The Peclet number for this case is zero. The solutions are well known (Crank 1956) and give the concentration distribution in space and time. The solution is time-dependent since the particles must be drawn from increasing distances from the collector as the surrounding layers are depleted. It is of limited interest in the understanding of benthic suspension feeding.

A more relevant problem is that of diffusion with flow past the collector, and the mechanism can be analyzed as a combination of diffusion and advection by the passing fluid (Pich 1966). If it is assumed that the particles are present at sufficient concentration that they can be treated as a continuum, their mass balance over an infinitesimal fluid element is given by

$$\frac{\partial c}{\partial t} + u \cdot \nabla c = D\nabla^2 c \tag{5.18}$$

where c is the concentration of particles at this point, D is the diffusion coefficient for the particles, and u is the velocity vector at the position of the fluid element (Crank 1956). This equation may be understood as a combination of Fick's second law of diffusion (the first and last terms) and a term expressing the advection of material by the local velocity.

For this equation to be solved the velocity field around the cylinder must be known. Solutions have been developed for low Reynolds number flows and for potential flows (see Pich 1966; Shimeta and Jumars 1991). The separation efficiency for the case of low Reynolds number is

$$E_D = 2.92(2 - \ln \mathrm{Re}_c)^{-1/3}\left(\frac{Ud_c}{D}\right)^{-2/3} \tag{5.19}$$

There is some uncertainty about the value of the coefficient (Pich 1966).

Shimeta (1993) gives the following expression for the encounter rate for a cylinder in low Reynolds number flow:

$$F_D = 1.17D^{2/3}cl_c(Ur_c)^{1/3}(2 - \ln \mathrm{Re}_c)^{-1/3} \tag{5.20}$$

This result predicts that the encounter rate due to Brownian diffusion increases with the diffusion coefficient, as would be expected, but also with the velocity, which is less obvious. The efficiency of collection is inversely related to velocity, but since increased velocity delivers more particles to the region of the cylinder the net effect is to increase the encounter rate.

Electrostatic attraction

If there is an attractive electrostatic force between a particle and a collecting surface, the particle will move across streamlines and is more likely to be collected. This is an important mechanism in gas cleaning, which is exploited in electrostatic precipitators and in fabric filtration. Particles in seawater, however, have a "surface charge that is neutralized locally by counterions that concentrate on the solution side to form an electrical double layer" (Spielman 1977). The electrical fields extend to only about 0.02 μm from the surface, and it is reasonable to expect that in marine conditions, electrostatic forces will determine the adhesion of the particle to the collecting surface but that other collecting mechanisms will be of more importance in getting the particle in close proximity to the collecting surface.

Shimeta and Jumars (1991) present equations for the encounter rate and the efficiency index for electrostatic attraction, which are given here in order to complete the set of relationships for the aerosol mechanisms. If Q is the charge per unit length of collector and q is the charge on a particle of diameter d_p, the encounter rate for an infinite circular cylinder is given by

$$F_E = \frac{4cQql_c}{(3\mu d_p)} \tag{5.21}$$

and the efficiency index by

$$N_E = \frac{4Qq}{(3\pi\mu d_p d_c U)} \tag{5.22}$$

These equations were derived for particles and collecting surfaces which are smooth, in the sense of having irregularities much smaller than 0.02 μm, conditions which are not commonly encountered in benthic suspension feeding. Although the equations may not be directly applicable in many cases, the electrostatic mechanism may still be important in the final stages of collection as the particle approaches the surface. How this mechanism interacts with direct interception will be discussed in the next section.

In freshwater applications, because of the much lower ionic strengths involved, electrostatic attraction should not be dismissed without careful consideration (Shimeta and Jumars, 1991). Gerritsen and Porter (1982) showed that relative ingestion of three sizes of polystyrene spheres by *Daphnia magna* could be modified if the surface charges on the particles were neutralized or if a non-ionic surfactant were added to reduce wettability.

Combined mechanisms

Each of the mechanisms of capture has been analyzed on the assumption that the other mechanisms are of negligible importance. Particles which are massive enough to separate by gravity, for example, will often also have enough inertia to separate from curvilinear streamlines and so be removed by inertial impaction simultaneously. In many, if not most, cases of practical interest, more than one mechanism may contribute. For some solutions which have been derived for the aerosol case, see Pich (1966). In the case of liquids, these solutions may not always be useful because of the differences between liquids and gaseous systems in the final stages of capture, viz. the complications which arise from the drainage of fluid from the gap between the particle and the collecting surface and the van der Waals electrostatic forces which counter this resistance.

Particles which are not of neutral buoyancy will be influenced by gravity and inertial effects. Yoshioka and Kansoka (1972) have analyzed the combined effects of gravitation, interception, and inertial separation in gas–solid systems. Theoretical equations for collection efficiencies at $Re_c < 1$ are presented and their experimental measurements are consistent with the theory. If particles are of neutral buoyancy, gravitational capture and inertial separation are not factors. Direct interception may be influenced by electrostatic forces and by Brownian diffusion.

The theoretical analyses of interception in liquids have been extended to cover some of the cases where two or more processes contribute to the collection of particles. We will briefly look at approaches to the solutions of this problem for low Reynolds number flow around a cylinder. We can approach the problem of combined Brownian diffusion and direct interception it in at least two ways. The fundamental analysis of particle migration can be carried out with more than one mechanism in mind. For instance, Langmuir (as cited by Dorman 1966) has modified his equation for diffusion of a particle to allow for direct interception by assuming that the particles diffuse, not to the surface of the collector, but to an imaginary shell at a distance of one particle radius from the collector. Another approach is to assume that each mechanism is operating independently and that, in the case of the total collection efficiency, it is the linear sum of the efficiencies of each mechanism. Fuchs (1964) and others have pointed out that in fact considerable error can arise from this view of the problem when each mechanism makes a significant contribution. It is, however, useful in identifying the ranges of variables in which one or the other mechanism is dominant.

The second approach was used by Shimeta (1993) in order to quantify the dependence of encounter rate on the size of non-motile particles. In the circumstances of interest to us, particle size may range widely. Potential prey may be as small as colloids and viruses, that is under 0.1 μm, with successively larger bodies as indicated in Fig. 5.1. If it is assumed that the encounter rates for direct interception and Brownian diffusion are addi-

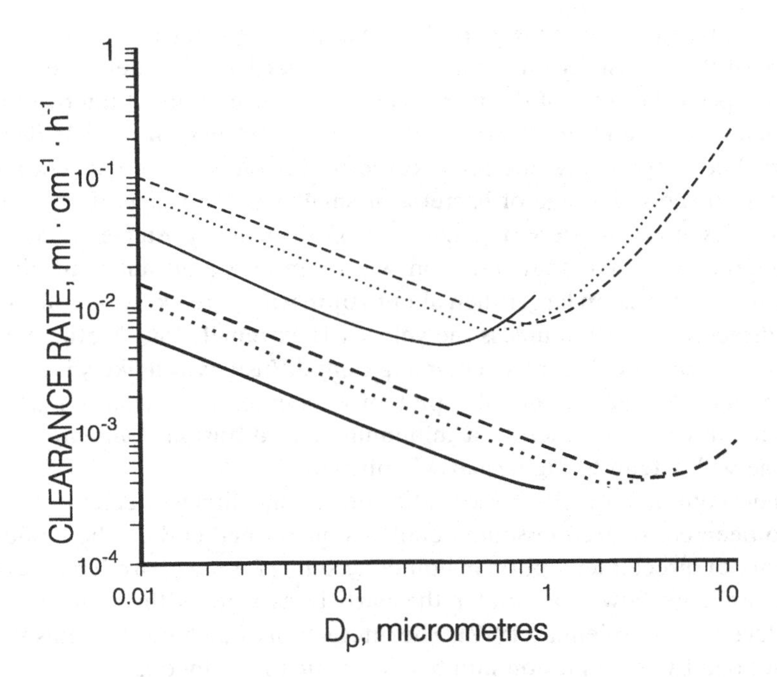

Figure 5.3 Combined collection by direct interception and Brownian movement. Clearance rates for a right circular cylinder in milliliters cleared per hour per centimeter of length versus particle diameter. Velocity of approaching stream: lower three curves, $0.01\,cm \cdot s^{-1}$; upper three curves, $1.0\,cm \cdot s^{-1}$, solid, dotted, and dashed lines, cylinder diameters of 10, 50, and 100 µm; $d_p < 0.1\ d_c$; $Re_c < 1.0$.

tive, equations (5.6) and (5.20) can be added to give the combined encounter rate. Following the procedure of Shimeta (1993), this calculation has been done for cylinders of three diameters at velocities of 0.01 and $1.0\,cm \cdot s^{-1}$ (Fig. 5.3). The Reynolds number of the flow around the largest cylinder at the highest velocity is unity. The ordinate is the clearance rate per unit of length of the cylinder, equivalent to the encounter rate divided by concentration of particles and collector length. This definition of clearance rate, based on a unit length of the collector, is different from the clearance rate, *R*, defined in Chapter 2, which is based on a single animal, or *S*, the seston removed by a unit weight of a suspension feeder. The diffusion coefficient has been calculated from equation (5.15) at 20°C. The curves have been terminated where the particle diameter exceeds 10% of the collector diameter.

In the Brownian diffusion-dominated regime, the predicted clearance rate is inversely related to particle size as a consequence of the dependence of the diffusivity on the inverse of the particle diameter. The minimums occur because of the increasing importance of direct interception at higher velocities and its second-power dependency on particle diameter. The curves show the importance of diffusion to the collection of seston of the size range of bacteria or smaller in the range of sizes and velocities used in preparing the plot. For smaller obstacles at lower velocities it is clear that diffusion will be more significant than direct interception. Seston larger than about 10 μm will more likely be collected by direct interception unless the velocity is extremely low. Motile particles will behave differently. Their apparent diffusivity will likely increase with particle size so that the plot of encounter rate against particle diameter will not generate a minimum. The diffusion-dominated size range will extend to higher particle diameters.

The combination of electrostatic forces and direct interception has also been considered in some detail. As mentioned earlier, the modern theory of direct interception in liquids accounts for the hydrodynamics of the drainage flow from under the particle as well as the influence of surface forces (Spielman 1977). The interaction of the two mechanisms is described by the adhesion number N_{ad}, which is defined as

$$N_{ad} = \frac{Q r_c^2}{9\pi\mu r_p^4 A_F U}$$

where

Q = Hamaker constant
r_c = radius of the collector
r_p = radius of particle

The term A_F describes the flow field around the collector, equation (5.7b). The term Q is a constant of proportionality in Hamaker's equation for the force between a spherical particle and a much larger collector. It depends on the properties of the molecules of the two surfaces and of the liquid medium (Spielman 1977). Monger and Landry (1991) have used a value of 3.7×10^{-14} erg for general lipid bilayer systems in seawater. The adhesion number can be interpreted as the ratio of van der Waals attraction to hydrodynamic forces.

Spielman and Fitzpatrick (1973) solved the equation for the limiting particle trajectory around a cylinder in the presence of an attractive force. By knowing this trajectory, the collection efficiency can be calcu-

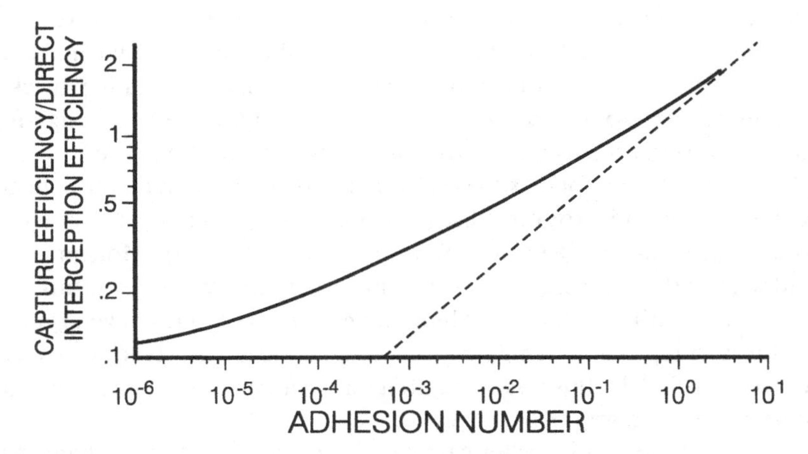

Figure 5.4 Normalized capture efficiency versus adhesion number for capture of neutrally buoyant particles by a cylinder: direct interception (Eq. 5.7) and van der Waals attraction. Dashed line (Eq. 5.21), capture by van der Waals attraction only (Spielman 1977).

lated. Their results were presented in the form of a dimensionless plot of the normalized capture efficiency vs. the adhesion number (Fig 5.4). The normalized capture efficiency is the predicted capture efficiency divided by that of the classical theory equation (5.5). It is interesting to note that the classical theory does not agree with the more advanced theory except coincidentally at one value of the adhesion number: i.e. there is no asymptotic convergence of the two treatments. At low adhesion numbers the capture efficiency is lower than predicted by the classical theory, primarily because of hydrodynamic drainage from under the particle. At high adhesion numbers, the electrostatic effects become dominant and the capture efficiency is given by

$$E_E = \left(\frac{3\pi N_{\text{ad}}}{4}\right)^{.333} \tag{5.23}$$

These theoretical results for capture by cylinders were compared with experimental measurements by Schrijver, Vreeken, and Wesselingh (1981), who used a glass cylinder of 250-μm diameter and latex particles ranging in size from 0.8 to 25.7 μm. The working fluid was an aqueous electrolyte solution so that the only important surface force was the London–van der Waals force. These particles were not of neutral buoyancy. The cylinder was mounted in a vertical position to prevent gravitational collection for measurements of collection efficiency. The results fell within a factor of 2 from the theoretical prediction and confirmed the general form of the relationship derived by Spielman and Fitzpatrick, although the interpretation of the results was complicated by a significant contribution of Brownian diffusion of the smallest particles.

Spielman and co-workers have also derived similar results for collection by spheres and a rotating disc (Spielman 1977). The effect of gravity has also been included (Spielman and Fitzpatrick 1973). Monger and Landry (1990) have used a similar approach to analyze the capture of bacteria by marine flagellates. Shimeta and Jumars (1991) have argued that the assumptions of smooth collecting surfaces and steady velocity are not satisfied in this case and that this refinement of interception theory is unnecessary.

Clearance rates were earlier calculated for the combined mechanisms of direct interception and Brownian diffusion for some typical conditions (Fig. 5.3). A similar calculation has been made by using Spielman and Fitzpatrick's treatment for direct interception with an allowance for surface forces (Fig. 5.5), using $Q = 3.7 \times 10^{-14}$. The plots are for two velocities, 0.01 and $1.0\,\mathrm{cm \cdot s^{-1}}$, and two cylinder diameters, 10 and $100\,\mu\mathrm{m}$. One of the curves from Fig. 5.3 has been superimposed on the plot to show the relative importance of Brownian diffusion. If the plots are compared it is seen that particles of the typical size of bacteria (say diameters about $0.5\,\mu\mathrm{m}$) or smaller are cleared more rapidly by Brownian diffusion (curve on the upper left of Fig. 5.5) than by direct interception and electrostatic attraction, particularly at lower velocities. The curves on the right hand side of both graphs represent collection dominated by direct interception, but these have been calculated differently and the clearance rates are not consistent for Figs. 5.3 and 5.5. The classical theory, which does not incorporate surface forces, was used for Fig. 5.3, and the clearance rates of Fig. 5.5 should be considered to be the more accurate estimates.

There is one important caveat to be raised regarding these results. The implicit assumption in the treatment of surface forces is that the surfaces of the cylinder and of the spherical particles are smooth. Any surface roughness or asperities of the order of a few hundred angstroms or more will change the rate at which the particle approaches the surface, as will any departures from sphericity of the particle. Collecting surfaces and seston particles are more likely to violate these geometric constraints than not.

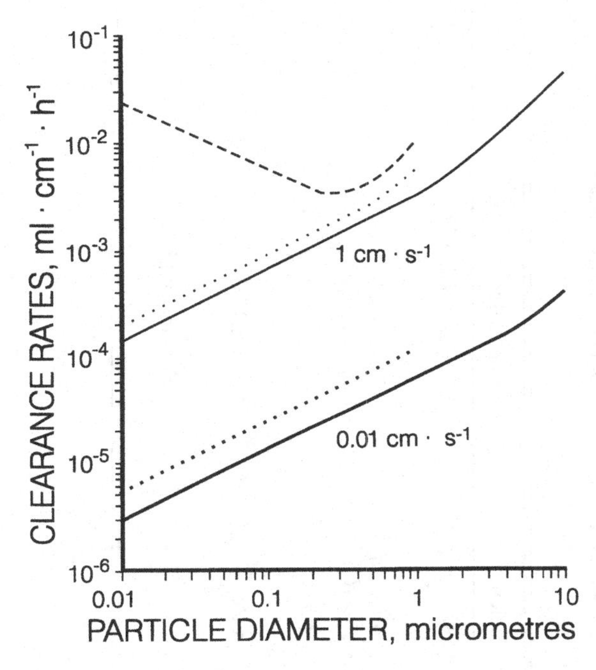

Figure 5.5 Clearance rates for right circular cylinders of diameter 10 μm (solid) and 100 μm (dotted) at velocities of 0.01 and $1.0\,\mathrm{cm \cdot s^{-1}}$, attributable to direct interception and electrostatic attraction. The dashed line is taken from Fig. 5.4 for direct interception, Brownian diffusion for $U = 1\,\mathrm{cm \cdot s^{-1}}$; $d_c = 10\,\mu\mathrm{m}$.

Assessing the relative importance of capture mechanisms

Aerosol theory suggests a simple method of assessing the relative importance of direct interception, Brownian diffusion, and electrostatic attraction in the collection process. The efficiency indices which have been defined for each mechanism provide a simpler means of estimating which is likely to be the predominant mechanism in a specific case (Ranz and Wong 1952; Shimeta and Jumars 1991). These have been collected in Table 5.2. In many practical situations, substitution of reasonable values of the key parameters into these ratios can quickly identify the relative importance of each of the collection mechanisms. As the preceding discussion has suggested, however, more than one mechanism may be important. In addition, relative encounter rates are not necessarily reflected well by the efficiency indices, as has been discussed by Shimeta and Jumars (1991), and the ratios should be interpreted with caution.

Table 5.2. *The five mechanisms of the aerosol theory: efficiency indices, engineering collection efficiencies, and encounter rates for infinite right circular cylinder at low Reynolds number.*[a]

Mechanism	Efficiency index	Engineering efficiency	Encounter rate
Direct interception (classical theory)	$N_R = d_p d_c^{-1}$	$E_R = 2A_F d_p^2 d_c^{-2}$	$F_R = 2A_F cUl_c\, d_p^2 d_c^{-1}$
Inertial impaction	$N_I = l_s d_c^{-1}$ [b]	$E_I = 2l_s d_c^{-1}$ if $l_s << d_c$ $E_I = 1.0$ if $l_s >> d_c$	$F_I = 2cUl_s l_c$ $F_I = cUd_c l_c$
Gravitational deposition	$N_G = w_s U^{-1}$	$E_G = w_s U^{-1}\left(1 + d_p d_c^{-1}\right)$	$F_G = cw_s l_c\left(d_c + d_p\right)$
Brownian diffusional deposition	$N_D = DU^{-1}\, d_c^{-1}$	$E_D = 2.92\left(2 - \ln \mathrm{Re}_c\right)^{-1/3}\left(U d_c / D\right)^{-2/3}$	$F_D = 1.17 D^{2/3} c l_c \left(U r_c\right)^{1/3}\left(2 - \ln \mathrm{Re}_c\right)^{-1/3}$
Electrostatic attraction	$N_E = 4Qq\left(3\pi\mu\, d_p d_c U\right)^{-1}$	$E_E = N_E$	$F_E = 4cQql_c\left(3\mu\, d_p\right)^{-1}$

[a] Symbols defined in text.

[b] $l_s = U\left(\rho_p - \rho\right)\left(18 d_p^2 \rho\mu\right)^{-1}$.

Additional insight into this question has been provided by Shimeta and Jumars (1991), who have calculated encounter rates and efficiency indices for a number of particles of importance to suspension feeders, which range in size from 1 μm (medium clay) to 2500 μm (marine snow). Collecting surfaces are circular cylinders of sizes of practical interest. They have presented their results in the form of plots of clearance rates and of encounter rates and efficiency indices normalized by those for direct interception alone as a function of ambient velocity. These plots are restricted to $Re_c < 1$. The rate of direct interception has been calculated using equation (5.5).

One of their plots has been reproduced here (Fig. 5.6). Clearance rates per centimetre of cylinder are plotted against ambient velocities from 0.1 to 50 cm·s^{-1}. In plot A, the clearance rates attributable to direct interception for nine particles of relevance to benthic suspension feeders (Table 5.2) are shown. The clearance rate is directly dependent on the size of the particles and the velocity but is independent of collector radius for this mechanism. The plot is for $Re_c < 1$, which implies that at the highest velocity of 50 cm·s^{-1} the plot should not be used if the diameter of the collector exceeds 2 μm. These plots were derived on the assumption that the particles are much smaller than the diameter of the collector. In the case of medium silt, for instance, the particle diameter is 20 μm. At 5 cm·s^{-1}, the maximum acceptable cylinder diameter, according to the Reynolds number limit, is also 20 μm so that the theory, devised for particles small with respect to the cylinder, is really only applicable to velocities well under 5 cm·s^{-1}. Plot B shows clearance rates attributable to inertial impaction on a cylinder of diameter of 50 μm. The dashed portions of the lines denote violation of the low Reynolds number condition. As would be expected the largest clearance rates are seen for the particles with the largest settling velocities. The smaller particles are little affected by inertial forces. At a velocity of 1 cm·s^{-1} the clearance rate for direct interception of small phytoplankton cells is almost two orders of magnitude greater than for inertial interception. The combined clearance rates due to inertial impaction and gravitational deposition are shown on plot C ($d_c = 50$ μm) and plot D ($d_c = 1000$ μm). The horizontal sections of the curves show where gravitational deposition is dominant. Particles with the highest settling rates have the highest clearance rates, usually comparable with the direct interception rates, especially at the higher velocities. Clearance rates due to Brownian diffusional deposition are shown on plot E for the smallest particles and collectors of various diameters. They are not neg-

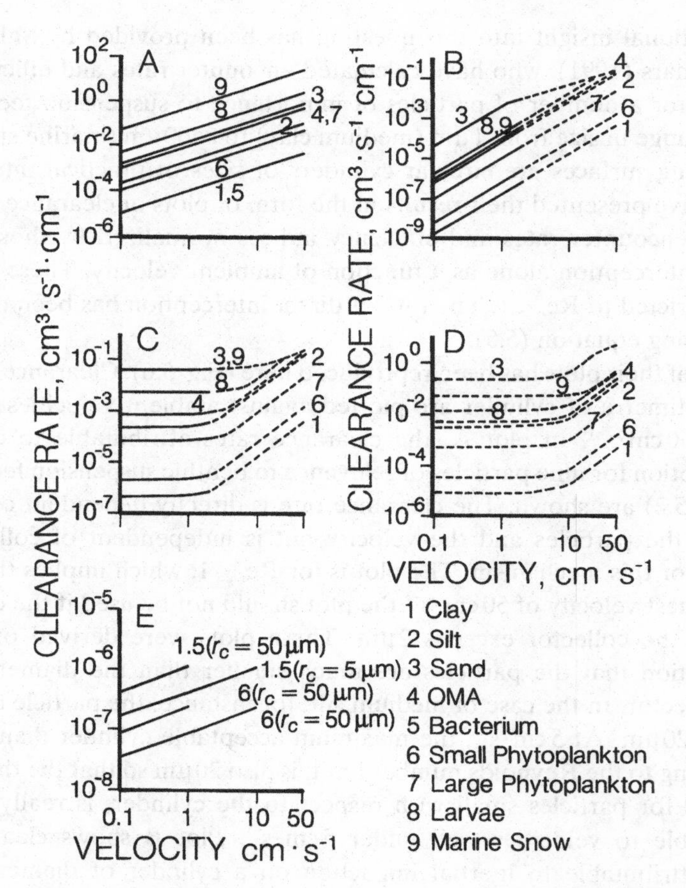

Figure 5.6 Capture of typical seston particles by a circular cylinder (expressed as clearance rates per unit length against velocity) for the mechanisms of the aerosol theory: A, direct interception; B, inertial impaction; C and D, inertial impaction and gravitational deposition with $d_c = 50\,\mu m$ (C) and $1000\,\mu m$ (D); E, Brownian diffusional deposition (r_c = radius of cylinder); OMA, organic matter aggregate. Lines are either ended or dotted if $Re_c > 1$ (Shimeta and Jumars 1991).

ligible with respect to direct interception, especially at the lowest velocities.

Although these plots are not applicable to all situations, a methodology for assessing the relative importance of the collection mechanisms for a particular case is clear from Shimata and Jumars's presentation. This approach is more illuminating than a simple use of indices. Note that a more accurate method of calculating the direct interception rates could have been used, as has been shown by Shimeta (1993), who used equation (5.7) rather than equation (5.5).

Higher Reynolds numbers

The primary collecting elements of suspension feeders are typically small and of cylindrical shape. It has been argued that in a majority of cases the velocities are such that they operate in the creeping flow regime (LaBarbera 1984). This appears to be so for many active benthic suspension feeders and for planktonic suspension feeders. Many of the passive benthic suspension feeders, however, can be shown to have primary collecting elements which commonly experience Reynolds numbers in excess of unity (Shimeta and Jumars 1991). For example, the pinnules of the soft coral *Alcyonium siderium* are about 380 μm long and 60 μm in diameter spaced at gaps of about 230 μm (Sebens and Koehl 1984). The velocity past the pinnule need only exceed $1.67\,\text{cm}\cdot\text{s}^{-1}$ for Re_c to be over unity, which is well within the range of velocities observed in nature.

The relationships developed for encounter rates for the creeping flow case cannot be extended to Re_c greater than about 0.1 without some error. These relationships are based on assumptions about the form of the streamlines of the flow around a cylinder which are derived from Lamb's solution. Experimental drag force measurements, which reflect the shape of the streamlines, show that at $Re_c = 1.0$ the Lamb solution overpredicts drag by about 25%. The streamlines, symmetrical about the cylinder in creeping flow, become compressed on the upstream side and relaxed downstream. The critical trajectory for direct interception will be different, consistent with a greater collection efficiency. At Re_c of about 10, a pair of attached vortices form on the downstream side of the cylinder. The form of the streamlines around a long circular cylinder at $Re_c > 1$ can be estimated (see, for instance, Davies and Peetz (1955) for a calculation at $Re_c = 10$). At $Re_c > 40$, vortices form alternately on each side of the downstream face and are shed periodically. At higher velocities the wake becomes turbulent.

From this description of the hydrodynamics of this flow, it is clear that the encounter rate equations lose their predictive value at higher Reynolds numbers and that the functional dependencies of encounter rate on collector size and velocity, used to deduce the relative importance of mechanisms, are not reliable. For a full discussion of the details of particle collection at higher Reynolds numbers, see Shimeta and Jumars (1991).

Other geometries

Encounter rate predictions for spherical collectors under low Reynolds number conditions can be found in Shimeta and Jumars (1991). Cylinders of non-circular cross sections are not amenable to simple analysis, except in the case of gravitational collection, because of the lack of hydrodynamic solutions for the flow around them, or because of their complexity if they exist.

Assemblages of collectors

Many passive suspension feeders do not present single collecting elements to the flow but instead use arrays in a more or less parallel configuration. For example, consider the pinnules on the tentacles of the sea pen (Best 1988), the octocorals (Patterson 1991a), the comatulid crinoids (Mayer 1979), and the stalkless crinoid *Antedon mediterranea* (Leonard 1989). Active suspension feeders may also use similar arrays, as is the case with bivalve molluscs. The first effect of introducing additional collecting elements is to modify the velocity field near the surface of the collectors. The term A_F, which was defined in equation (5.7a) for the case of flow around an isolated circular cylinder, is modified for arrays of cylinders (Spielman 1977). Streamlines will be compressed and direct interception efficiency will be improved. The second effect is that the bulk flow will be influenced. In the case of passive collectors presenting an array of collecting elements to a stream, the flow between the collecting elements will be reduced and more of the approaching fluid will be diverted around the array if the elements become more closely spaced.

Active suspension feeders will be required to supply pumping energy to move fluid through the filtering apparatus at a rate proportional to the superficial velocity and the area of collecting surface presented under low Reynolds number conditions.

Approximate solutions to the problem of the filtering of aerosols by beds of cylinders have been developed for low Reynolds number flows. The approach has been to simplify the problem to one in which the particles are assumed to be small with respect to the diameter of the collecting cylinders so that the velocity distribution need only be known close to the surface of the cylinder. The influence of the neighboring cylinders is reflected in the value of the term A_F introduced earlier for isolated cylinder flows in equation (5.7b). This term can be related to the volume fraction of fibres in the filter bed. The equations developed for capture by an isolated fibre can then be applied with the new value of A_F (Spielman 1977). The functional dependencies of encounter rates by the aerosol mechanisms on parameters such as velocity, particle diameter, and collector diameter are unchanged.

Interpreting mechanisms from parameter sensitivity

It is sometimes argued that the dominant collecting mechanism can be identified if the dependency of collection efficiency or encounter rate on changes of certain independent variables is observed. For instance, if an experiment is carried out using particles of different sizes with all other variables fixed, the efficiency of collection should increase as d_p^2, equation (5.6), if direct interception is the dominant mechanism. If Brownian diffusion, on the other hand, is dominant, the dependency is to the –2/3 power, equation (5.19). In principle, this should provide a means of identifying dominant mechanisms, but inspection of Table 5.2 shows that relationships are not that straightforward in many instances. The dependence of efficiencies on velocity, in particular, is complicated by the term A_F, which must be known for the flow past the collecting elements and which may not be easily determined in either laboratory or field experiments. In practice, the indices should be used with caution and calculations of encounter rates for model systems similar to the organism under study are a preferred method of analysis.

Ambient flow affects particle capture

Our treatment has been based on the assumption that collecting elements are rigid cylinders in a steady flow, conditions which are by no means universal. Benthic suspension feeders often live in wave-driven and turbulent environments, and the collectors experience periodic and randomly varying velocities. They may also be in shear zones, as opposed to uniform flow zones. The scale of the turbulent motions may be larger or smaller than the characteristic size of the collector, depending on the energy viscous dissipation rate of the turbulent flow and the collector size. Collection in wakes of either whole organism or single collecting elements may also be important. All of these hydrodynamic factors will affect the rate of collection of particles. In addition, collecting elements are often not rigid, but either move with the motion of the water, periodically sweep, or respond behaviorly to the presence of seston or velocity (see Chap. 4).

The adhesion or retention coefficient has been assumed to be unity in this discussion. In reality little is known of its magnitude, which depends

Table 5.3. *Passive suspension feeders for which aerosol theory seston encounter efficiency index calculations are available.*

Scientific name	Major taxa	Reference
Alcyonium siderium	Soft coral	Sebens and Koehl (1984)
A. siderium	Soft coral	Patterson (1984, 1991a,b)
Briarium abestimum	Soft coral	Lasker (1981)
Pseudoplexaura porosa	Soft coral	Lasker (1981)
Pseudopterogorgia americana	Soft coral	Lasker (1981)
?*Alcyonium* sp.	Soft coral	McFadden (1986)
Meandrina meandrite	Scleractinian coral	Johnson and Sebens (1993)
Ophiopholis aculeata	Brittle star	LaBarbera (1987, 1984)
Ptilosarcus gurneyi	Sea pen	Best (1988)
Oligometra serripinnia	Feather star	Holland et al. (1986)
Florometra serratissima	Crinoid	Byrne and Fontaine (1981)

on the balance of hydrodynamic forces and electrostatic forces. It is then dependent on the flow, the particle and collector geometry, and the chemistry of the surfaces. It is not necessarily constant for any given combination of seston and collecting element and can be expected to be velocity-sensitive.

Seston capture mechanisms

Five major categories of suspension feeders were recognized in Chapter 4: passive, active (inclusive of sieving and non-sieving forms), facultative–active, combined passive–active, and deposit–suspension switching species. Of these, only the first two groups have been cursorily examined as to which of the aerosol theory seston capture mechanisms was predominant. These include passive and a few active suspension feeders.

Passive suspension feeders

A list of the species for which data involving seston encounter efficiency calculations have been made is shown in Table 5.3. The species involved are either corals or echinoderms. A review of much of this work is presented by Shimeta and Jumars (1991). They point out that LaBarbera (1978) used an inappropriate way of calculating the efficiency of seston capture by direct interception. Further analysis by LaBarbera (1984), incorporating the hydrodynamic retardation model of Spielman (1977) and allowance for surface forces in seston capture by the brittle star, was also criticized by Shimeta and Jumars (1991) on several grounds. Among those grounds was that the hydrodynamics of the flow near the collector was not fully defined, that simple measures of mainstream flow are insufficient to characterise the appropriate velocity to the collector, and that streamlines around the collector depend on the Reynolds number, which, in the case of the brittle star, was >1 and therefore strictly not covered by aerosol theory.

Most of the seston encounter efficiency results obtained by authors shown in Table 5.3 suggest that direct interception was the predominant seston capture mechanism. Patterson (1984, 1991a), for example, found that the details of the capture process in *Alcyonium siderium* were more complicated than might be visualised by a simple application of aerosol mechanisms. Capture sites were not evenly distributed about the colony, and the frequency and location of capture were influenced by the velocity and turbulence of the ambient flow. Patterson also observed that cysts which were caught on the aboral side of the tentacle were usually lost before transfer to the mouth, leading him to suggest that large particles are more likely to be caught effectively by the tentacles on the downstream side of a polyp. The efficiency indices were calculated by using the tentacle width as a characteristic dimension, typical velocities near the tentacle, and a settling velocity of cysts of $0.08\,cm \cdot s^{-1}$. The indices for a velocity of $30\,cm \cdot s^{-1}$ were as follows:

Gravitational deposition	0.003
Direct interception	0.67
Inertial impaction	4×10^{-5}
Diffusive deposition	8×10^{-12}

For a velocity of $3\,cm \cdot s^{-1}$, the direct interception index was unchanged, the gravitational and diffusive deposition indices were increased by a factor of 10, and the inertial impaction index was reduced by a factor of 10. The indices suggest that direct interception was the dominant mechanism at velocities commonly encountered in nature. It should be emphasised, however, that some of the assumptions used in the development of the indices are violated. The efficiency of collection of cysts was also directly measured at flow velocities ranging from 2.7 to $19.8\,cm \cdot s^{-1}$ and found to decrease with velocity, qualitatively consistent with a dominant

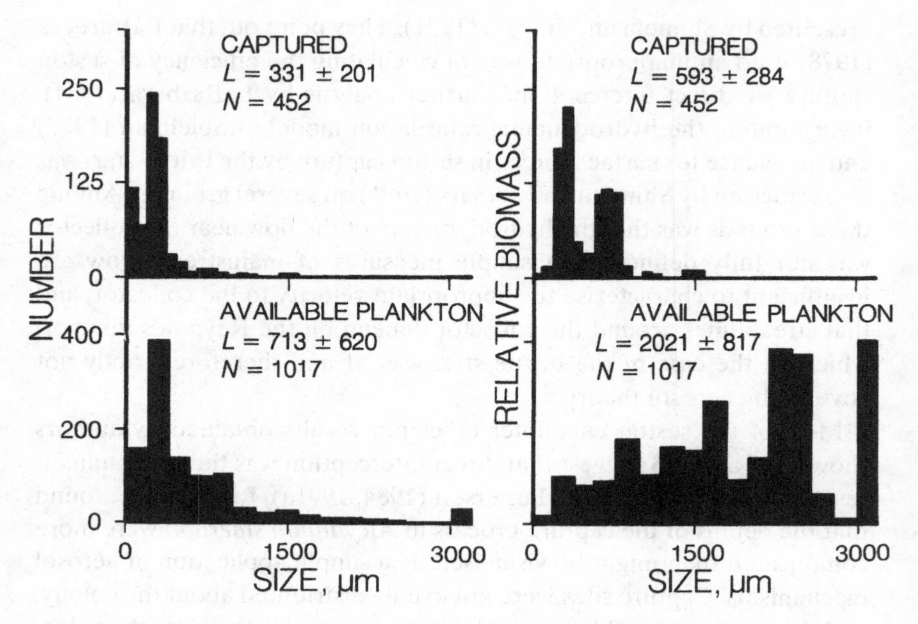

Figure 5.7 Size distributions of particles captured by *Alcyonium siderium* and of the available plankton: L, length ± 1 SD; plots on left, based on number of items; plots on right, based on biomass (Sebens and Koehl 1984).

mechanism of direct interception (equation [5.6], the weak inverse dependence on velocity appearing in the A_F term). Sebens and Koehl (1984) showed that the same species of soft coral was a benthic zooplanktivore selecting smaller prey, similar in size to the tentacles (width of 300 μm), but much larger than the pinnule diameter of 60 μm (Fig. 5.7). A regression of collection efficiency against the Reynolds number of polyps (based on the diameter of the oral dise and the velocity 1 cm above it) gave a dependence on Reynolds number to the power −1.55. This result suggests that in this case particle capture is influenced by factors in addition to those of the aerosol theory. Perhaps most important is the observation that many particles which struck the surface of the tentacles were not retained; Patterson (1991a) noted the need for additional study of this process. Changes in velocity also change the orientation of the polyps in the stream, resulting in differences in collection efficiency between upstream and downstream polyps.

Other possible aerosol mechanisms reported by the authors listed in Table 5.3 include gravitational deposition at low velocities in the coral *Meandrina meandrites* (Johnson and Sebens 1993) and strong adhesion of particles in feather stars (Holland et al. 1986).

Active suspension feeders

The active suspension feeders which have been examined to determine the mechanism of seston capture include only two categories: sieving suspension feeders and suspension feeding bivalve molluscs.

Sieving suspension feeders

The polychaete *Chaetopterous variopedatus* generates a flow through its U-shaped tube by muscular contractions of its fan-shaped notopodia (see Fig. 4.10). A mucus net used for feeding is generated continuously within its tube at the twelfth segment and is rolled up and eaten at 15-min intervals (Jørgensen et al. 1984). The net is made of longitudinal filaments estimated to be about 0.1 μm thick and transverse filaments of smaller diameter. The mesh size is 0.76 ± 0.96 by 0.46 ± 0.12 μm (Flood and Fiala-Medioni 1982). Velocity through the mesh was estimated to be $3\,mm \cdot s^{-1}$. Particle retention efficiency was measured by Jørgensen et al. by two different makes of electronic particle counters, which gave different results in the size range of interest. Retention of 90% was seen at a particle size of 1.2–1.7 μm, depending on the measurement method used. Bacteria of about 1 μm were retained at about one-third efficiency, as measured by fluorescence microscopy. The dominant collection mechanism is clearly sieving, but there is no evidence of how, if at all, smaller bacteria (say about 0.5 μm) are captured. These particles will be captured by direct interception and Brownian diffusion. As the seston is 5–10 times the filament diameter size, the aerosol theory is strictly inapplicable.

Similar measurements were made of particle retention by the gastropod *Crepidula fornicata*, which generates a current through the mantle cavity by the lateral cilia on the gill filaments (Jørgensen et al. 1984). Two mucus nets, one coarse and at the entrance to the mantle cavity, the other fine and transported across the gill surface, collect particles. The morphology of the fine net is not known but it is better than 90% efficient at removing particles larger than 1.2 μm. The pressure drops across the nets of both of these mucus net feeders were estimated and seemed to be consistent with the capacity of the pumping mechanisms.

Jørgensen et al. (1984) also described particle capture by mucus nets of the solitary ascidians *Ascidia obliqua*, *A. virginia*, and *Ciona intestinalis*. Seawater flow in the trophic fluid transport system of ascidians is described in Chapter 4, the section, Sieving suspension feeders; also see Fig. 4.8. A mucus net is continuously formed and drawn across the pharynx wall. These nets become leaky to particles in the 1- to 2-μm size

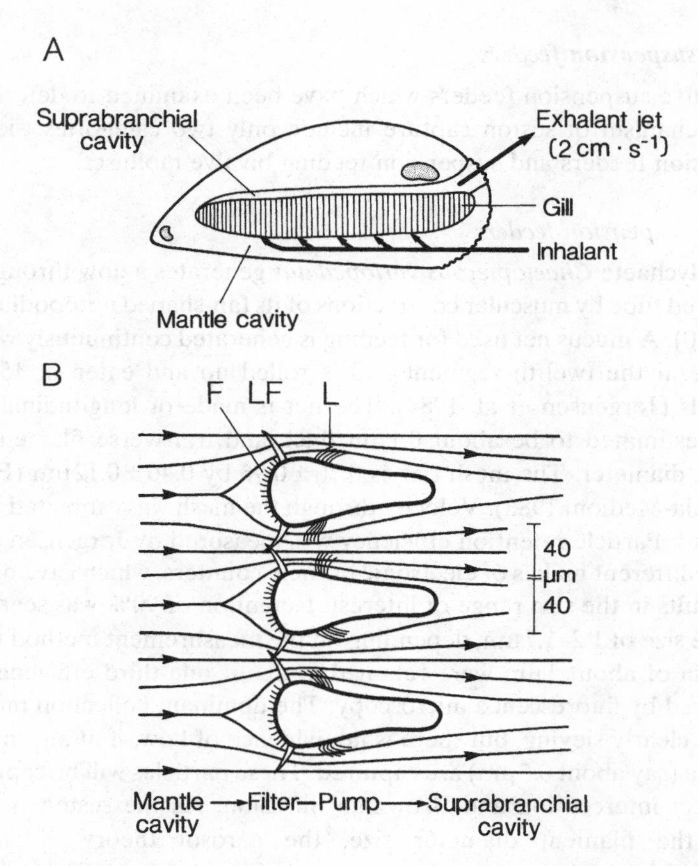

Figure 5.8 A, schematic of water flow through *Mytilus edulis*; B, mean flow streamlines through three gill filaments: F, frontal cilia (beating to the plane of the diagram; LF, laterofrontal cirri (effective stroke to the left); L, lateral cilia (effective stroke to the right) (Silvester and Sleigh 1984).

range, depending on the species, as is consistent with a sieving mechanism from what is known of mesh sizes (Flood and Fiala-Médioni 1981).

Bivalves

Although there is a long history of scientific study of the mechanism of seston capture in suspension-feeding bivalve molluscs, there is still no consensus on how it is achieved. The earliest view was that seston was removed from inhalant water by passing it through sievelike structures or gills. Mucus was thought to contribute to this mechanism by removing the already captured seston and/or providing the actual sieving net;

hence, the *mucociliary sieving theory*. The ciliary pump in active suspension feeders consists of ciliary bands with seston collected by another set of cilia beating in the direction of the mouth (Jørgensen et al. 1984). The seston may be collected upstream of the ciliary pump, as in bivalves, or downstream, as in sabellid polychaetes (Jørgensen et al. 1984). The mechanisms by which sestonic particles leave the viscous stream through pinnules and gill filaments to enter the surface flow to the mouth are still a matter of contention. The trophic fluid transport system of the blue mussel *Mytilus edulis* is shown schematically in Fig. 5.8A.

The sieving hypothesis generally considered the laterofrontal cirri the location where sieving occurred (Fig. 5.8B). The Reynolds number of the flow, based on the size of the laterofrontal cirri, is of the order of 10^{-4} (Jørgensen 1983). Each laterofrontal cirrus is in fact a compound structure consisting of 22–26 pairs of cilia arranged in two parallel but alternating rows (Moore 1971; Owen 1974). The spacing of the cirri is about 2.0–3.5 μm, and the branched cilia are separated by about 0.6 μm. Silvester and Sleigh (1984), Silvester (1988), Jones et al. (1992), and Jørgensen et al. (1988) provide a discussion of the dimensions of the cirri and the interfilamentary channel. The cirri are arranged in alternate rows so that the branched cilia overlap slightly to form a barrier to the passage of particles. The alternate rows have an effective stroke of 90° directed towards the frontal cilia, one row being in position to screen water coming through the ostium, while the other, a half cycle out of phase, is in the recovery stroke (Owen 1974). The gills of *Mytilus edulis* have been seen to remove particles larger than about 5 μm with close to 100% efficiency. Particles of 1–2 μm are removed with about 50% efficiency and particles smaller than 1 μm are removed with about 10% efficiency (Jørgensen 1975).

Jørgensen (1975) was among the first to doubt the sieving hypothesis in bivalve molluscs. He pointed out that observations on the dependence of retention efficiency on particle size were not consistent with the mesh openings provided by the laterofrontal cirri. In addition, the pressure drop expected across the laterofrontal cirri in a position which blocked the flow in the interfilamentary passage was too high to be supplied by the ciliary pump (Jørgensen 1981a). Silvester and Sleigh (1984) challenged the pressure loss argument on two grounds: that the cilia formed a more complicated organ than the series of parallel cylinders assumed by Jørgensen (1975), leaving larger openings for the passage of inhalant water, and that the velocities calculated by Jørgensen in the interfilamentary canals were too high. Observed particle retentions were reconciled by a sieve model having openings of a range of dimensions and separation by sieving, without contributions from aerosol theory mechanisms.

Because of his doubts with regard to the sieving theory in mussels, Jørgensen (1981a,b) proposed an alternative one: *the hydromechanical shear force theory*. According to this proposal, the laterofrontal cirri function as flow modifiers rather than filters. This is consistent with the theoretical analysis by Cheer and Koehl (1987) that in viscous flow, bristled appendages act more as paddles than as rakes, with the cirri moving water rather than sieving it. The seston particles are separated from the main trophic fluid flow by hydrodynamic forces which move particles across shear zones (see also Jørgensen 1983, 1990) and which are well known in fluid mechanics (Leal 1980). The three-dimensional nature of the flows in front of the gills (Jørgensen 1990) and, in particular, the interactions between the pulsating flows generated by the laterofrontal cirri and frontal cilia are not clear. There is very little guidance available from established fluid mechanics to confirm or refute the hydromechanical shear force theory. Silvester and Sleigh (1984) argued against it on the basis that the order of magnitude of the lateral forces available for diverting particles was too low to explain the particle separation achieved by blue mussels.

Work on the blue mussel pump by Jørgensen (1982) and Jørgensen et al. (1986, 1988) (and see Chap. 4) has provided further evidence against the sieving hypothesis in bivalve suspension feeders. The pressure drop across the laterofrontal cirri cannot be estimated correctly if they are assumed to be arrays of cylinders blocking the interfilamentary canals, as proposed by Silvester and Sleigh (1984). The 2-μm spaces between alternating cirri beating in phase would suggest that a fraction of the water bypasses the cirri and is not filtered. The mussel pump model suggests that only the distal half of the lattice is passable and the forward stroke of the cirri will produce an additional loss component. The pressure drop across the laterofrontal cirri cannot be measured directly, but the resistance the cirri represent can be removed experimentally by adding serotonin to the water in low concentrations. Serotonin is a nerve transmitter in the bivalve gill, which, when added to ambient water, causes the laterofrontal cirri to remain more or less fixed at the end of the stroke, thus eliminating their flow resistance (Jørgensen 1990). Comparing the pump characteristic for the system with and without serotonin, it was concluded that the frictional resistance was not measurably affected by the removal of the laterofrontal cirri. This result would support the contention that most of the flow is not sieved by the cirri. When the laterofrontal cirri are immobilized by serotonin, the retention efficiency of algal particles smaller than 6 μm is drastically decreased but that of 14-μm particles is little changed. Jørgensen et al. (1988) suggest that the laterofrontal cirri do not act as sieves of the through current, but rather as modulators in the process of particle retention, improving its efficiency for the smaller particles. Recent work by Nielsen et al. (1993) has lent support to this view. Videographs of particle tracks in the vicinity of an isolated gill filament showed a large fraction of sestonic particles passing from the through to the frontal currents without any contact with cilia. This diversion of particles was largely absent when the laterofrontal cirri were immobilized. In bivalve taxa other than mussels, the laterofrontal cilia may be absent or less well developed, and this is considered by Riisgård (1988) and Nielson et al. (1993) to be an adaptation concerned with improving the capture of smaller particles.

Summary

It is clear that it is usually possible to identify the dominant aerosol mechanism(s) operating for a specified passive benthic suspension feeder after a relatively cursory examination of the parameters of the problem. The more ambitious aim of calculating the efficiency of collection or the encounter rate for a whole animal from observations of animal morphology, fluid flow conditions, and type and quality of seston collected has rarely been successfully completed in the same animal. Although of practical interest, further work of the latter kind may not be very helpful in advancing our understanding of the mechanisms of suspension feeding, which are the focus of this chapter.

Some of the details of the mechanisms of suspension feeding have been clarified by the use of the aerosol theory, since their introduction to marine biology in 1977, but there have not been as many advances as might have been reasonably expected. Some of the reasons for limited success of the aerosol theory are described later. One limitation is that it is only well developed for the case of capture by long circular cylinders of smooth surfaces at low Re_c. Although the low Re_c condition is often satisfied, the collecting elements are often not circular in cross section and the surfaces may have roughnesses and protuberances. Nor are they always oriented normal to the flow vector and they may be flexible rather than rigid. In addition, the theoretical efficiency predictions are developed on the assumption that the seston diameter is much smaller than the collector diameter. In most of the field and laboratory cases studied, seston sizes were widely distributed and frequently overlapped the diameter of the collector. This may not be a difficulty in identifying dominant mechanisms, however, since it is inevitable that seston particles close in size to that of the collector will be captured mainly by interception.

Real flow situations are seldom as simple as the idealized flows treated by the theory. Nearby structures may alter the streamlines about the collector, although this effect may be compensated for by the analyses developed for fibre filter beds. Oscillating flows typical of wave environments or the stochastic flow caused by turbulence may alter the encounter rates and have not been fully studied. In passive suspension feeders, the collecting elements may see only a portion of the flow which approaches the whole animal. Much of the flow may be diverted around the animal. Some will flow through but not encounter the filtering elements, passing between the tentacles, for instance. This means that the flux of seston based on the approaching flow overestimates what the encounter rate, summed over all the collecting elements, actually is.

Passive suspension feeders have proved generally easier to study than active suspension feeders. The following conclusions can be drawn:

- Much of the work that has been reported on benthic suspension feeders up to 1995 had dealt with animals with primary collecting elements close to cylindrical in shape (though there are many exceptions) and with a flow around them of Re_c order of 1 or less. On the other hand, the preferred seston usually is not significantly smaller (say by an order of magnitude) than the diameter of the collectors so that this constraint on the theory is not satisfied.
- Direct interception is an important mechanism for most non-motile organic particles captured by passive suspension feeders, but if the modernized theory of interception is applied, surface properties must be known. This difficulty may be more real in theory than in practice in view of fact that surface irregularities of typical collection elements are often larger than the characteristic distance over which the surface forces are important (order of $0.02\,\mu m$). It is unfortunate that the theory for direct interception does not reach an asymptotic state at low values of the adhesion number so there is no limiting expression, as there is for the other extreme of very large intersurface forces. The "classical" expression for collection efficiency, equation (5.7), should give a reasonable estimate for many situations.
- Gravitational separations may be important for certain particles which have a significant sinking rate, but few reports are found in the literature.
- Brownian diffusion is an important mechanism for collection of particles of the size of the order of $1\,\mu m$, e.g. bacteria.
- At the typical scales and Reynolds numbers encountered, inertial impaction is probably not a significant mechanism for living organic seston. Inorganic particles are often much more dense and, although not common in the literature reviewed, some inertial effects might be expected under favorable circumstances of particles with large stopping distances in rapid flows.

The ciliary, flagellate, and muscular pumps in suspension feeders characteristically operate in trophic fluid transport systems where viscous flow is present. Clearance measurements show that many benthic active suspension feeders efficiently remove particles smaller than the gaps between the filter elements so that other mechanisms than sieving are probably involved. Detailed study of the blue mussel has shown that the bivalve gill is not a static filter, but is characterized by pulsating flows and moving cilia and cirri at very low Reynolds numbers. A mechanism other than that included in the aerosol theory may be involved, as has been described in our brief account of the controversy between mucociliary and particle migration by shear force theories. Few other active benthic suspension feeders have been studied in enough detail to determine their mechanisms of seston capture.

Finally, the theory associated with sieving, or aerosol mechanisms, has not yielded experimentally testable hypotheses by benthic biologists. This applies also to the complex theory of hydromechanical shear force proposed by Jørgensen (1981a) in mussels. Consequently, we have not found in the work concerned with seston capture in benthic suspension feeders any key hypotheses around which this chapter could be constructed. Thus, this chapter is different from Chapters 3 through 8. Perhaps the way forward depends on further development of a technology concerned with observation and manipulation of living microscale collection surfaces, including local viscous flows throughout the trophic fluid transport system. Then the framing of key hypotheses, particularly for active suspension feeders, could lead to the determination of the mechanisms of seston capture.

References

Best, B. A. 1988. Passive suspension feeding in a sea pen: effects of ambient flow on volume flow rate and filtering efficiency. Biol. Bull. 175: 332–342.

Brun, R. J., W. Lewis, P. Perkins, and J. Serafini. 1955. Impingement of cloud droplets on a cylinder and procedure for measuring liquid–water content and droplet sizes in supercooled clouds by rotating multicylinder method. Nat. Adv. Comm. Aero. Report 1215. NACA, Washington, D.C.

Byrne, M., and A. R. Fontaine. 1981. The feeding behaviour of *Florometra serratissima* (Echinodermata: Crinoida). Can. J. Zool. 59: 11–18.

Cheer, A. Y. L., and M. A. R. Koehl. 1987. Paddles and rakes: fluid flow through bristled appendages of small organisms. J. Theor. Biol. 129: 17–39.

Crank, J. 1956. The mathematics of diffusion. Clarendon Press, Oxford.

Davies, C. N., and C. V. Peetz. 1955. Impingement of particles on a transverse cylinder. Proc. R. Soc. A. 234: 269–295.

Dorman, R. G. 1966. Filtration, p. 195–222. In C. N. Davies (ed.) Aerosol science. Academic Press, London.

Flood, P. R., and A. Fiala-Médoni. 1981. Ultrastructure and histochemistry of the food trapping mucous film in benthic filter-feeders (Ascidians). Acta. Zool. (Stockholm) 62: 53–5.

1982. Structure of the mucous feeding filter of *Chaetopterous variopedatus* (Polychaeta). Mar. Biol. 72: 27–33.

Fuchs, N. A. 1964. The mechanics of aerosols. Pergamon Press, New York.

1986. Methods for determining aerosol concentration. Aerosol Sci. Tech. 5(2): 123–143.

Gerritsen, J., and K. G. Porter. 1982. The role of surface chemistry in filter feeding by zooplankton. Science 216: 1225–1227.

Harvell, C. D., and M. LaBarbera. 1985. Flexibility: a mechanism for control of local velocities in hydroid colonies. Biol. Bull. 168: 312–320.

Holland, N. D., J. R. Strickler, and A. B. Leonard. 1986. Particle interception, transport and rejection by the feather star *Oligometra serripinna* (Echinodermata: Crinoidia), studied by frame analysis of videotapes. Mar. Biol. 93: 111–126.

Johnson, A. S., and K. P. Sebens. 1993. Consequences of a flattened morphology: effects of flow on feeding rates of the scleractinian coral *Meandrina meandrites*. Mar. Ecol. Prog. Ser. 99: 99–114.

Jones, H. D., O. G. Richards, and T. A. Southern. 1992. Gill dimensions, water pumping rate and body size in the mussel *Mytilus edulis* L. J. Exp. Mar. Biol. Ecol. 155: 213–237.

Jørgensen, C. B. 1975. On gill function in the mussel *Mytilus edulis* L. Ophelia 13: 187–232.

1981a. A hydromechanical principle for particle retention in *Mytilus edulis* and other ciliary suspension feeders. Mar. Biol. 61: 277–282.

1981b. Feeding and cleaning mechanisms in the suspension feeding bivalve *Mytilus edulis.* Mar. Biol. 65: 159–163.

1982. Fluid mechanics of the mussel gill: the lateral cilia. Mar. Biol. 70: 275–281.

1983. Fluid mechanical aspects of suspension feeding. Mar. Ecol. Prog. Ser. 11: 89–103.

1990. Bivalve filter feeding: hydrodynamics, bioenergetics, physiology, and ecology. Olsen and Olsen Press. Fredensborg, Denmark.

Jørgensen, C. B., P. Famme, H. S. Kristensen, P. S. Larsen, F. Mohlenberg, and H. U. Riisgård. 1986. The bivalve pump. Mar. Ecol. Prog. Ser. 34: 69–77.

Jørgensen, C. B., T. Kiorboe, F. Mohlenberg, and H. U. Riisgård. 1984. Ciliary and mucus-net filter feeding, with special reference to fluid mechanical characteristics. Mar. Ecol. Prog. Ser. 15: 283–292.

Jørgensen, C. B., P. S. Larsen, F. Mohlenberg, and H. U. Riisgård. 1988. The bivalve pump: properties and modelling. Mar. Ecol. Prog. Ser. 45: 205–216.

LaBarbera, M. 1978. Particle capture by a Pacific brittle star: experimental test of the aerosol suspension feeding model. Science 201: 1147–1149.

1984. Feeding currents and particle capture mechanisms in suspension feeding animals. Am. Zool. 24: 71–84.

Lamb, H. 1962. Hydrodynamics. 6th ed. Cambridge University Press, Cambridge.

Langmuir, I., and K. B. Blodgett. 1946. U.S. Army Air Forces Tech. Rept. 5418. U.S. Office of Technical Services. PB 27565.

Lasker, H. L. 1981. A comparison of the particulate feeding abilities of three species of gorgonian soft coral. Mar. Ecol. Prog. Ser. 5: 61–67.
Leal, L. G. 1980. Particle motion in a viscous fluid. Annu. Rev. Fluid Mech. 12: 435–476.
Leonard, A. B. 1989. Functional response in *Antedon mediterranea* (Lamarck) (Echinodermata: Crinoidea): the interaction of prey concentration and current velocity on a passive suspension-feeder. J. Exp. Mar. Ecol. 127: 81–103.
Leonard, A. B., J. R. Strickler, and N. D. Holland. 1988. Effects of current speed on filtration during suspension feeding in *Oligometra serripinna* (Echinodermata: Crinoidea). Mar. Biol. 97: 111–112.
Licht, W. 1980. Air pollution control engineering: basic calculations for particulate collection. Marcel Dekker, New York.
Loudon, C. 1990. Empirical test of filtering theory: particle capture by rectangular-mesh nets. Limnol. Oceanogr. 35: 143–148.
Loudon, C., and D. N. Alstad. 1990. Theoretical mechanics of particle capture: predictions for hydropsychid caddisfly distributional ecology. Am. Nat. 135: 360–381.
May, K. R., and R. Clifford. 1967. The impaction of aerosol particles on cylinders, spheres, ribbons and discs. Ann. Occup. Hyg. 10: 83–95.
Mayer, D. L. 1979. Length and spacing of the tube feet in crinoids (Echinodermata) and their role in suspension feeding. Mar. Biol. 51: 361–369.
McFadden, C. S. 1986. Colony fission increases particle capture rates of a soft coral: advantages of being a small colony. J. Exp. Mar. Biol. Ecol. 103: 1–20.
Monger, B. C., and M. R. Landry. 1990. Direct interception feeding by marine zooflagellates: the importance of surface and hydrodynamic forces. Mar. Ecol. Prog. Ser. 65: 123–140.
Moore, H. J. 1971. The structure of the latero-frontal cirri on the gills of certain lamellibranch molluscs and their role in suspension feeding. Mar. Biol. 11: 23–27.
Morris, C. C., and D. Diebel. 1993. Flow rate and particle concentration within the house of the pelagic tunicate *Oikopleura vanhoefferi*. Mar. Biol. 115: 445–452.
Nielsen, N. F., P. S. Larsen, H. U. Riisgård, and C. B. Jørgensen. 1993. Fluid motion and particle retention in the gill of *Mytilus edulis*: video recordings and numerical modelling. Mar. Biol. 116: 61–71.
Owen, G. 1974. Studies on the gill of *Mytilus edulis*: the eu-latero–frontal cirri. Proc. R. Soc. London (B) 187: 83–91.
Parsons, T. R., M. Takahashi, and B. Hargrave. 1977. Biological oceanographic processes. 2nd ed. Pergamon Press, Oxford.
Patterson, M. R. 1984. Patterns of whole coral prey capture in the octocoral *Alcyonium siderium*. Biol. Bull. 167: 613–629.
1991a. The effects of flow on polyp-level prey capture in an octocoral, *Alcyonium siderium*. Biol. Bull. 180: 93–102.
1991b. Passive suspension feeding by an octocoral in plankton patches: empirical test of a mathematical model. Biol. Bull. 180: 81–92.
Pich, J. 1966. Theory of aerosol filtration by fibrous and membrane filters, p. 223–285. *In* C. N. Davies (ed.) Aerosol science. Academic Press, London.
Ranz, W. E., and J. B. Wong. 1952. Impaction of dust and smoke particles on surface and body collectors. Ind. Eng. Chem. 44: 1371–1381.
Riisgård, H. U. 1988. Efficiency of particle retention and filtration rate in 6

species of northeast American bivalves. Mar. Ecol. Prog. Ser. 45: 217–223.
Rubenstein, D. I., and M. A. R. Koehl. 1977. The mechanisms of filter feeding: some theoretical considerations. Am. Nat. 111: 981–994.
Schrijver, J. H. M., C. Vreeken, and J. A. Wesselingh. 1981. Deposition of particles on a cylindrical collector. J. Coll. Interf. Sci. 81: 249–256.
Sebens, K. P., and M. A. R. Koehl. 1984. Predation on zooplankton by the benthic anthozoans *Alcyonium siderium* (Alcyonacea) and *Metridium senile* (Actiniaria) in the New England subtidal. Mar. Biol. 81: 255–271.
Shimeta, J. 1993. Diffusional encounter of submicrometer particles and small cells by suspension feeders. Limnol. Oceanogr. 38: 456–465.
Shimeta, J., and P. A. Jumars. 1991. Physical mechanisms and rates of particle capture by suspension feeders. Oceanogr. Mar. Biol. Ann. Rev. 29: 191–257.
Silvester, N. R. 1983. Some hydrodynamic aspects of filter feeding with rectangular-mesh nets. J. Theor. Biol. 103: 265–286.
1988. Hydrodynamics of flow in *Mytilus* gills. J. Exp. Mar. Biol. Ecol. 120: 171–182.
Silvester, N. R., and M. A. Sleigh. 1984. Hydrodynamic aspects of particle capture by *Mytilus*. J. Mar. Biol. Assoc. U.K. 65: 859–879.
Spielman, L. 1977. Particle capture from low-speed laminar flows. Annu. Rev. Fluid Mech. 9: 297–319.
Spielman, L. A., and J. A. Fitzpatrick. 1973. Theory for particle collection under London and gravity forces. J. Coll. Interf. Sci. 42: 607–623.
Trager, G. C., J.-S. Hwang, and J. R. Strickler. 1990. Barnacle suspension feeding in variable flow. Mar. Biol. 105: 117–127.
Wildish, D. J., and D. D. Kristmanson. 1993. Hydrodynamic control of bivalve filter feeders: a conceptual view, p. 299–324. *In* R. F. Dame (ed.) Bivalve filter feeders in estuarine and coastal ecosystem processes. Springer-Verlag, Berlin.
Yoshioka, N., and C. Kansoka. 1972. Collection efficiency of an aerosol by an isolated cylinder: gravity and inertia predominant region. Chem. Eng. (Japan) 36: 313–319.

6

Behavioral responses to flow

The life form of benthic suspension feeders includes both infauna and epifauna, and as Table 6.1 suggests, the majority are epifauna. The central problem for both sessile and tube-living epifauna is that they must grow upwards into the benthic boundary layer to reach good quality, or higher quantities of, seston, both of which may be height or velocity dependent (Muschenheim 1987a,b). Yet, these suspension feeding epifauna may be restricted by the increasing velocity encountered higher in the benthic boundary layer, because it increases drag and the likelihood of dislodgement for them. By contrast, infauna are less susceptible to dislodgement, which does not occur unless a substantial part of the sediment is swept away.

Behavioral responses to flow include *rheotaxis*, which is a directed response to flow direction involving locomotion or muscular turning of body parts, and *rheokinesis*, which is a non-directed response causing random movement proportional to flow velocity. *Rheotropisms* are debatably behavioral responses but fit the definition of behavior in Carthy (1958); that is they involve flow-induced differential growth perceived as twisting stresses by the axial support system, e.g. of gorgonian corals which have semi-conductor properties that provide electrical stimuli for skeletal secreting cells (Wainwright and Dillon 1969). The result is greater skeletal growth on the stimulated side.

This chapter is organized around two general responses of suspension feeders to hydrodynamics. The first involves only non-radially symmetrical suspension feeders which respond behaviorally so as to position themselves to flow in order to optimize filtration and, hence, growth. The second is suspension feeding epifauna, which respond behaviorally to flow in order to minimize lift and drag on their body parts. In addition to this, we present a preliminary discussion of the phenomenon of aggrega-

Table 6.1. *Life form categories of benthic suspension feeders.*

Life form	Examples
Infauna	Some Polychaeta, e.g. *Chaetopterus variopedatus* Some Bivalvia, e.g. soft shell clam *Mya arenaria*, cockle *Cardium edule* Phoronidea – *Phoronopsis virida*
Sessile epifauna	Cnidaria: Hydrozoa – hydroids Cnidaria: Anthozoa – gorgonian corals Cnidaria: Antipatharia – octocorals Bryozoa – bryozoans Brachiopoda – lampshells Echinodermata: Crinoidea – stalkless crinoids Porifera – sponges Ascidiacea – sea squirts Cirripedia – barnacles
Tube-living epifauna	Some Polychaeta and Crustacea (see Table 6.3)
Free-living epifauna	Some Echinodermata, e.g. feather stars (Crinoidea), sand dollars (Echinoidea) Some decapod Crustacea, e.g. sand crabs (Hippidae), porcelain crabs (Porcellanidae) Some Bivalvia, e.g. scallops (Pectinidae), mussels (Mytilidae)

tion among benthic suspension feeders, inclusive of a consideration of the proximate causes of the behavior. Although only a few free-living epifaunal suspension feeders have developed any swimming ability, we have also considered this behavior in relation to flow.

Optimizing growth by orientation to flow

Infauna

Suspension feeding burrowing polychaetes and bivalves are intimately connected with water movement because they depend on moving seawater for distributing sexual products and for obtaining food, by drawing it into either their burrows, e.g. *Chaetopterus variopedatus*, or their siphon tubes, e.g. the soft-shell clam. They also depend on water movement to replenish the otherwise seston-depleted seawater which would build up around them in static conditions (Creutzberg 1975). Many suspension-feeding infaunal polychaetes have radially symmetri-

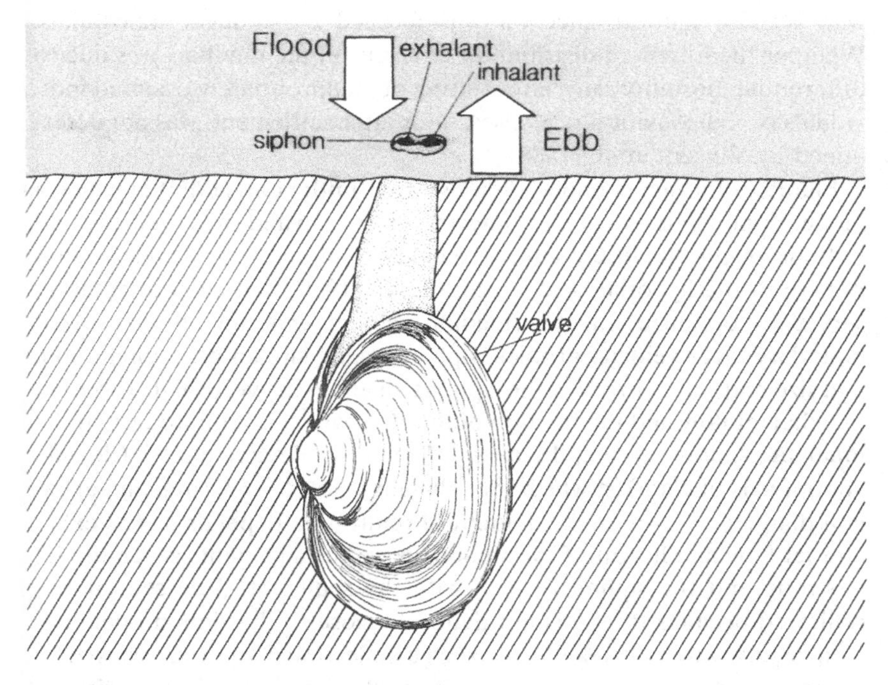

Figure 6.1 The clam *Mya arenaria*: in life the valves are buried with only the tips of the siphon reaching the sediment surface.

cal feeding structures, e.g. many sabellids, and so orientation with respect to a particular current direction is not usually of concern. This is not the case in bilaterally symmetrical infauna, e.g. bivalves, where individual orientation to current direction may be of more importance.

It was noticed by Vincent, Desrosiers, and Gratton (1988) that the deep-burrowing soft-shell clam *Mya arenaria* L. was oriented on tidal mudflats in the Gulf of St. Lawrence such that the clam's sagittal plane was perpendicular to the current direction (Fig. 6.1). Tidal currents over the clam flats here were predominantly bidirectional, flood and ebb, and it was hypothesized that this orientation was a means of optimizing filtration. In any other orientation with respect to bidirectional currents, some degree of exhalant mixing with the inhalant, and therefore seston dilution, would occur, resulting in availability of a lower seston load for filtration. No field evidence to support this view could be found by Vincent et al. (1988) when individual orientation was related to growth and reproductive measures in older clams, and it was suggested that this was because natural selection was focused on younger individuals. Whether the observed distribution of clams on the mudflats was due to differential mortality rates after setting of inappropriately oriented individuals or to behavioral responses, at or after settlement, was not determined by Vincent et al. (1988).

Flume experiments undertaken by Monismith et al. (1990) in which a model siphon pair of a clam connected to a pump and shear velocity and bottom roughness closely simulated field conditions, provide results supportive of the hypothesis of Vincent et al. (1988) that clam orientation serves to optimize filtration/feeding. Model experiments show that boundary layer velocity profiles, siphon height, siphon pair orientation, and size and exit velocity of the siphons all affect the vertical distribution of exhalant flow and the fraction of this flow which is refiltered when the inhalant is downstream. In an inhalant downstream model (dorsal–ventral axis of clam) in which the inhalant diameter = 3.2 mm and exhalant diameter = 6.4 mm, the amount of refiltration depends on the exhalant jet flow and height of the siphon above the sediment. When the jet is strong and the siphon is flush with the sediment, little seawater is refiltered because the jet is not deflected close to the siphon openings. Conversely, when the jet is weak, such a deflection does occur, caused by the cross-flow, and results in up to 35% refiltration. In similar conditions, except that the siphon height is adjusted to be 10 mm above the sediment, much less of the exhalant is refiltered. Monismith et al. (1990) concluded that the optimum conditions for clams to take advantage of bidirectional tidal flows was with the siphons oriented with their centres on a line normal to the flow.

In further flume studies in which siphons of a bivalve reef were modelled, O'Riordan, Monismith, and Koseff (1993, 1995) presented flume data simulating either *Tapes japonica* or *Potamocorbula amurensis* from San Francisco Bay. The experiment involved simulated siphons of each bivalve jetting or withdrawing realistic amounts of seawater in a flume boundary layer. At realistic siphon diameters for each species, simulated by appropriately sized plastic tubing and at field observed densities of bivalves, the amount of exhaled seawater which was refiltered could be determined. For *T. japonica* at a density of $625 \cdot m^{-2}$ the maximum refiltration ratio was 18%, and for *P. amurensis* at $11{,}025 \cdot m^{-2}$, this value reached 48%. The major factors influencing refiltration were bivalve density and increases in relative shear velocity. Two lesser factors which influenced refiltration were siphon height and cross-flow Reynolds number (O'Riordan et al. 1995).

In growth experiments with the hard clam *Mercenaria mercenaria* in a laboratory-induced oscillating flow, Turner and Miller (1991a) found that the clam siphons were randomly oriented in the flume sediment, thus not supporting the optimum orientation theory of Vincent et al. (1988) and Monismith (1990). However, in Turner and Miller's (1991a) work, the maximum growth period was 12 d and the hard clams burrowed into the sediment while the flume was being filled, i.e. when no directional cues were available to guide behavior. Field observations and a more intensive experiment directed to individual hard clam orientation in relationship to flow direction are required. Dolmer, Karlsson, and Svane (1994) also found that the blue mussel *Mytilus edulis* L. did not undertake rheotactic behavior. These experiments involved underwater stereophotography of undisturbed mussels, as well as laboratory experiments in a recirculating flume, and showed that foot and byssus thread movements were unrelated to the current direction at velocities of 0–20 $cm \cdot s^{-1}$.

The phoronidean *Phoronopsis psammophila* from sublittoral sandy sediments in the Gulf of Marseilles was studied by Emig and Becherini (1970) in a recirculating raceway. When subjected to a unidirectional flow, the lophophore, mouth, and surrounding cone of tentacles of this phoronidean are turned to oppose the flow, and if the flow is reversed, the lophophore orientation is also reversed. In this case, the response is a rheotactic one, probably associated with the optimum orientation for filter feeding, although feeding success was not determined. Johnson (1988) studied another phoronidean, *P. viridis*, present on intertidal sands of Bodega Bay, California. Phoronideans live in an infaunal tube and, during feeding, extend the anterior part of their body, which carries the terminal lophophore. The latter is a ciliated crown of tentacles which creates an active feeding current when *P. viridis* is feeding in still seawater. Both live animals and models were studied in a recirculating flume at velocities measured 4 cm above bottom at up to 10 $cm \cdot s^{-1}$. Since the flume boundary layer was not characterized in this study, it was not clear whether or not the measured velocities were in the free-stream flow or within the flume benthic boundary layer. Flow around the phoronideans was visualized with fluorescein dye and photography. It was shown by Johnson (1988) that *P. viridis* models involved sediment scouring around their anterior body, where it emerged from the sediment, causing entrainment upwards towards the lophophore because the body was increasingly deformed in the downstream direction as flow was increased. As was pointed out, this is similar to the results for tube-living

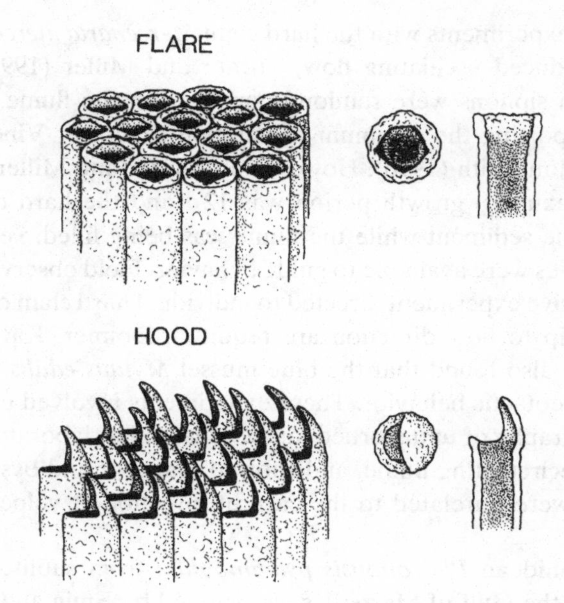

Figure 6.2 Structure of aggregations of the upper part of *Phragmatopoma californica* tubes (Thomas 1994); note that flare and hood are ~7 mm above the sediment–water interface. End-on and sectional views of an individual tube on right.

polychaetes found by Eckman, Nowell, and Jumars (1981) and Carey (1983) discussed later in the section, Tube normal to flow. However, Johnson (1988) did not demonstrate enhanced feeding in live *P. viridis* or adjustment of its body to optimize feeding, as demonstrated by *P. psammophila* (Emig and Becherini 1970).

The sabellarid polychaete *Phragmatopoma californica* is an infaunal species of Californian shores which extends its tentacles into the water column for passive filter feeding. These worms occur in dense aggregations at suitable locations, and Thomas (1994) noticed that *P. californica* built two types of tube extensions, flares and hoods, which protruded ~7 mm above the sediment–water interface (Fig. 6.2). Field observations suggested that tubes with flare extensions were associated with wave-exposed, whereas hood tubes were associated with wave-protected localities. The open face of the hood (Fig. 6.2) was shown by field observations to face the dominant field current direction. Tube extensions were also rapidly replaced after damage, suggesting their importance to the worm, although repair was slower in the case of flare extensions (Thomas 1994). Flume studies showed that hoods and flares resulted in a decrease of fluid exchange with seawater above the colony; thus the extensions were not linked to possible food depletion above the worms. In flume experiments hood orientation with respect to unidirectional flows or complete removal of the hood showed that the hood acted to reduce tentacle deflection at higher flows when present at its normal field orientation. Since deflection of filtering elements by flow in other suspension feeders (Chap. 4) results in a reduction in filtering efficiency, it is reasonable to suppose that the extensions assist the worms to filter feed at higher limiting velocities. Further experiments involving filtration or growth are necessary to test this hypothesis experimentally.

Sessile epifauna

The coral sea fans (Gorgonacea) are frequently variable in growth form with a particular morphology dependent on water movement. Théodor (1963) reported water movement–related morphology in the Mediterranean gorgonian *Eunicella stricta*. This coral has a concave form in strong wave surges, a fan shape in moderate flows, and a whip form in lesser flows. Leversee (1976) reported two growth forms of the sea whip *Leptogorgia virgulata* in North Carolina waters: a bushy form where directionally unpredictable, turbulent currents were present, and fan-shaped colonies where predictable, bidirectional currents were found. The Caribbean reef coral *Agaricia agaricites*, studied by Helmuth and Sebens (1993), grew in three forms: bifacial, horizontal unifacial, and vertical unifacial (Fig. 6.3). These growth forms were related to water movement, which was inversely related to depth. Thus, the bifacial form was found at the greatest depths, where the water movement was least, and appears to be locally adapted by opposing rather than being parallel to the flow, as in the unifacial forms. Other authors who have studied corals report that the fan is generally opposed, or perpendicular to, the major bidirectional flow occurring at a particular field location.

Leversee (1976) was the first to demonstrate that the orientation of fan-shaped gorgonian corals with respect to flow direction was important in determining the rates of prey capture in this passive suspension feeder. Experimental tests were made in a 60-L recirculating flume in which the fan-shaped form of *Leptogorgia virgulata* was presented to the unidirectional flow (at a velocity of $4.0 \pm 0.05\,cm \cdot s^{-1}$), either opposed or parallel to it. *Artemia nauplii* at a concentration of $20 \pm 3 \cdot L^{-1}$ were used as food, and the rate of depletion of the nauplii monitored over a 3-h period. Of

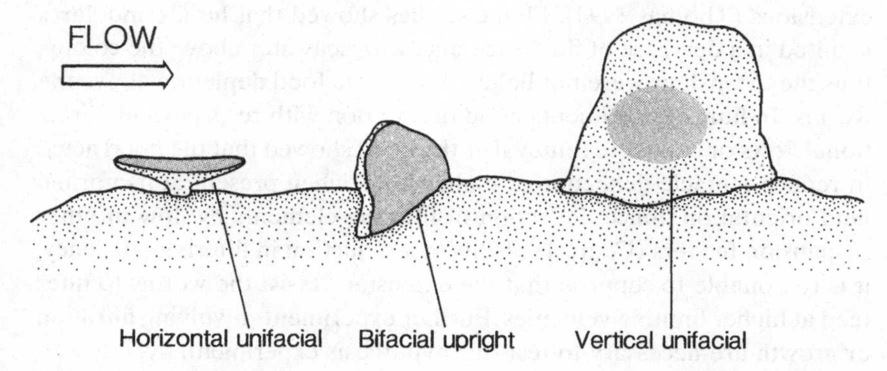

Figure 6.3 Colony form of *Agaricia agaricities* in the Caribbean (Helmuth and Sebens 1993). Amorphous encrusting forms may also be present, but are not shown in the diagram.

six colonies tested, four showed significantly greater feeding when opposed versus when parallel to the flow. Similar results were obtained by Helmuth and Sebens (1993) with the scleractinian coral *Agaricia agaricites* in unidirectional flume flows of 3 and $6\,cm \cdot s^{-1}$ and food supplied as *Artemia* cysts at 1000 cysts $\cdot L^{-1}$. Results suggest that at $3\,cm \cdot s^{-1}$, significantly more cysts were captured by the opposed coral versus that with the fan parallel to the flow. Although more than twice the number of *Artemia* cysts were captured by opposed compared to parallel corals at a speed of $6\,cm \cdot s^{-1}$, this difference was not statistically significant. A review of the shapes produced by growth of some passive suspension feeders in different locations is presented by Warner (1977).

Helmuth and Sebens (1993) compared the feeding rate of three growth forms of *Agaricia agaricites* found in Caribbean waters (Fig. 6.3). Their results reveal that for a given unit area of feeding surface of this colonial coral, more polyps were present in unifacial (horizontal 12.0 ± 3.3, vertical 13.7 ± 3.2 calices $\cdot cm^{-2}$) than in bifacial (8.1 ± 2.0 calices $\cdot cm^{-2}$) forms, and that this led to higher feeding rates when expressed either on a per polyp or unit area basis. We have reanalyzed their results in terms of additional data (total number of cysts captured for the whole colony) supplied by Helmuth (pers. commun. 1995). The additional results (Table 6.2) suggest that the bifacial upright colony feeds best at lower ambient velocities with a nominal feeding peak $\sim 18\,cm \cdot s^{-1}$, compared to the vertical unifacial colonies, which feed best at higher velocities and with a peak at $>30\,cm \cdot s^{-1}$. The latter results are consistent with the known field distribution of the water movement–related

Table 6.2. *Food capture rates of* Agaricia agaricites.

	Bifacial		Horizontal unifacial		Vertical unifacial	
Velocity $cm \cdot s^{-1}$	Cysts/cm^2	Whole colony no. of cysts	Cysts/cm^2	Whole colony no. of cysts	Cysts/cm^2	Whole colony no. of cysts
3	2.5	171	6.5	284	3.0	144
6	3.0	242	11.0	371	5.0	224
18	8.0	508	15.0	514	15.0	615
30	3.0	293	15.0	623	38.0	1320
50	1.0	87	16.0	739	6.0	297

Results are adjusted to the same ambient cyst concentration, 1 $cyst \cdot ml^{-1}$.
Source: Estimates from (1) mean, unit surface area values as given in Fig. 4 of Helmuth and Sebens (1993), and (2) total number of cysts captured per colony, based on unpublished results (Helmuth, pers. commun.).

morphotypes: bifacial, deep, low velocity or unifacial, shallow, higher velocity in Caribbean waters down to 20m (Helmuth and Sebens 1993). Similar to the conclusions reached by Helmuth and Sebens (1993) based on unit area, analysis based on whole colony feeding suggests that the bifacial colony performs better at intermediate velocities. At $18\,cm \cdot s^{-1}$, the bifacial colony can feed with efficiency equal to that of the horizontal unifacial colony and nearly as well as the vertical unifacial form.

The hypothesis tested by Abelson, Miloh, and Loya (1993) was that the shape of passive suspension feeders was critical in determining the quality and quantity of sestonic particles available as food at different levels within the benthic boundary layer. Two general categories, based on the height to width ratios (= slenderness ratio [SR]), were recognized by Abelson et al. (1993): high SR, high branching corals, and low SR, with flattened or massive corals or sponges. High SR corals protrude to a greater height within the benthic boundary layer than low SR corals. The latter encounter increased seston concentration due to wake resuspension from the sediment. Because low SR corals encounter high fluxes of bedload material, they may be specialized in this type of feeding. On the basis of a hydrodynamic model constructed during this study, sestonic particle concentration and particle settling velocity were determined around model passive suspension feeders placed in the field and these measurements supported the model predictions.

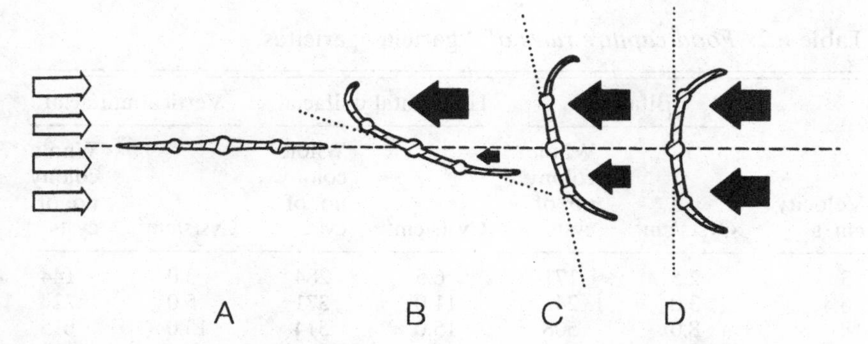

Figure 6.4 Orientation of a sea fan to ambient flow from left to right: A, fan blade parallel to the flow; B, blade at a low angle; C, blade at a larger angle to the flow; D, blade perpendicular to the flow; B and C, unequal twisting stresses, indicated by filled arrows, are present (Wainwright and Dillon 1969).

How then can these growth patterns be stimulated by flow? Wainwright and Dillon (1969) studied two sea fans, *Gorgonia ventalina* and *G. flabellum*, which lived in patches at depths of 2–10 m on the seaward reefs of the Florida Keys, United States. Like the authors mentioned earlier, they found that the majority of larger fans were oriented in the same plane (although how this was related to water movements was not determined). They observed a large number of fans of different size – from 10 to 90 cm in height – and showed that, whereas the smallest colonies were randomly oriented, the largest ones were aligned in the same plane. Wainwright and Dillon (1969) suggested that as they grow in size and become more influenced by the greater velocities that they reach higher in the benthic boundary layer, twisting (= torque) stresses (Fig. 6.4) due to bidirectional flows are used by the colony to bring about colony orientation opposed to this flow. Twisting stresses occur when an unequal area on one side of the colony is presented to the flow, causing greater force to be applied there. If the colony was seeking to minimize drag, then Fig. 6.4A would be the optimum orientation strategy, but, since D in Fig. 6.4 is the morphology realized in larger sea fans, then drag must normally be of little concern. Both A and D in Fig. 6.4 are stable orientations for which unidirectional flows can cause no colony twisting. The flexible sea fan blade of *Gorgonia* consists of a vertical, thickened axial support that extends from a holdfast to the blade. The stiffened axial support is thicker at its base and tapers to the top of the blade. As a result of twisting forces in B and C (Fig. 6.4), the axial support displays creep. As a result of the secretion of new skeletal material around that already present, the sea fan is held in its new position. Evidence of changes in fan orientation which occurred during growth were seen by Wainwright and Dillon (1969) on sectioning the blade. The axial fan of gorgonian sea fans consists of calcite and fibrous gorgonin containing a collagenlike material, probably supplemented by soluble organic matter. These materials have piezoelectric and semi-conductor properties that can provide electrical stimuli to cells to cause them to secrete the skeletal supporting material.

Brachiopods from Washington on the U.S. Pacific coast were used in laboratory experiments in 15-L recirculating water tunnels by LaBarbera (1977). Experiments with orientation of dead shells in flow and tubing carrying a dye which was released from a model lophophore within the shell confirmed the flow pathways of ambient seawater shown in Fig. 6.5. In life, these flows are generated by ciliary pumping of the lophophore with seawater being drawn in via the right and left lateral margins and exiting at the anterior shell gape. Of the four orientations shown in Fig. 6.5, D is the hydrodynamically optimum position – that is, when the hinge axis is opposed to the flow direction. With this orientation bidirectional flows, which are a common occurrence in the field, can be accommodated because the ambient flow simply changes from the left to the right lateral margin. Some brachiopods are able to adjust their position with respect to flow direction, and the rheotactic behavior observed may be linked to ambient flow enhancement of filtration, as proposed by LaBarbera (1977). Alternatively, we suggest that brachiopods do not experience ambient flow enhancement and may respond to dynamic pressure differences between inhalant and exhalant, as do scallops (see the sections, Bivalves, and Lamphells, Chap. 4). According to this view, dynamic pressure differences between inhalant and exhalant cause them either to reduce pumping, to work harder to effect the same rate, or to change their position behaviorally with respect to flow – all in order to optimize feeding rates.

LaBarbera (1977) found that of two species of articulate brachiopods he could observe in the field off San Juan Island, Washington, both preferred the D orientation with approximately half facing one way and the other half facing the opposite way, consistent with the flows at the site being bidirectional. Experiments with live brachiopods placed in the water tunnel so that the initial orientations were either as A or as B in Fig. 6.5, demonstrate that two species, *Laqueus californianus* and *Terebratulina unquicula*, actively reoriented to position D. In the proc-

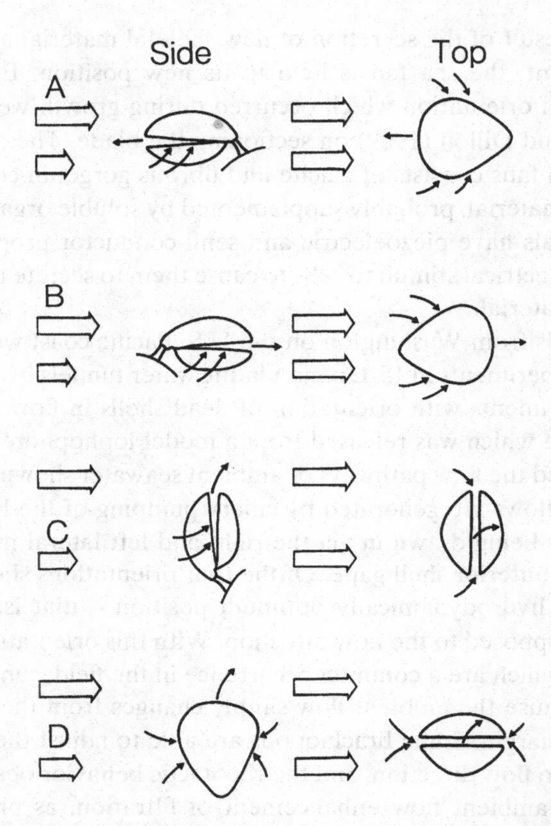

Figure 6.5 Possible orientations indicated by A, B, C, or D of a brachiopod relative to ambient flow as seen from the side and top of the animal: small arrows, activity by the ciliary pump; heavy arrows, ambient flow with the length suggesting the velocity (redrawn from LaBarbera 1977).

ess, both could traverse an arc as great as 120° to achieve their preferred orientation. Although the mechanism was not studied by LaBarbera, he suggested that the pedicle adjustor muscles were involved although the sensory organs and nervous pathways were not identified. In *L. californianus*, movement of the pedicle was preceded by three to five coughs (a rapid sharp closure followed by a full valve opening), whereas the rheotactic behavior of *T. unquicula* did not include coughing but rapid rotation through 60° to 90° with the valves closed or slightly opened or with the valves open and a slower rotation, about 1 s per 30° through a final arc of 20° to 40°. One other species, *Hemithryris psittacea*, showed similar rheotactic behavior but could only adjust through 45°, whereas a fourth species, *Terebratulina transversa*, never showed readjustments of body positions during the flume experiments. So possibly a non-behavioral mechanism is involved in achieving the preferred D position with respect to field flow in *T. transversa*.

Threshold current speeds for rheotaxis in *L. californianus* (~$1.5\,cm \cdot s^{-1}$) and *T. unquicula* (~$0.75\,cm \cdot s^{-1}$) were similar to the excurrent velocity measured with a thermistor flow meter, suggesting that dynamic pressure is the factor responded to in this process. Eshelman and Wilkens (1979) observed that *T. transversa* in Pacific waters of the British Columbia, Canada, coast were ~70% in position D and ~30% in position C, in line with LaBarbera's (1977) preferred orientation hypothesis. *T. transversa* was absent from sites without a current flow or with currents that were too extreme (>$133\,cm \cdot s^{-1}$).

In a study of the mechanism of brachiopod orientation to ambient flows (LaBarbera 1978), active rheotactic behavior was considered to be achieved by muscular activity. A structural study of the pedicle musculature suggested that the dorsal adjustors were primarily involved in generating torque around the long axis of the pedicle to resist the torque applied by the ambient flows. The latter was measured in model brachiopods at flows up to $14\,cm \cdot s^{-1}$. LaBarbera (1978) found that the torque around the pedicle was low and that the dorsal pedicle adjustor muscles were easily capable of resisting this force, although sufficient data were not obtained to estimate the energetic cost of this behavior.

A common North Atlantic barnacle, *Semibalanus balanoides*, preferentially settles in fine grooves on hard substrates and orients positively to light. Immediately after settlement and metamorphosis, orientation may be changed by rheotropic response (Crisp and Stubbings 1957) where flows are predominantly unidirectional. This brings the rostrum to oppose the major flow direction and involves torsional growth resulting from mechanical strain unequally distributed on the barnacles' body. Crisp (1960a) showed that *S. balanoides* oriented with the rostrum to the flow grew better in wet weight than those in which the orientation was experimentally varied by 90° or 180°. A whale barnacle, *Cornula diadema*, is present as an ectocommensal with the rostrum opposed to the flow as the whale swims forward (Crisp and Stubbings 1957) similar to the orientation of *S. balanoides*. According to Crisp (1960b), *S. balanoides* can move up to 2 cm away from the original settlement site under direct competitive pressure from faster growing, larger barnacles of the same or different species. Trager, Hwang, and Strickler (1990) and Trager et al. (1992) have shown that flow provides important cues in the

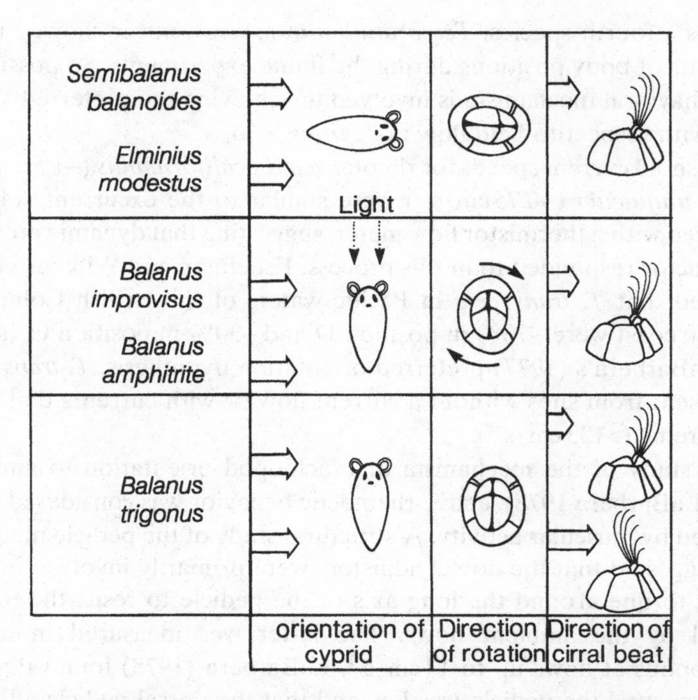

Figure 6.6 Diagrammatic representation of cyprid orientation, direction of rotation, and direction of adult cirral beat for five species of barnacles (based on Crisp 1953; Ayling 1976).

feeding behavior of barnacles, such as *S. balanoides* (see the section, Barnacles, Chap. 4). This species switches from active to passive feeding at ~3.1 $cm \cdot s^{-1}$, and in oscillating flows, seawater deceleration is the cue for switching the cirral fan orientation through 180° so that the barnacle can anticipate the maximum wave surge energy on the concave surface of the fan during passive feeding.

The common barnacle of New Zealand, *Balanus trigonus*, shows some important differences from *S. balanoides*. Whereas the latter is a facultatively passive/active suspension feeder, *B. trigonus* is an obligate active suspension feeder with cirral beats of 2–3 beats $\cdot s^{-1}$, whatever the ambient flow velocity (Ayling 1976). At low flows, *B. trigonus* cirral beating is opposed to the flow, but at higher flows (~30 $cm \cdot s^{-1}$), the cirri are reversed and beat in the same direction as the flow. A further difference is that the orientation of the body, defined by the rostrocarinal axis, is initially at right angles to the flow, whereas in *S. balanoides*, the axis is aligned in the same direction as the flow. Thus, in *B. trigonus* the cirral fan must swivel through only 90° (Fig. 6.6) to oppose flows from either direction. For *S. balanoides* in oscillating flows, the cirral fan must be swivelled through 180° to oppose the flow when it approaches the barnacle from the carina end. There appear to be no comparative growth studies for *B. trigonus* at different orientations to prove the contention of Ayling (1976) that the preferred orientation to flow allows this barnacle to minimize energetic costs involved in suspension feeding.

Otway and Underwood (1987), working with an Australian intertidal barnacle, *Tesseropora rosea*, showed that its preferred orientation was with the rostrum and cirral fan opposed to the main flow direction, as in *S. balanoides*. Major flow in this habitat was the wave backwash. In growth experiments on plates cut out of the solid rock, the plates were rotated through 90° or 180° and compared to controls which were undisturbed or prepared exactly as the treatments. Results showed that significant depressions of wet weight occurred in rotated plates in 15- to 19-mo-long experiments, but there was no effect on the rate of shell growth, mortality, or egg mass production.

Preliminary descriptions of the behavior of stalked barnacles *Trilasmis fissum hawaiense* (family Poecilasmatidae), which are epizoic ecto-commensals on the mouth parts of spiny lobsters from Hawaii, have been given by Bowers (1968). The behavior includes movement of the cirral net and other movements of the peduncle in response to water movements.

Tube-living epifauna

Macrofauna which build tubes that project into the benthic boundary layer are usually, although not exclusively, suspension feeders. On the basis of their adult tube dimensions, *H*, height above the sediment–water interface, and tube diameter, *D*, opposing the major flow vector, the ratio *H/D* (using consistent units for *H* and *D*) can be used to characterize the main types (Table 6.3) as follows:

Tube normal to flow	$H/D = 2–10$
Truncated cone tube	$H/D \leq 2$
Spar buoy tube	$H/D \geq 20$

In addition, some tubes have one, or both, open ends opposed to bidirectional flow; the ratio *H/D* is not applicable to them. The spar buoy tube is so named because it is articulated near its base and can passively move

Table 6.3. *Tube height,* H, *cm, above the sediment–water interface and maximum diameter,* D, *cm.*[a]

Species	Higher taxa	Tube dimensions			Reference
		H	*D*	*H/D*	
Tube normal to flow					
Lanice conchilega	Polychaeta, Terebellidae	2.0	0.5	4	Carey (1983)
Pseudopolydora paucibranchiata	Polychaeta, Spionidae	3.5	0.5	7	Levin (1982)
Streblospio benedicti	Polychaeta, Spionidae	3.5	0.5	7	Levin (1982)
Eudistylia vancouveri	Polychaeta, Sabellidae	12.5	1.7	7	Merz (1984)
Ampelisca abdita	Amphipoda, Ampeliscidae	0.5–1.0	0.2–0.3	3–5	Mills (1967)
Ampelisca vadorum	Amphipoda, Ampeliscidae	0.4–1.0	0.2–0.5	2–5	Mills (1967)
Haploops fundiensis	Amphipoda, Ampeliscidae	1.4	0.5	3	Wildish and Lobsiger (1987)
Tube opposed to flow					
Net-spinning caddis larvae	Trichoptera, Hydropsychidae	—	—	—	Edington (1968)
Ampithoe valida	Amphipoda, Ampithoidea	—	—	—	Borowosky (1983)
Tanais cavolinni	Tanaidacea	—	—	—	Borowosky (1983)
Truncated cone tube					
Spio setosa	Polychaeta, Spionidae	5.0	5.0	~1	Muschenheim (1987a)
Fabricia limnicola	Polychaeta, Sabellidae	0.6	0.4	~1	Levin (1982)
Mesochaetopterus sagittarius	Polychaeta, Chaetopteridae	0.9	0.6	~1	Bailey-Brock (1979)
Phyllochaetopterus verrilli	Polychaeta, Chaetopteridae	0.9	0.6	~1	Bailey-Brock (1970)
Spar buoy tube					
Potamilla neglecta	Polychaeta, Sabellidae	1.5–2.5	0.08	19–31	Wildish and Lobsiger (1987)
Complex tube shape					
Diopatra cuprea	Polychaeta, Onuphidae	2–5	0.5–0.8	2–10	Myers (1972)

[a] In some cases, tube dimensions were estimated from the original figures.

in whatever direction the flow dictates. In this respect, it is like the navigational buoy of this name, which is common where a considerable tidal prism is available.

Anyone who has watched snow being scoured around the base of a tree, or seawater flow scouring sediments at the base of a pier piling, is familiar with the general effects to be expected from flow around animal tubes. Isolated tubes which project into the logarithmic part of the benthic boundary layer cause momentum transfer, sediment deposition, and sediment entrainment. A theoretical and experimental flume study using a rigid circular cylinder as a tube mimic (Eckman and Nowell 1984) provides a good basis for those considering flume experiments with live tube-living epifauna. The range of tube dimensions tested by Eckman and Nowell (1984) was $H/D \simeq 1.0\text{–}6.7$, so their results are applicable to most species listed in Table 6.3, although not to the thin tube of *Potamilla neglecta*. Boundary shear stresses at tube height, which vary with ambient velocities, and to a lesser extent tube diameter are the critical factors involved in the small-scale geophysical effects seen. An interesting prediction from the Eckman and Nowell model is that if the tube is squat and $H \leq 100$ times the viscous sublayer thickness, a net deposition of sediment may occur if the boundary shear stresses are low enough to prevent scour in the downstream vortex.

Tube normal to flow

Laboratory studies in a recirculating flume by Eckman et al. (1981) with the polychaete *Owenia fusiformis* collected from False Bay, Washington, showed that isolated tubes caused sediment scouring around their base, which was assisted by tube bending. The scoured pits became traps for bedload transport on which *Owenia* could feed. In a flume study of *Lanice conchilega* collected from the Tay estuary, Scotland, Carey (1983) showed that the tube wakes (Fig. 6.7) resuspended material in such a way that the low density particles were available for downstream tentacle capture. The tube top was extended into two semi-circular plates which were arranged normal to the major flow direction and supported a branching fringe constructed of sand particles which was presumed to support the tentacular crown of *Lanice* during feeding. It is likely that these structures have importance in protecting and regulating the local flow environment near the tentacles.

Tube reconstruction of the sabellid polychaete *Sabellastarte magnifica* has been observed in the laboratory after removal of the worm from its old tube (Fitzsimons 1965). Beginning with the worm horizontal on the

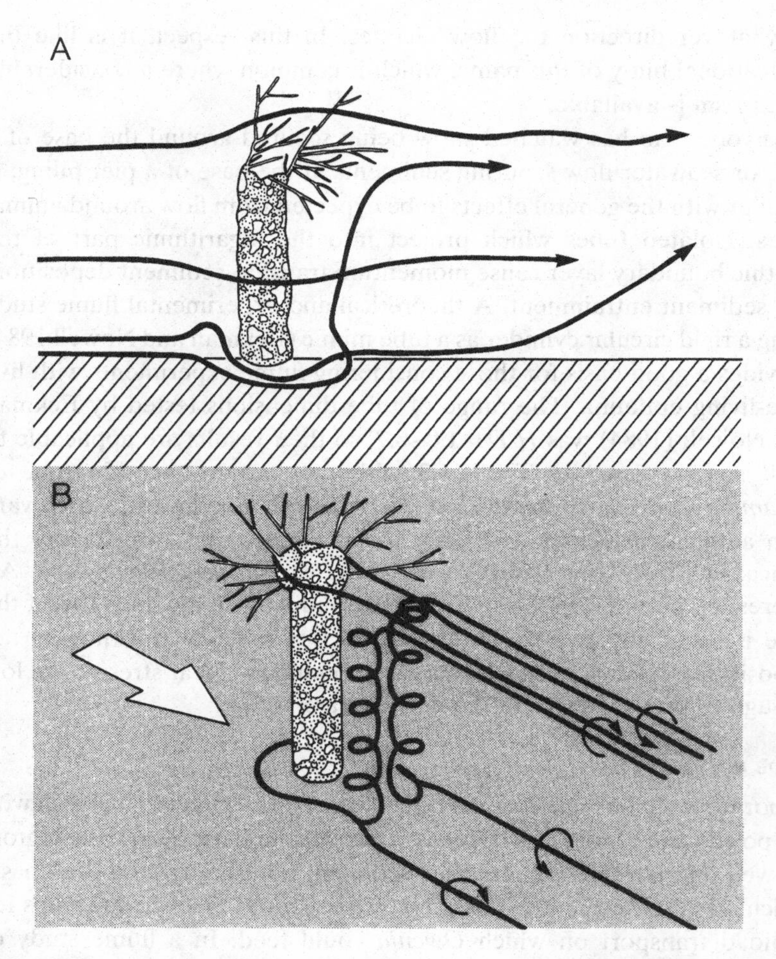

Figure 6.7 *Lanice conchilega* tubes; flow patterns and sediment scouring near the tube are based on time-lapse photography of egg white particles (Carey 1983); A, side elevation; B, from downstream; large arrow, flow direction.

substrate, the tube is eventually bent up at 90° to the substrate so that the finished tube is normal to ambient flows. Tube building occurs at the rate of 3–4 $mm \cdot d^{-1}$ in laboratory conditions.

No laboratory flume studies have yet been made of ampeliscid amphipods (Fig. 6.8), which have a tube with an elliptical cross section. The study by Eckman and Nowell (1984) considered only rigid right circular cylinders. Enequist (1949) and Mills (1967) have described ampeliscid tube building and feeding, aided by observations in aquaria

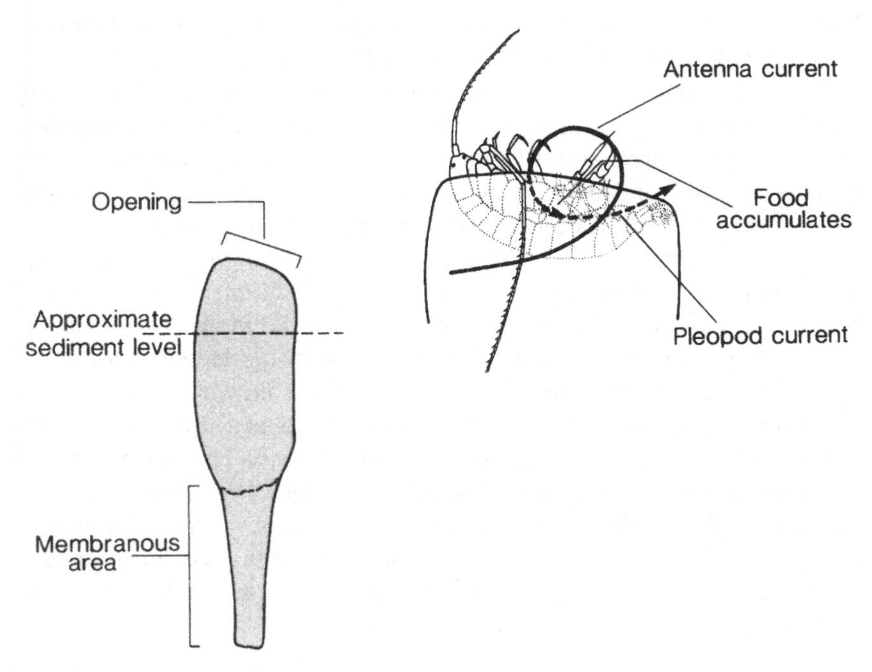

Figure 6.8 *Ampelisca abdita* amphipod: position adopted for feeding and complete tube (Mills 1967).

but without simulation of water movement. Tube building began after a freshly disturbed *Ampelisca abdita* had begun to burrow into the substrate, when it turned ventral side up and began secreting threads from glandular tissue on peraeopods 1 and 2 through ducts on the dactyls (Mills 1967). The threads were stretched around the body by the urosome and uropods and covered by sand particles placed there by the gnathopods. The inner lining of the finished tube consisted of smooth, whitish glandular secretion to which fine particles were stuck on the outside of the tube. It was noticed by Mills (1967) that both *A. abdita* and *A. vadorum* tubes were oriented so that the widest part of the tube opposed the major flow direction on intertidal sands. Tube orientation of *Haploops fundiensis* found at 80-m depth in the Bay of Fundy brought up in undisturbed grabs or indicated in underwater photographs, showed that the tubes were all oriented in the same plane and that the flat surface opposed the flow (Wildish and Lobsiger 1987). The elliptical tube section of *H. fundiensis*, 0.5 cm wide × 0.2 cm thick, has openings at the top and bottom and is quite flexible. No flows are induced in the tube, and the end inserted in the substrate is plugged with sediment.

During feeding, ampeliscid amphipods sit ventral side up at the top of their tube (Fig. 6.8), with the body along the long axis of it, braced by the posterior peraeopods and with the first two pairs of peraeopod dactyls hooked over the tube lip. If the animal moves lower into the tube, the lips close after it, thus preventing outside access. Three feeding methods are possible (Enequist 1949; Mills 1967): *passive filtration*, holding long setose first and second antennae into the flow; *whirling*, movements of both second antennae beating in synchrony, which creates a flow across the gnathopods where particles are filtered, with the feeding flow augmented by pleopod beating, which also drives the flow from anterior to posterior; and *deposit feeding*, which is possible when the long second antennae are scraped across the sediment surface. The specificity of these feeding methods is not clear, although Mills (1967) considers *A. vadorum* and *A. abdita* as mainly "whirling" feeders, whereas *Haploops fundiensis* must be a passive filterer or whirler since the second antennae are not long enough to reach the sediment for deposit feeding from its position in its tube.

Tubes opposed to the flow

Although some freshwater organisms may orient their feeding structures to take advantage of ambient flow in passive filtration, e.g. the net spinning caddis larvae (Table 6.3), whose pocket-shaped nets on stones are opposed to freshwater stream flow, we know of no marine tube builders which oppose the flow to filter passively. Perhaps the amphipod *Ampithoe valida* and tanaid *Tanais cavolinii*, which both have open-ended tubes, oriented horizontally on the substrate surface (Borowsky 1983), are examples of this type.

Truncated cone tubes

Short, squat tubes are built by cementing particles to the rim of the tubes, which grow upwards as sedimentation proceeds. Polychaetes with truncated cone tubes (Table 6.3) include *Spio setosa*, which have been studied by Muschenheim (1987a, b) in unidirectional laboratory flume flows. In nature, tubes are often broken off to the sediment surface level by wind–wave effects. By comparing the suspension feeding of broken tube worms, feeding at 0–2 cm above the bed, with that of intact tube worms feeding at 4–6 cm above the bed, Muschenheim (1987a) showed that the latter were feeding on better quality food, as evidenced by a greater

Table 6.4. *Effect of ambient velocity on helical coiling in* Pseudopolydora kempi japonica *palps.*[a]

Velocity ($cm \cdot s^{-1}$)	Number of observations	Median coils per palp
5	10	1.00
12	8	1.25
19	6	2.25
26	9	2.00

[a] Velocities are measured at palp height.
Source: Data from Taghon et al. (1980).

proportion of enriched organic mineral aggregates in gut contents. Spionid polychaetes are considered to be deposit-suspension feeders, using a pair of long palps for feeding. Taghon et al. (1980) conducted flume studies with a spionid, *Pseudopolydora kempi japonica*, over a unidirectional velocity range of 0 to ~16 $cm \cdot s^{-1}$, measured at 0.4 cm above the bed. Behavior of the worms was observed under low power magnification. As ambient velocity increased, the proportion of worms deposit feeding steadily decreased. Worms suspension feeding held their palps in a downstream helical coil with the number of turns increasing with ambient velocity (Table 6.4). The rheotactic behavior of palp coiling is directly proportional to velocity – at least in the range up to 20 $cm \cdot s^{-1}$ – and can be shown to increase the surface area of the tentacle facing the flow. Further experiments with narcotized worms showed that the behavior was absent when the muscular and nervous systems were inactivated, suggesting that a coordinated behavioral response was involved.

Turner and Miller (1991b) have provided a laboratory flume study of the chaetopterid polychaete *Spiochaetopterus oculatus* subjected to oscillatory flow. The water tunnel used permitted maximum oscillatory flows of 5–40 $cm \cdot s^{-1}$ with amplitudes of 8–32 cm and periods of 4–25 s. This chaetopterid is a facultative deposit–suspension feeder like the spionids mentioned. It has a pair of long palps which touch the sediment during deposit feeding and transfer the particles collected along the ciliated groove to the mouth. During suspension feeding, the palps become helically coiled, the degree of coiling increasing with higher ambient velocities. Exposure to oscillating flows showed that the worm rotated the palps at flow reversal so as to keep the ciliated food groove on the palps opposed to the flow. The time spent suspension feeding was related to ambient velocity so that it predominated at maximum flows >15 cm·s^{-1}. Deposit feeding predominated at maximum flows <10 cm·s^{-1}, and a change to suspension feeding appeared to be related to the inability to keep the palp tip on the sediment surface at velocities >10 cm·s^{-1}. The relevant hydrodynamic variable to characterize the transition could be the [bed] sediment shear stress. In Barnes's (1964) study of the same species, the worms also spent time building their tubes, which involved withdrawing their palps and exposing the fourth parapodia for this purpose or tube cutting as velocities increased by means of the fourth setae. Tube building in *Spiochaetopterus* was maximal in motionless seawater, and within 15 min of initiating oscillating flow, tube cutting also started (Barnes 1964; Turner and Miller 1991b). Reducing the height of the tube is presumably related to adjusting the worm's feeding ambit to one where good quality seston is available.

Spar buoy tubes

A sabellid polychaete, probably *Potamilla neglecta*, was the first benthic suspension feeder recognized as belonging to the group of spar buoy tubes. It was studied by underwater time-lapse photography at an 80-m-deep station in the Bay of Fundy (Wildish and Lobsiger 1987). Because these sabellids have long, very thin tubes, conventional grabs are inadequate for sampling and underestimates their density. Time-lapse photography (10- to 12-min intervals) over a 3-d period showed that the tube positions passively followed the current direction indicated by a current meter attached to the camera frame. This suggests that the tubes are articulated near their base and that the conical, branchial crowns are always able to feed in the same downstream position, whatever the flow direction.

The active suspension feeding sabellid polychaete *Sabella penicillus* studied experimentally by Riisgård and Ivarsson (1990) may also have a spar buoy tube, although it has a much greater tube height (≈7.5 cm) than *Potamilla* spp. *Sabella* has a crown-filament pump in which the laterofrontal cilia on pinnules borne from the filaments represent the active pump.

Complex tube shapes

The polychaete *Diopatra cuprea*, which is found intertidally and subtidally from Cape Cod to Florida, has an inverted J-shaped tube

(Myers 1972). Its tube extends 2 (1–5 range) cm above the sediment surface, depending on local sediment erosion/deposition near the tube. *Diopatra* can reduce the tube height by cutting if erosion is marked and build the tube upwards if considerable deposition of sediment occurs. Where currents are high, greater worm densities are present and the worms can orient the tubes behaviorally to within a 90° arc, so that the long axis of the J, seen from above, is at right angles to the flow. The worms may also rebuild the tubes so that a better orientation to flow is obtained (Myers 1972). Although it was suspected that the tube orientation was connected with feeding, no experimental results confirming this were presented.

A flume study of *Diopatra cuprea* was undertaken by Luckenbach (1986) to determine the erodibility of sediments affected by tubes in flow or by associated deposit–feeding macrofauna. During these experiments, it was not stated how the tubes were oriented with respect to the unidirectional flume flow. Those sediments with high macrofaunal densities, with or without *Diopatra* tubes, were more erodible than consolidated sediments with fewer macrofauna, suggesting that the tubes were unimportant in sediment stabilization, at least at friction velocity values employed during these experiments (Luckenbach 1986).

Free-living epifauna

The scallops (Pectinidae) are mobile epifauna with some swimming capabilities, which are more fully discussed later in the section, Swimming. Because their body plan is bilaterally symmetrical with inhalant and exhalant openings at specific positions on the body (Fig. 6.9) – as in the brachiopods observed earlier – it is plausible to propose that rheotaxis can optimize their orientation with respect to ambient flows.

Observations by SCUBA of natural populations of *Pecten maximus* in Strongford Loch, Northern Ireland (Hartnoll 1967), showed that the majority of scallops opposed the predominantly unidirectional flow (59% of individuals at 45° to 135°; see Fig. 6.9C). Mathers (1976) also observed *P. maximus* on the west coast of Ireland, at two locations where the tidal flows reversed on flood and ebb tides. Here it was found that approximately half of the scallops opposed the ebb and half opposed the flood currents with the ventral margin. The digestive histology of flood and ebb subpopulations showed that digestion was offset by 6 h, suggesting that each group fed on either the flood or ebb tide only, but not on

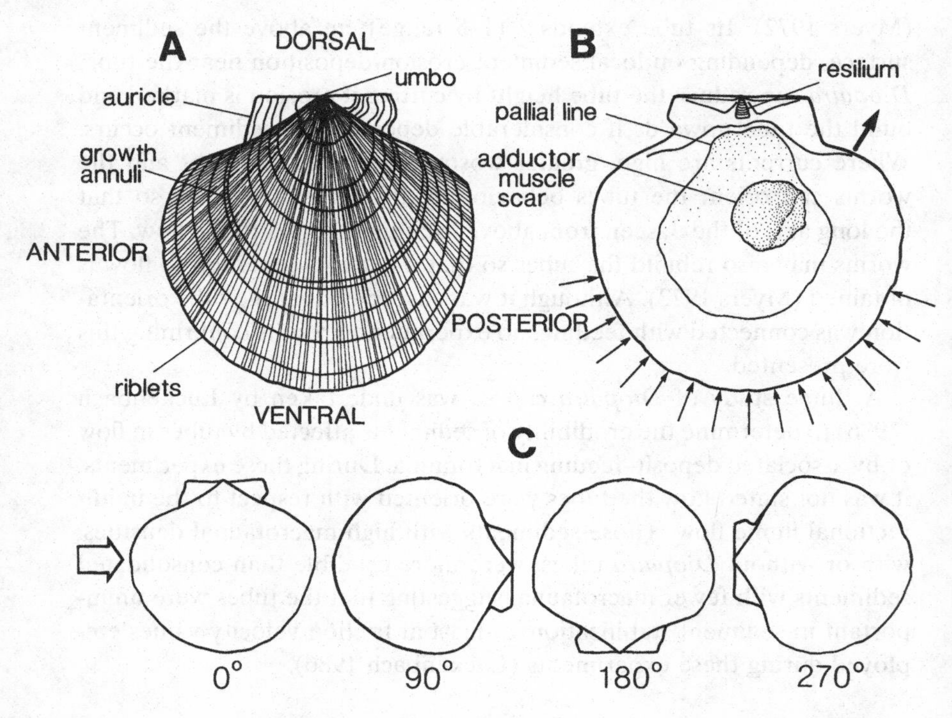

Figure 6.9 Scallop valve orientation and morphology: A, left upper valve from above; B, right, lower valve, inside view; small arrows, inhalant; large arrow, exhalant; C, valve orientation with respect to flow direction (large arrow).

both (Mathers 1976). Giant scallop (*Placopecten magellanicus*) beds in the Gulf of St. Lawrence studied by Caddy (1968) had most individual scallops opposing the largely unidirectional flow.

Evidence that scallop orientation to flow does affect filtration rate, and hence feeding and growth, is provided by Wildish et al. (1987) and Wildish and Saulnier (1992). Experiments with scallops glued in place in a Blažka respirometer at three flow velocities (Table 6.5) showed that giant scallops with their exhalant opening directly facing the ambient flow at 225° had significantly lower seston uptake rates at both 10 and 18 $cm \cdot s^{-1}$ than in the other treatments. Filtration rates for all orientations at 31 $cm \cdot s^{-1}$ were significantly less than at 10 $cm \cdot s^{-1}$. Similar results are obtained if giant scallop growth is measured. Multiple-channel flume growth experiments (Table 6.6), in which four concurrent treatments are possible with natural seston concentrations which are the same in each channel, show that at high velocities (22.5 $cm \cdot s^{-1}$) growth is significantly

Table 6.5. *Giant scallop seston uptake rates, μg Chla·g wet^{-1}·h^{-1}, mean of two replicate experiments at each treatment.*[a]

Scallop orientation (see Fig. 6.9C)	Velocity, cm·s^{-1}					
	10		18		31	
	μg·g^{-1}·h^{-1}	%	μg·g^{-1}·h^{-1}	%	μg·g^{-1}·h^{-1}	%
45°	0.77	100	0.78	101	0.24	31
135°	0.96	100	0.76	80	0.49	51
225°	0.54	100	0.22	51	0.07	13
315°	0.77	100	1.44	188	0.53	70

[a] Conditions: temperature = 9.5 ± 0.2°C (±SE); initial food concentration = 10^4 cells·ml^{-1}.
Source: Wildish et al. (1987).

Table 6.6. *Multiple-channel flume growth experiments in which individual giant scallops were glued to the flume floor with the orientation shown.*[a]

Treatment		Percentage daily growth[b] ± SE	
Speed (cm·s^{-1})	Scallop orientation (°)	Valve height (h)	Wet weight (W)
22.5	225	0.103 ± 0.007	0.263 ± 0.020
22.5	45	0.125 ± 0.004	0.344 ± 0.015
3.6	225	0.149 ± 0.005	0.321 ± 0.011
3.6	45	0.135 ± 0.006	0.336 ± 0.014

[a] Conditions: 28-d duration starting on November 5, 1986, mean temperature = 7.6°C with range of 6.1°–9.5°; mean Chla = 0.65 (range 0.48–0.99) μg/L, 30 juvenile giant scallops per treatment.
[b] $100 \cdot (h_1 - h_0)/h_0N$, $100 \cdot (W_1 - W_0)/W_0N$, where N = time in days.
Source: Wildish and Saulnier (1992).

limited when the exhalant faces the flow (225°), but this effect is absent at low ambient velocities (3.6 cm·s^{-1}) for the same orientation.

Behavioral cues which must be involved in scallop orientation behavior are not well known, although rheotaxis may be involved. The ability of scallops to spin around by directing a jet of water from only one side of the animal (Caddy 1968) may help to explain how it is done. An alternative hypothesis that scallop orientation to flow is a passive process brought about by ambient water movement has been proposed by Grant et al. (1993).

Natural populations of feather stars (Echinodermata, Crinoidea) found on the Great Barrier Reef were observed by time-lapse cinematography (Meyer et al. 1984). It was observed that the current regime caused changes in body posture (rheotaxis). Feather stars such as *Comanthus bennetti* and *Himerometra robustipinna* were able to crawl for periods of ~10 min at a very slow speed (1 arm length$\cdot$min^{-1}) to achieve a favoured position for passive filter feeding. The body position adopted at different flows was predominantly fan-shaped when flows were maximal, but transitional, meridional, and hidden positions were also seen. No experimental studies related the body position to feeding success, however, and the exact purpose of the rheotaxic behavior thus remains conjectural.

The feeding behavior of porcelain crabs – *Petrolisthes leptochelys* from the Red Sea and *P. oschimai* from the Indian Ocean – was shown by Trager et al. (1992) to be controlled by oscillating flows. Setae on the third maxillipeds are used by porcelain crabs as a filter. In very low flows, the paired third maxillipeds are rapidly swept through seawater, but, at higher flows, they are used passively much as they are in other facultative active suspension feeders such as barnacles. Also as in barnacles, it is the concave face of the filter presented to the flow. In low oscillating flows (0.16 Hz), both third maxillipeds acted in concert during passive feeding. At higher oscillating flows (>0.58 Hz), the third maxillipeds were extended alternately so the left and right sampled the flow on alternate pulses (Trager et al. 1992).

Avoiding flow dislodgement

Macrobenthic suspension feeders can be dislodged by wind–wave action and tidal flows. Experimental or observational studies which seek to understand the mechanisms and behavioral responses which are concerned with minimizing dislodgement are considered here.

Sessile epifauna

The greatest hydrodynamic forces exerted on macrobenthos occur on intertidal rocky shores. The reef coral *Acropora reticulata* was shown by Vosburgh (1977) to be limited to depths >7.7 m in Eniwetak Atoll, where it escaped the most violent breaking waves. Drag and mechanical tests

Table 6.7. *Force coefficients for intertidal rocky shore suspension-feeding organisms.*[a]

Species	Re	C_D	C_L	C_M
Semibalanus cariosus	10^5	0.52	N/A	1.31
Mytilus californianus – end-on	10^5	0.20	—	1.20
– broadside	10^5	0.80	—	2.00
Models				
Sphere	10^5	0.47	0.4	1.67
Cylinder	10^5	0.73	N/A	2.00

[a] C_D = drag coefficient, C_L = lift coefficient, C_M = inertia coefficient. (See the section, Free-living epifauna).
Source: Denny (1985).

were performed in a tow tank on air-dried specimens to determine the forces required for coral breakage. The extreme hydrodynamic forces predicted at these depths on the windward side of Eniwetak Atoll were less than the critical values for coral breakage.

Denny (1982, 1985) constructed a wave force telemetry system suitable for field operations and made observations on individual wave forces through many tidal cycles at one exposed location on the coast of Washington. Denny (1985) found that 2- to 4-m-breaking-height waves at this location produced maximum velocities of at least $8\,m \cdot s^{-1}$ and accelerations of at least $400\,m \cdot s^{-2}$ near the substratum. Denny (1985) has interpreted the field data obtained in terms of the force which will be experienced by a few representative intertidal organisms for which data are available (Table 6.7) where the Re values are in the range 10^4–10^6. The forces are of three types: drag, wave acceleration reaction, and lift (see Vogel 1981a, b). For example, exposure to shear forces in breaking waves has been estimated by Denny (1985) by calculating the force on the basis of drag alone and ignoring the acceleration reaction, which contributes <15% to the total force. For waves up to a breaking height of 10 m, *Semibalanus* experiences lower shear forces than do mussels. It is of interest that the bilaterally symmetrical mussel has a much greater exposure to shear force if it is oriented broadside rather than end on to the flow. From Table 6.7, it can be seen that drag coefficients are high for the intertidal organisms measured in comparison to those of streamlined shapes, e.g. of scallops, for which precise data are unavailable. Streamlining is only effective in reducing drag of an animal when it is optimally oriented with respect to the flow. On the wave-swept rocky shore, flow directions are unpredictable and therefore streamlining is probably not a realistic option. As Denny (1985) points out, mussels may be present in densely packed reefs where the forces are shared by all rather than by an individual mussel. Some intertidal organisms, e.g. *Semibalanus cariosus*, are firmly cemented to the substrate by a special adhesive to resist the lift and drag forces involved.

Adhesive forces involved in *Semibalanus balanoides* attachment have been studied by Yule and Crisp (1983) and Yule and Walker (1987) by using strain gauges. Eckman, Savidge, and Gross (1990) used an indirect method of determining drag and detachment forces for *Balanus amphitrite* from shear forces in a small flume. The latter study suggests that the drag and detachment forces associated with *Balanus* are lower than those determined directly with strain gauges in *Semibalanus*. Whether this is a real difference consistent with the more exposed lifestyle of *Semibalanus* or is due to differences in measurement technique awaits further standardization of both methods.

In terms of the preceding introduction to forces commonly present on wave-swept rocky shores, we must ask whether behavioral responses to flow are involved in helping the organism to avoid flow dislodgement. Adhesive attachment of benthic organisms on a wave-swept rocky shore is a common solution to dislodgement, e.g. in barnacles (see earlier discussion), or mussel byssus threads which are glued to the substrate (Waite 1983). Skeletal stiffening by spicule formation, e.g. in sponges and cnidarians (Koehl 1982), and strengthening by coral formation or development of various forms of body elasticity which allow form changes proportional to velocity, resulted in reduced drag (Koehl 1977a). Most of these must be considered as structural adaptations to resist the forces generated by flow.

Size may also represent a refuge from the worst effects of orbital wind–wave action, e.g. for the sea anemone *Metridium senile* (Anthony and Svane 1994). Large numbers of small sea anemones of pedal disc diameter of 0.3 cm are found where wave exposure is strong versus fewer, much larger, 2.5- to 3.5-cm-pedal-disc-diameter individuals in subtidal lower wave-exposed habitats.

Behavioral responses may be involved (Koehl 1984) in withstanding moving water; they include adjusting the orientation of the individual to minimize lift and drag. One of the characteristics shown by the sea anemone *Anthopleura xanthogrammica* which live in exposed intertidal surge channels of rocky shores on the U.S. West Coast was found by

Koehl (1977b) to involve a behavioral response to flow. In the field, Koehl showed that there was a significant height difference between individuals from more exposed (2.1-cm) and those from protected (5.6-cm) sites. Subsamples of exposed and protected anemones were placed in a laboratory sea table where the flows were maintained at a constant low level of $<5\,cm \cdot s^{-1}$. After 8 d, the exposed individuals had significantly increased in height to 4.3 cm, whereas those from the protected site showed no significant change. This suggests that the response was rheotropic and that excessive drag is prevented by maintaining a low profile in the benthic boundary layer. This was possible because this anemone is not a suspension feeder but feeds on dislodged mussels which fall onto the oral disc. Drag forces measured on live and model anemones showed that drag was greatest on the tallest ones. A retracted sea anemone also had a streamlined shape consistent with waves coming from any direction.

Changes of body shape or orientation of sessile epifauna with respect to ambient flow which involve muscular movement or locomotion, e.g. withdrawal of the tentacles of scleractinian coral polyps (Hubbard 1974) or locomotion by normally sessile bivalves such as mussels, may be related to the flow forces they experience. We could find no other detailed studies of rheotactic behavior by sessile epifauna associated with avoiding flow dislodgement.

Free-living epifauna

Arnold and Weihs (1978) developed a hydrodynamic model for plaice rheotaxis which is applicable to most benthic macrofauna where $Re = 10{,}000 - 1{,}000{,}000$ (where Re is defined by multiplying body length times free-stream velocity divided by the kinematic viscosity of seawater) as a means of analyzing the effect of flow forces in dislodging free-living epifauna. In the model, the fish is assumed to be dislodged by the flow when the drag force on it is just balanced by the static friction force between the fish and the bottom. The static friction force is proportional to the difference between the fish's buoyant weight and the lift force generated by the flow over the fish.

An experimental study of mussel dislodgement of both *Mytilus californianus* and *M. edulis* was made by Witman and Suchanek (1984), using a direct drag force measurement system. The force that was required to dislodge the mussel by breaking the bysuss threads was recorded. It was found that attachment strengths increased with shell area and varied with position in the mussel reef and with the degree of wave exposure of the habitat. The measured attachment strength of the byssus threads was found to be proportional to the degree of wave exposure of the mussel's habitat. The Re under flume flow conditions and based on the valve length of the mussels used by Witman and Suchanek (1984) ranged from 9000 to 21,000, but with simulated epifloral attachments of considerable length, this increased to 53,000–27,000,000. Thus, mussels that had been overgrown by kelp experienced flow-induced drag forces that were two to six times greater than on the mussels alone, and this explains why, during storms, mussels could be dislodged. Laboratory experiments by Young (1985) were designed to determine the effect of various environmental variables on the rate of byssus formation by *M. edulis*. The most important factor was found to be periodic agitation, simulating wave surges, which caused 15.8 threads $\cdot$ d^{-1} for each mussel to be produced. This rate was twice that of any other factor tested such as temperature, salinity, tidal submergence, or season, although all of those affected byssus thread formation. In agreement with results of Witman and Suchanek (1984), Young (1985) found that wave-exposed mussels had more byssus threads (87 per mussel) than those from sheltered sites (48 per mussel).

Preliminary studies using the Arnold and Weihs (1978) model with giant scallops *Placopecten magellanicus* have been presented by Wildish and Saulnier (1993). In these experiments, the scallops were placed on the smooth bottom of a flume and the velocities at which they were first dislodged observed.

In order to study the hydrodynamics of the scallop, it is necessary to study their body structures. The geometry of giant scallop valves (Fig. 6.10) shows that the valve length is less than the valve height at <10 cm, but is greater at >10-cm valve height. In natural conditions, the giant scallop opposes the flow with its ventral edge. The frontal area A_f presented to this flow (Fig. 6.10) is irregularly ellipsoidal in cross section, and the lower, right valve is flatter than the upper, left valve. Because of this irregularity, A_f was determined empirically by Wildish and Saulnier (1993) from photographic cutouts calibrated to area (rather than by calculation of πlw). Seen from the posterior edge (Fig. 6.10), the maximum valve width does not occur centrally but more dorsally as shown. Because of the choice by the scallop to oppose the flow with the ventral edge first, i.e. with the "trailing edge" (Fig. 6.10) rather than the "blunt" end, it is not optimally streamlined for lift (like the aeroplane wing). The potential hydrodynamic instability of this orientation has been discussed

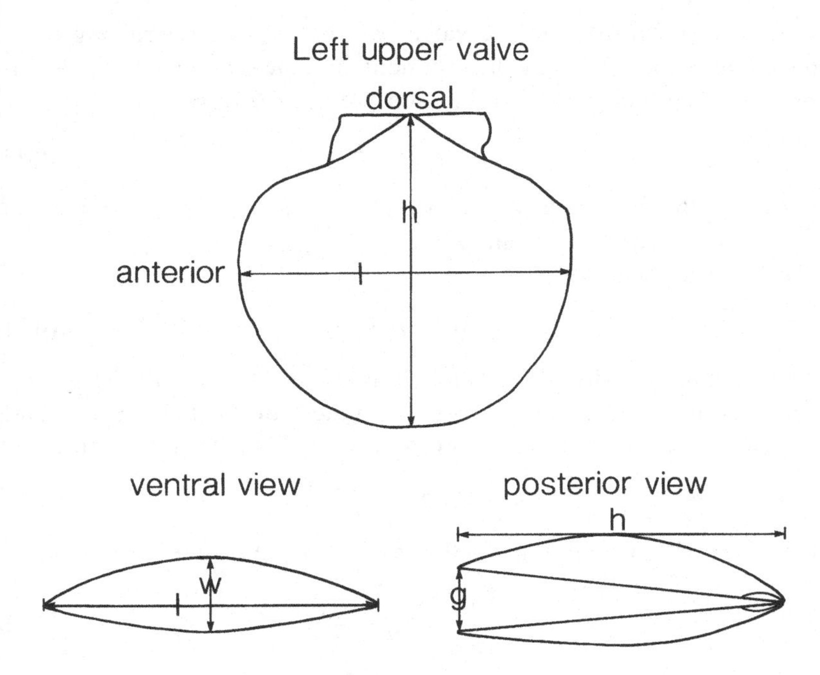

Figure 6.10 Geometry of the adult giant scallop valves showing the measuring origins adopted: *h*, valve height; *l*, valve length; *w*, maximum valve width or thickness; *g*, maximum valve gape.

by Grant, Emerson, and Shumway (1993). Its common occurrence in the field seems unrelated to optimum feeding orientation because Wildish et al. (1987) showed that this was at 315° (exhalant 90° to flow) and in a field test at high flows (>10 $cm \cdot s^{-1}$) scallops grew better at 270° (exhalant 45° to flow) than opposing the flow at 90°.

During flow over scallop valves, according to Bernoulli's principle, a decrease in pressure occurs above the left valve as a result of the faster flows there. The pressure on the right valve, which may be partly recessed in the sediment, is higher, resulting in lift. Lift forces act in the geometric centre of the scallop valve and tend to push the ventral edge of the valve upwards into the flow. When the negative buoyancy is balanced by the lift force (F_L), the scallop will be lifted off the sediment. The velocity at which this occurs (V_L) was called the "lift-off speed" by Arnold and Weihs.

It is possible that at velocities less than required to lift the scallop, viscous and form drag on the valves will exceed the restraining force exerted on the left valve by the sediment, and the scallop will slip downstream. At this slip speed, Arnold and Weihs (1978) show

$$F_D = \kappa(W - F_L) \tag{6.1}$$

where F_D is the drag force, κ is the static friction coefficient, and W is the submerged weight of the scallop.

The lift is defined as

$$F_L = 0.5\rho_s A_f C_L U^2 \tag{6.2}$$

where ρ_s is the density of seawater, A_f is the frontal area of the scallop seen from the ventral edge, C_L is a non-dimensional lift coefficient, and U is the ambient free-stream velocity. The drag is determined from

$$D = 0.5\rho_s A_f C_D U^2 \tag{6.3}$$

where C_D is the frontal drag coefficient. Substituting and rearranging,

$$\frac{2W}{\rho A_f U_s^2} = C_L + \frac{C_D}{\chi} \tag{6.4}$$

The term on the left of the equation where U_s is the slip speed, $cm \cdot s^{-1}$, determined experimentally, should be nearly constant for individuals with valves of geometrically similar shapes and sizes yielding $Re > 10^4$. Re is defined here by valve height and local free-stream velocity. The ratio on the left side of (6.4) can be referred to as Arnold and Weihs's ratio and should be useful in correlating slip velocities with size of benthic animals in which $Re > 10^4$.

For live, unrestrained giant scallops ($n = 17$), a $U_s = 11$–$37\,cm \cdot s^{-1}$ of 2.7 to 12.0 cm valve height of animals was determined by Grant et al. (1993) on a smooth acrylic floor. Independent data for the same species of live scallop, but with the valves restrained with a rubber band at the anterior end so that the band did not touch the smooth acrylic floor of the Saunders and Hubbard recirculating flume were obtained by Wildish and Saulnier (1993). The scallops ($n = 26$) were of 2.6- to 10.4-cm valve height and gave a $U_s = 9$–$32\,cm \cdot s^{-1}$ (Wildish and Saulnier 1993), similar to the results obtained by Grant et al. (1993). In addition to live giant scallops, Wildish and Saulnier (1993) undertook experiments on model scallops consisting of the valves of live specimens used in the previous experiments, but with all wet tissue removed and the valves dried. Each set of

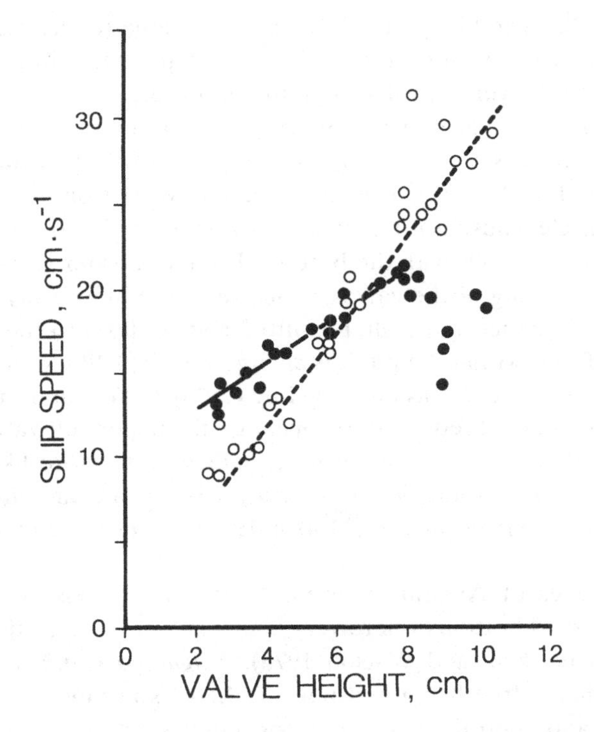

Figure 6.11 Valve height of the giant scallop as a function of slip speed: open circles, live, shut valves, filled circles, dead, open valves (Wildish and Saulnier 1993).

hinged valves was then adjusted to a buoyant weight equal to that in life by the addition of plasticine and the valves propped open to simulate the observed maximal gaping in life – *g* in Fig. 6.10.

Results suggest that the slopes of U_s as a function of valve height (Fig. 6.11) for live and model scallops are significantly different. Larger individuals (>8 cm valve height) in which the valve is maximally gaped have anonymously low slip speeds. This is linked to the increased frontal area, which increases drag. In scallops of sizes >8 cm valve height, frictional resistance with the bottom and buoyant weight is less efficient in overcoming drag and lift. Possibly the behavioral phenomenon of recessing in older scallops may be linked to this physical effect resulting from allometric changes in the animal (Wildish and Saulnier 1993). Recessed scallops are subject to lower drag forces because much of the body may be beneath the benthic boundary layer. Additionally, the sides of the recessed pit will prevent slipping although not affect the lift forces sufficient to cause the ventral valve edge to flip upwards.

Recessing by *Pecten maximus* when placed in a large seawater tank (flow characteristics not specified) on an 8-cm-deep bed of sand has been described by Baird (1958). The scallop jetted water from the right side near the auricle, causing the animal to swing around and the valves to become inclined at 60° with the bottom. The jet of water is repeated by opening and closing of the valves, which serves to blow away the sand. By an extra large jet, the scallop is lifted and settles into the pit it has excavated. From within the pit, it is enlarged by expelling water forcibly around the mantle edge, as described by Caddy (1968), and this causes, typically, a coating of sediment to appear on the upper, left valve. Older, larger individuals of *Placopecten magellanicus* were more likely to be recessed than smaller individuals. Recessing is also a common phenomenon in *Amusium pleuronectes* (Morton 1980) and *Pecten ziczac* (Waller 1975).

Some species of Australian terebratellid brachiopods are specially adapted for sand bottoms where very high wave surge velocities may be present (Richardson and Watson 1975). *Magadina cumingi* is such a species, which is free living and not attached, as in most brachiopods. The dorsal and ventral pedicle adjustor muscles act antagonistically to extend and retract the pedicle, which is used as a pogo stick to raise the animal off the bottom. Few details of the behavior of this interesting brachiopod were given.

Although most sand dollars (Echinodermata, Echinoidea) are deposit feeders (Telford and Mooi 1987) and thus do not warrant inclusion here, they have been studied by hydrodynamic methods and concepts similar to those presented here. Telford (1983) believed that in the sand dollar *Mellita quinquiesperforata*, the lunules present act as lift spoilers, that is reduce lift by inducing flow through the lunules, which tends to reduce the pressure differential between the top and underside of the animal. Lift spoilers increase the ambient velocity, which induces slip and critical lift-off in this sand dollar. *Dendraster excentricus* (deposit–suspension feeder), which lives in wave surge channels of the western U.S. coast, avoids lift behaviorally by partially burrowing into the sand with the result that its thin, flat body is upright in the sediment with the result that all individuals are oriented parallel to the flow direction (Merrill and Hobson 1970). They recognized bay, tidal channel, and outer coast populations and the primary importance of water movement in their distribution. At moderate levels of tidal flow and wave surge, the highest densities were found in the "inclined" position – that is with the anterior end of the test inserted in the sand and the thin test profile parallel to ambient flows. In lesser flows, they opposed the flow or assumed a horizontal position on the sand. During storm wave conditions, *Dendraster* burrowed into the sediment. The "inclined" behavior of this species would serve both to optimize feeding and to prevent washout by waves. O'Neill (1978) showed that dense aggregations of the sand dollar also benefitted feeding efficiency when they were passively suspension feeding. At flows $>10\,cm \cdot s^{-1}$, the inclined position improves the feeding success of the oral surface tube feet compared to that of the normal horizontal position. This is because the cambered shape of the test causes the streamlines between adjacent individuals to compress. Individuals which are packed closely take advantage of local streamline curvature to enhance particle capture for each other (O'Neill 1978).

Epifauna and aggregation

Evidence is presented here that aggregation among benthic suspension feeders does commonly occur. In investigating the proximate causes of aggregation, a number of general hypotheses should be considered (Table 6.8). For those suspension feeders which are sessile after larval attachment, the factors influencing larval recruitment are critically important (see Chap. 3) and presumably linked to the flow-related advantages in numbers 1 and 2 of Table 6.8. Other species can move away and reattach at a new and perhaps better location, e.g. juvenile mussels. Yet others are colonial: the original larval settler undergoes a sexual reproduction to create a colony at the original settlement site, e.g. some sponges, corals, and ascidians. Because larval recruitment and its possible link to improved reproductive success (number 3) and larval recruitment linked to predator protection (number 4, Table 6.8) are not directly related to flow, they are not considered in detail. Nevertheless, the two latter general hypotheses must be part of the synthetic phase of thinking when the topic of aggregation is considered.

According to the first hypothesis of Table 6.8, aggregated groups of individuals occupy areas where a specialized feature of the hydrodynamic regime provides a locally higher flux of seston past the group of suspension feeders. The spatial extent of the area occupied by the group, or patch, is determined by the spatial limits dictated by the hydrodynamics and the absolute seasonal seston flux values, which also dictate the

Table 6.8. *Hypotheses concerned with aggregation in suspension-feeding macrofauna.*

No.	Hypothesis		Reference
1	H_0	Larval recruitment to a group is not related to the location's being optimal for suspension feeding	Fager (1964) Okamura (1986)
	H_1	Larval recruitment to a group is related to the location's being optimal for suspension feeding	
2	H_0	Larval recruitment to a group is not related to providing protection from energetic water movement	Barkai (1991)
	H_1	Larval recruitment to a group allows protection for some individuals from energetic water movement	
3	H_0	Larval recruitment to a group is unconnected with reproductive success	Crisp (1979) Okamura (1986)
	H_1	Larval recruitment to a group results from the group's being optimal for reproductive success	
4	H_0	Larval recruitment to a group is not related to protection from predators	Okamura (1986)
	H_1	Larval recruitment to a group results in better protection from predators	

carrying capacity (by addition or death of individuals) for the patch. At high carrying capacities where densities may reach very high levels, behavioral mechanisms may serve to limit densities, as shown by Johnson (1959) in *Phoronopsis viridis*. In densely packed bivalve mollusc reefs the nearest neighbor distance may be so small that the valves are not able to open completely, thereby imposing spatial limits to growth (Bertness and Grosholz 1985).

Scallops *Placopecten magellanicus* in intensive culture (Wildish et al. 1988) interact behaviorally ("biting syndrome") at high densities by locking of valves, which can prevent feeding or involve adductor muscle damage. Coté, Himmelman, and Clarereboudt (1994) grew juvenile scallops in pearl nets with and without dummy scallops made by gluing valves of dead scallops together. The pearl nets were suspended in the southwestern Gulf of St. Lawrence and left for a summer's growth. The results obtained by Coté et al. (1994) clearly showed that at up to a maximum density of 250 scallops per net (≡1563 scallops·m^{-2}) tested, the dummy scallops did not decrease the growth of 25 scallops per net as did living scallops up to this density. This suggests that the major effect was caused by either food depletion or behavioral interaction, e.g. by interference competition in cages, but not directly by space limitation.

There are some suggestions in the literature that special and localized hydrodynamic conditions controlling seston flux are linked to aggregations of sessile suspension feeders. Fager (1964) described a reef of the polychaete tube-dwelling *Owenia fusiformis*, which was 60 × 150 m long with its long axis to the beach and at a depth of 6 m. The reef was associated with a rip current which could serve to concentrate both the seston supply and *Owenia* larvae at this site. In studies undertaken by Bosence (1979) in an Irish loch, the serpulid polychaete *Serpula vermicularis* was substrate limited. The larvae initially settled on rare rock outcrops, but after this substrate was covered, social settlement occurred, resulting in the building of extensive reefs. The reefs were formed by subsequent waves of larvae which settled on the adult tubes of serpulids, forming bushlike aggregations of tubes. In Narragansett Bay blue mussel reefs were described by Nixon et al. (1971) as developing across narrow inlets on bars near the mouths of tidal channels and on muddy sandflats where energetic tidal currents rather than wind–wave energy were the major source of water movement.

From the preceding and similar further examples available in the literature, we conclude that the evidence that location is important for the development of patches of suspension feeders is circumstantial and weakly supportive of hypothesis 1 in Table 6.8. However, very little information on the behavioral mechanisms which must be involved is available.

One form of aggregation among sessile suspension feeders involves coloniality, e.g. in some hydroids, sponges, bryozoans, and ascidians. In these cases, it is probably difficult to determine a proximate cause, or causes, of coloniality. Two examples will be given to illustrate this difficulty. The larvae of the colonial ascidian *Botryllus schlosseri* can recognize genetically similar forms and settle gregariously so as to increase the chances of fusion as adults (Grosberg and Quinn 1986). This forms a larger, more competitive type of colony. Genetically dissimilar forms which meet in this way are unable to fuse and instead act as spatial competitors. The second example is an interesting type of coloniality discussed by Hughes (1979) among vermetid prosobranch gastropods. Vermetids are sedentary forms with a loosely coiled (= vermiform) shell which is cemented to a hard substratum. Many vermetids form loose aggregations which may not qualify as true colonies, although the South African *Dendropoma corallinaceum* seems to be a true colonial form, producing dense sheetlike aggregations on rock surfaces which are potent competitors for space, although it appears to tolerate mutualistically the presence of other encrusting forms such as the calcareous alga *Lithothamnion* sp. (Hughes 1979). *D. corallinaceum* is found in a narrow zone near mean low water springs and suspension feeds by means of a mucous net. In this species, the nets overlap and coalesce to form a communal net, so that individuals, when they haul in the net, may be feeding on part of a net secreted by their nearest neighbor. Hughes (1979) considered that vermetid aggregation was linked to successful fertilization by waterborne sperm and/or the use by juveniles of the presence of adults to indicate a suitable niche for settlement.

Evidence that many epifaunal suspension feeders do live together in aggregated groups or reefs comes from standard grab or core sampling (Table 6.9), where the area sampled is usually $0.1\,m^2$ and maximally $1.0\,m^2$. The examples included in Table 6.9 were selected because they had high densities or biomasses and it can be assumed that the grab sampled within an aggregated group. In no individual reports in Table 6.9 was the sample plan area varied, and, therefore, the type of spatial distribution (random, regular, or contagious [Elliot 1977; Thrush 1991]) at all scales remains unknown. Another difficulty with the data shown in Table 6.9 is that the densities given may include a preponderance of very young, small individuals. This becomes important if there are no accompanying biomass data, as proved to be the case in many of the studies cited.

Despite the obvious sampling problems, some impressively high numbers and biomasses are shown in Table 6.9. Thus, suspension-feeding bivalves may reach a dry weight standing crop of $21\,kg \cdot m^{-2}$. We conclude that for a range of species, available evidence is consistent with their living closely together as groups. Tube-living amphipods have high densities, but because each individual is small, the population biomass is modest. All the amphipods of Table 6.9 are probably deposit–suspension feeders (Enequist 1949), suspension feeding when water movement is sufficiently energetic by means of rigidly held, feathery antennae for passive suspension feeding. The antennae can, in some cases, also be used to surface deposit feed if flows slacken. Only the suspension-feeding crinoids and ophuroids (Warner 1979) appear to be present at significant biomasses (unfortunately no data are available in this reference) because the adults link arms and passively filter feed by extending their arms into the boundary layer. This has the effect of anchoring them to resist energetic water movements and to create turbulence in the benthic boundary layer and hence increase the feeding potential.

An example of clumping behavior, which provides protection from wave effects able to dislodge sessile epifauna, comes from a study of sublittoral holothurians on the west coast of South Africa (Barkai 1991). The holothurians involved were all passive suspension feeders. In locations where wave exposure was high, *Thyone aurea* (maximum density $377 \cdot m^{-2}$) and *Pentacta doliolum* (maximum density $464 \cdot m^{-2}$) occur together in dense aggregations. *Thyone*, which has degenerate tube feet, insinuates its body beneath *Pentacta*, utilizing it as a means of attachment. In more protected sites, *Thyone* creates separate clumps but *Pentacta* is absent there. Flume studies showed that the strength of attachment and resistance to increasing velocity as lift-off speed, U_s – the velocity at which a holothurian was swept away – were much higher in *P. doliolum* than *T. aurea*. Barkai (1991) considered that *Thyone* commensally used the *Pentacta* colony to extend its range from protected sites to more exposed ones. Model sea cucumber experiments in flumes in which grouped and singular models were compared showed that U_s of grouped models was often two times higher than that of singular models. The precise behavioral mechanisms involved in aggregation of these two holothurians are unknown.

Usually, clumps are of individuals of a single species, e.g. polychaete or amphipod tubes. Nowell and Church (1979) provide tube model experiments in a flume that predicts tube density conditions which permit reduction of drag within the clump. When tube density is sufficient to give approximately one-twelfth the plan area of the bottom, the flow changes to a "skimming flow" regime. In skimming flow over tube clumps, much less kinetic energy reaches the bottom, and, under these conditions, the flow–sediment interaction is stabilized. As pointed out by Eckman et al. (1981), the Nowell and Church model does not accurately predict stabilized versus de-stabilizing tube–sediment relations in field conditions. They suggested that the discrepancy between field and laboratory results was due to the common occurrence in the field of sediment binding by mucus from bacteria or benthic diatoms. Eckman et al. (1981) point out that a purely hydrodynamic cause of reduced kinetic energy reaching the sediment as a result of tube clumping in field conditions has not been demonstrated.

Table 6.9. *Density and biomass among aggreging epifaunal suspension feeders.*[a]

Taxa	Location	Density: no·m^{-2} Mean	Density: no·m^{-2} Max	Biomass, g dry·m^{-2} Mean	Biomass, g dry·m^{-2} Max	Reference
Bivalve molluscs						
Modiolus modiolus	Glacial till, Bay of Fundy,	—	510[b]	—	3,038[b]	Peer et al. (1980)
	Canada	—	—	281	1,023	Wildish and Peer (1983)
Abra alba	Muddy sand, Consolidated	1,235	—	—	—	Warwick and Davies (1976)
	substrate, Mud, Bristol	121	—	—	—	
	Channel, UK	70	—	—	—	
Mesodesma deauratum	Fine–medium sand, Grand	488	1,550[b]	6,485	21,030[b]	Hutcheson and Stewart
	Banks, Nfld., Canada	3,010	12,000[b]	939	4,697[b]	(1994)
Crassostrea virginica	Estuaries of southeastern USA	—	4,077	—	214	Dame et al. (1984)
Mytilus edulis	Narragansett Bay, Rhode Island, USA	2,139		10,962		Nixon et al. (1971)
	Baltic Sea	36,000	158,000	101	—	Kautsky (1982)
	Newfoundland, Canada	—	—	596	1,101	Thompson (1984)
	Morecambe Bay, UK	—	—	1,234	—	Thompson (1984)
Perna perna	South Africa	—	—	826	1,284	Thompson (1984)
Echinoderms						
Amphiura filiformis	Silt–sand burrower of eastern Atlantic	—	2,200	—	—	Warner (1979)

Antedon bifida	Outcropping rock, Torbay, UK	—	1,200	—	—	Warner (1979)
Ophiothrix fragilis	High tidal current flows on various substrates	—	2,196	—	—	Warner (1979)
Polychaetes						
Owenia fusiformis	Fine sand–silt, La Jolla Bay, USA	500	15,000	—	—	Fager (1964)
Lanice conchilega	Fine–medium sand, Weser estuary, FRG	—	20,200	—	1,094	Buhr and Winter (1977)
Spio setosa	Sand beach, Halifax Harbour, Canada	408	2,002	61	109	Muschenheim (1987a)
Eudistylia vancouveri	Rock, San Juan Is., Pacific coast, USA	—	2,576	—	—	Merz (1984)
Phyllochaetopterus verrilli and Mesochaetopterus sagittarius	Fringing reef, Oahu, Hawaii, USA	—	62,400	—	—	Bailey-Brock (1979)
Amphipods						
Haploops fundiensis	Silt–clay soft sediment, 80-m depth, Bay of Fundy, Canada	376	923	0.35–0.53	0.96	Wildish (1984)
Ampelisca abdita	Fine sand, Atlantic coast	1,360	73,000	—	—	Mills (1967)
Ampelisca vadorum	Very coarse sand, Atlantic coast, N. Amer.	1,307	1,885	—	—	Mills (1967)

[a] Wet biomass converted to dry by dividing by 3.33.
[b] Maximum from single grab sample.

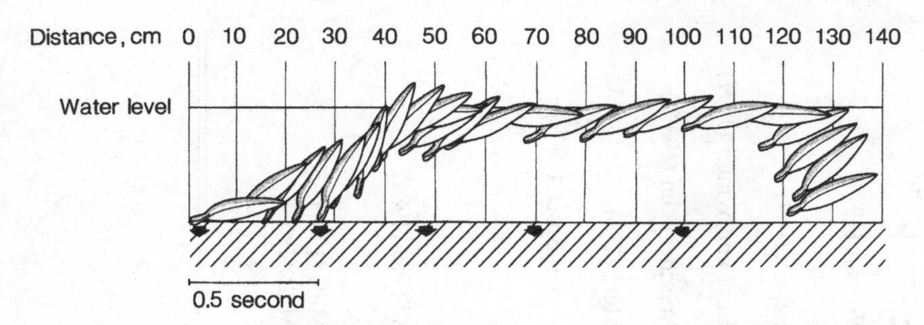

Figure 6.12 Swimming sequence of *Amusium pleuronectes* as recorded by a cine camera (Morton 1980). The arrows indicate the timing of muscle adductions.

Swimming

The pectinids are the bivalves with the best developed ability to swim, although Stanley (1970) mentions solenids, solemyids, and cardids as other bivalves with some locomotory abilities. Among scallops, two types of locomotion can be recognized (Moore and Trueman 1971): a predator escape reaction, in which a limited number of adductions of the valves moves the animal as a result of the rapid expulsion of pallial water which pushes the animal in the opposite direction from the point of predator attack, and a complex swimming behavior.

Field studies have described complex swimming sequences in *Placopecten magellanicus* (Caddy 1968) and *Amusium pleuronectes* (Morton 1980) as (1) a rise from the bottom, (2) level flight, and (3) passive sinking (Fig. 6.12). During the rise and level flight stages of swimming, periodic contractions of the adductor muscle occur, depending on the length of the flight, the species, and possibly the size of the individual. The timing of valve adduction is shown in the swimming sequence of *A. pleuronectes* shown in Fig. 6.12. The force powering swimming in scallops is provided by periodic jet propulsion. This involves valve gaping to take in pallial water, closing the mantle edge around the water, followed by rapid contraction of the adductor muscle, which forces it out in two vectored jets dorsally on either side of the auricles. Valve opening is usually described as being due to elasticity of the ligament or resilium (Demont 1990), although Vogel (1985) makes a good case for flow-induced forces augmenting the ligament in reopening the valves. According to this argument, dynamic pressure on the right valve ventral edge will be high and on the left valve low, thereby creating pressures which tend to open the valves. Calculations made by Vogel (1985) based on a gape angle of ~13° and at a swimming speed of $50\,cm \cdot s^{-1}$ suggest that the opening moment is ~22% of that due to ligament elasticity. Following this is a glide phase in which the valve margins are closed before initiation of another adduction jet propulsion sequence. It is local muscular contractions of the mantle which helps to produce the vectored jet propulsive forces, which are principally powered by adductor muscle contractions. Stephens (1978) has studied the nervous control mechanisms involved in swimming for *Argopecten irradians*. The jet propulsion rhythm is maintained during rise and level flight phases by neuronal feedback from stretch receptors in the adductor muscle which are activated during valve opening.

The pressure pulses involved in swimming of *Chlamys opercularis* have been recorded with a pressure transducer (Moore and Trueman 1971). The transducer consisted of plastic tubing and a hypodermic needle sealed into the valve so that it opened into the pallial cavity. Maximum pressures measured for normal swimming were 5–6 cm of water at the relatively low swimming speed (~$30\,cm \cdot s^{-1}$) of this species. Escape reaction movements involved periodic pulses of about half the pressure found in complex swimming by *Chlamys* (Moore and Trueman 1971).

Lift was determined experimentally in four scallops, *Chlamys islandica*, *C. opercularis*, *Placopecten magellanicus*, and *Pecten maximus*, by Gruffyd (1976). Measurements were made on model scallops of each species at zero flow, equal to the buoyant weight of the live animal, and at increasing velocities until the model became unstable as the lift generated by the flume flow approached the model weight. The results were dependent on scallop size and the angle of incidence, that is the angle of inclination of the ventral valve tips with the horizontal. Unfortunately, the frontal area, A_f, at different angles of incidence of the valves was not determined so that coefficients of drag could not be determined with these data. Re-examination of Gruffyd's results by Thorburn and Gruffyd (1979) when the angle of incidence was 0° showed that small amounts of lift were always generated, rather than the zero lift previously reported. The species examined by Gruffyd (1976) showed increased lift as this angle was increased. Maximum lift was produced at angles of incidence of ~20° in all species, except *P. magellanicus*, in which lift increased at angles at least up to 60°. During initiation of swimming behavior, the angle of attack of the body is important in achieving sufficient lift-off from the substrate. In a similar model study of *Chlamys opercularis*, Thorburn and Gruffyd (1979) showed that allometric changes in valve shape were insufficient to provide lift to counteract increasing weight with growth. This is because of the mass/surface area law, which states that mass increases according to h^3 while surface area increases according to h^2.

The lift and drag forces of plaster casts of six species of scallop have been determined in flume studies by Millward and Whyte (1992). New lift and drag coefficients using the plane form area of the valves, rather than frontal area, were determined. Typical C_L ranges from <0.1 to 0.8 for all species with stall angles of incidence of ~15° for planoconvex shells, e.g. *Pecten alba*, and ~25° for others (biconvex valves). Millward and Whyte (1992) point out that because of the use of an inconsistent set of units, the calculated values of Gruffyd (1976) and Thorburn and Gruffyd (1979) are two orders of magnitude greater than they should be. These authors also found a correlation between C_D and the thickness ratio of the valve – equivalent to A_f, as found also by Wildish and Saulnier (1993) in *Placopecten magellanicus*.

Flow patterns over the valves were studied (Thorburn and Gruffyd 1979) by visualizing natural particles present in seawater by dark field photography. The results suggest that a smooth attached flow over the left valve becomes a turbulent eddying wake at angles of attack >20° and at ambient velocities >25 cm · s^{-1}. The stability of *C. opercularis* during swimming was enhanced by the presence of auricles, which, when removed, caused instability. Auricles were also thought to increase lift at low swimming speeds.

Hydrodynamic characteristics of scallops vary as a size-related phenomenon because of body allometric changes which occur during ontogeny. Although this was recognized by Gould (1971), the first studies of swimming behavior which incorporated size as a variable were those of Morton (1980), Joll (1989), and Dadswell and Weihs (1990). Measures required to determine the hydrodynamic characteristics of a scallop valve (see Fig. 6.10) include valve height (h), maximum thickness or valve width (w), valve length (l), and valve area (a). From Dadswell and Weihs (1990), the Re of the scallop body can be expressed as

$$\mathrm{Re} = \frac{Uh}{\upsilon} \tag{6.5}$$

where U = ambient velocity and υ = kinematic viscosity of seawater at a given temperature. The fineness (F) of the scallop valves, which is related to drag encountered, is given by

Table 6.10. *Pectinid swimming speeds estimated from original scatter plots scaled against valve height.*[a]

Species	h, cm	Re $\times 10^4$	F	R_a	Swimming speed, $cm \cdot s^{-1}$ Mean	Max	Reference
Chlamys islandica	4.0	—	—	—	25	—	Gruffyd (1976)
	7.0				35	—	
C. opercularis	6.7	—	—	—	21	34	Thorburn and Gruffyd (1979)
Amusium pleuronectes	6.5	—	—	—	45	70	Morton (1980)
	10.0				37	40	
A. balloti	5.5	—	—	—	68	80	Joll (1989)
	10.0				90	100	
Placopecten magellanicus	5.5	2.7	4.4	1.19	49	68	Dadswell and Weihs (1990)
	10.0	4.0	3.3	1.32	40	53	

[a] h = valve height, F = fineness, R_a = aspect ratio.

$$F = \frac{h}{w} \tag{6.6}$$

The aspect ratio of the scallop valves is given by

$$R_a = \sqrt{\frac{l^2}{a}} \tag{6.7}$$

and the available lift, L, can be calculated from

$$L = \frac{1}{2}\rho V^2 C_L a \tag{6.8}$$

where C_L is the coefficient of lift for low aspect ratio wings obtained from Hoerner (1975), $0.4 < C_L < 0.8$.

Complete data are available only for *Placopecten magellanicus* (Table 6.10). The data for this species show that the fineness ratio fits a near-optimum shape, as does the relationship $x/h = 0.6$ in giant scallop adults (where x = distance from ventral edge to the widest point in cross section) for streamlining and fast swimming powers (Arnold and Weihs 1978). As pointed out in the section, Free-living epifauna, the blunt

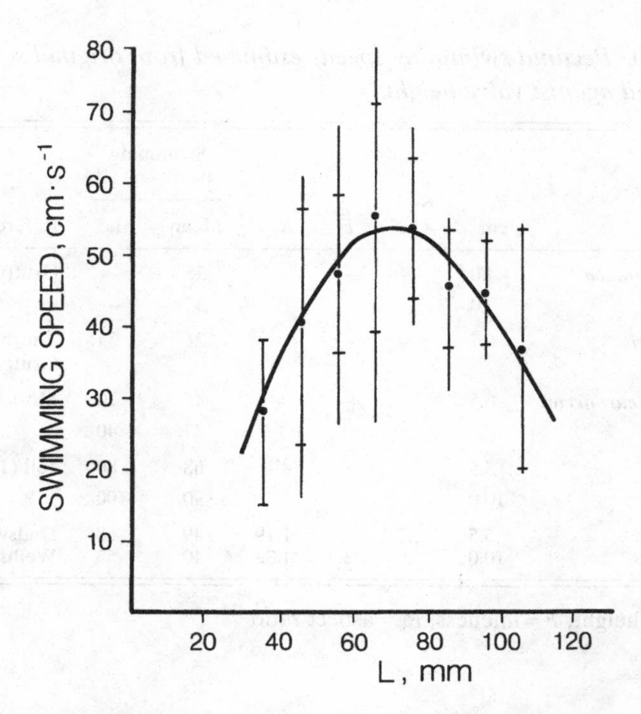

Figure 6.13 Swimming speed of *Placopecten magellanicus* as a function of valve height: vertical lines, sample range; points, means, horizontal lines, one standard deviation; curved line, fitted by eye (Dadswell and Weihs 1990).

trailing edge (dorsal side) opposed to the flow would be the optimum hydrodynamic shape. During the ontogeny of the giant scallop, the adductor grows toward the mid-centre point of the valve, a process which is complete by $h = 5.5$ cm (Dadswell and Weihs 1990). A similar ventral displacement of the adductor muscle in relation to the valves is not complete until $h = 8.0$ cm.

Swimming speeds shown in the table are approximate estimates made from scatter plots of the original data made under non-standard environmental conditions (e.g. temperature). Another difficulty in comparing the swimming speeds shown in Table 6.10 is that they were determined in different ways. Thus, Gruffyd (1976) and Morton (1980) measure the time taken over a set distance during level flight swimming, while Joll (1989) and Dadswell and Weihs (1990) use the total distance travelled divided by the time taken from the initiation of stage (1) to the termination of stage (2) – that is the rise and level flight stages – but exclude the passive sinking stage (3). Nevertheless, it is clear that of the swimming speeds of the species so far examined (Table 6.10), the fastest is the Australian saucer scallop *A. balloti*. The effect of scallop size on swimming speed is reported by Joll (1989) for *A. balloti* and by Dadswell and Weihs (1990) for *P. magellanicus* (Fig. 6.13) with a drop in performance of all larger individuals. In fact, Dadswell and Weihs (1990) calculated that the largest giant scallops, $h \geq 11.0$ cm, may be unable to generate enough lift to overcome their buoyant weight. Presumably, it is only higher ambient flows and increase in the angle of attack, both of which increase lift, which permit the larger scallops to initiate swimming behavior.

Other aspects of swimming performance, such as the distance travelled and the number and rapidity of adductions per swimming flight, are obviously important in determining the value of swimming to individuals. Although data for distance travelled per swimming flight and adduction rate are available for *Amusium* (Morton 1980; Joll 1989), there are insufficient data for other species to allow worthwhile comparison. Despite assertions by Morton (1980) that larger *Amusium pleuronectes* swim for a longer time (and hence distance), his own data of swimming time as a function of valve height have a non-significant *r* value. Similarly, for the relationship between the number of adductions per swimming flight, the relationship with valve height is non-significant. Data presented by Joll (1989) for *Amusium balloti* of swimming distance versus valve height were significantly correlated on log-transformed data. One group of scallops tested when ambient flow was zero showed significantly lower swim distances than when ambient flow was present.

The adduction rate of juvenile *Placopecten magellanicus* ($h = 0.04$–0.35 cm) during swimming is positively related to temperature with a $Q_{10} \simeq 3$ adductions$\cdot s^{-1}$ measured over the temperature range 9.0° to 14.5°C (Manuel and Dadswell 1991). These data suggest that scallop swimming performance comparisons must consider temperature as well as individual size and ambient flow conditions. Manuel and Dadswell (1993) studied swimming in relation to size of juvenile *P. magellanicus*. At a Reynolds number $<3 \times 10^3$, equivalent to a valve height <12–16 mm, the amount of swimming decreased with size. This critical size below which swimming was also inefficient hydrodynamically may explain the use of byssus attachment by the smallest juveniles of the giant scallop.

The type of sedimentary substrate on which adult bay scallops *Argopecten irradians* are placed determines the probability that they will swim (Winter and Hamilton 1985). Bay scallops were significantly more likely to swim when placed on clean sand than on seagrass beds, the preferred habitat. Distances swum by bay scallops depended inversely on the weight of fouling organisms, as simulated by adding 2–8 g of epoxy putty to the left valve. The predacieous gastropods *Fasciolaria tulipa* and *Murex pomium* always caused the bay scallop to swim when the tentacles came into contact with the snail's foot. The direction of swimming by the scallop was controlled by the location on the mantle edge contacted by the predator – the escape reaction was directly away from the source of stimulation (Winter and Hamilton 1980).

Summary

The foregoing shows that orientations to flow by a range of suspension feeders, e.g. infaunal clams, phoronidians, brachiopods, barnacles, and epifaunal bivalves, do take place. Research required to establish that suspension feeders orient rheokinetically or rheotactically to enhance their feeding includes observations of the behavior in field and lab and experimental confirmation that the response is to flow and that the individual derives trophic advantage from the preferred orientation. Other species grow with respect to flow direction – rheotropisms – thereby optimizing their feeding and growth. Research required to establish that rheotropic growth is occurring includes field observations which establish a preferred orientation with respect to flow and flume experiments which establish the effect of flow on growth. Although relatively few suspension feeders have locomotory capabilities, those that do may use them to increase their feeding opportunities. The research required to establish this mechanism is often difficult to complete satisfactorily. Steps involved in the process include field observations of suspension feeder movement; experimental confirmation of sensory cues, which frequently will not be flow-related ones; and some experimental verification that restrained animals grow more slowly than unrestrained ones.

Tube-living epifauna behavior has been little studied and so of the five types of tube recognized here (there may well be others), studies have only been started in one of the groups. The research required includes a careful field observation of normal behavior in the tube, hydrodynamic experiments to make clear the flow–tube interaction and the way ethology is linked to it and that improved feeding or growth results from the observed behavior. Abelson et al. (1993) used a physical model for passive suspension feeders such as corals and sponges to predict the quality and quantity of seston encountered. The "slenderness ratio" employed to characterise coral and sponge bodies uses the same height/diameter ratio used in this chapter for tube-living epifauna. The model predicts that high slenderness-ratio bodies (≃ high *H/D*) encounter higher fluxes of finer, good quality particles, whereas low slenderness ratio bodies (≃ low *H/D*) encounter coarser, poorer quality bedload particles. It is interesting to speculate why the ampeliscid amphipod tube is elliptical and oriented with the long axis opposed to the major flow vector. One possibility is that the amphipod feeds in the wake downstream of the tube to take advantage of resuspended sedimentary particles.

From the results discussed in this chapter, it will be clear to the reader that insufficient evidence is available for much of the behavior discussed to link it to one or another of the responses presented at the beginning of this chapter. In formulating future hypotheses about the behavior of suspension feeders and flow, optimizing feeding and avoiding flow dislodgement could be alternative hypotheses (H_0 and H_1) or competing multiple hypotheses. Okamura (1990) pointed out that many suspension feeders have alternative or parallel forms of nutrition. The importance of these, e.g. of uptake of dissolved organic matter directly from seawater, is still in doubt, although for some cases, e.g. photosynthetic microalgae in coral reefs and chemoautotrophic bacteria in deepsea pogonophorans, it is well established (Okamura 1990). All potential forms of nutrition need to be considered when constructing hypotheses regarding suspension feeding and flow.

Many ways of overcoming the flow dislodgement problem for epifauna do not involve behavior at all, but rather some form of morphological–biochemical solution, e.g. adhesives from antennular secretions of barnacle cyprids and byssus threads in bivalves. Some solutions do, however, involve behavioral adaptations, and the following were recognized: orientation of the body to minimize drag, clumping of individuals to reduce individual effects of drag, and recessing of scallops within the sediment to minimize drag. Research required for these studies includes a field or lab description of the flow avoidance behavior as well as a complete hydrodynamic model study to show which orientation is "best" in regard to feeding, slip or lift-off speed, individual drag within a clump versus an isolated individual, and slip or lift-off speeds of non-recessed versus recessed scallops. An example of the difficulty in deciding whether orientation of free-living epifauna is passive or active is provided by the scallop (Grant et al. 1993). If the response is active, it is not clear to what the scallop is responding – is it feeding optimization or flow avoidance? If the response is not concerned with feeding, as seems likely from the available data, an alternative possibility is that in bidirectional tidal flows the 90° to 270° axis is the best orientation to prevent being passively swept away. Another possibility is that the ventral edge of the valves opposing the flow may ensure the maximum lift available to initiate swimming, although this would only work for one tidal direction. These hypotheses deserve further experimental testing. That little research has been undertaken in the laboratory is surprising because much of it can be completed with the aid of model suspension feeders in flume studies, thereby simplifying it.

For many benthic suspension feeders, larvae represent the dispersal agent for the species. Yet, as we have shown (Chap. 3) for some suspension feeders, they may settle gregariously. Of four general hypotheses shown in Table 6.8 and proposed to determine the mechanisms involved in aggregation, none has received enough attention to determine their relative importance. Clearly, more observational and experimental work is required before general conclusions can be reached.

Swimming locomotion may be part of a predator avoidance reaction in some epifaunal bivalves, although many authors consider scallop swimming to have been refined for additional functions, e.g. in effecting reproductive dispersal, finding mates, choosing better feeding conditions. Predator avoidance behavior can readily be confused with flow-related behavior. Some idea of the sensory cues used to initiate the reflex responses involved in escaping predators is required, as well as the specificity of the stimulus and its threshold level. When behavioral studies involving swimming are contemplated, the alternate hypothesis of predator avoidance should be considered along with the flow-related one.

The complex behavior of swimming should ideally be studied in controlled environmental conditions (e.g. flow, temperature, salinity) as a function of body size, with swimming performance expressed as speed, distance travelled, and number of adductions per swimming flight. Relatively few studies have met these rigorous criteria, although it is clear that there is an optimum swimming performance by mid-sized individuals. The older and larger scallops swim at lower speeds than their smaller confrères. Morton (1980) thought that this was because of senescence in the adductor muscle with age and a consequent change in behavior. In contrast, Gould (1971) and later writers, particularly Dadswell and Weihs (1990), who linked ontogenic changes to hydrodynamic characteristics, consider that ontogenic changes are responsible for the size-related swimming speed differences. In particular, the weight, which scales according to h^3 during ontogeny, outstrips the lift available, which only increases as h^2.

References

Abelson, A., T. Miloh, and Y. Loya. 1993. Flow patterns induced by substrata and body morphologies of benthic organisms and their roles in determining availability of food particles. Limnol. Oceanogr. 38: 1116–1124.

Anthony, K. R. N., and I. Svane. 1994. Effect of flow-habitat on body size and reproductive patterns in the sea anemone, *Metridium senile*, in the Gullmarsfjord, Sweden. Mar. Ecol. Prog. Ser. 113: 257–269.

Arnold, G. P., and D. Weihs. 1978. The hydrodynamics of rheotaxis in the plaice (*Pleuronectes platessa* L.). J. Exp. Biol. 75: 147–169.

Ayling, A. M. 1976. The strategy of orientation in the barnacle *Balanus trigonus*. Mar. Biol. 36: 335–342.

Bailey-Brock, J. H. 1979. Sediment trapping by chaetopterid polychaetes on a Hawaiian fringing reef. J. Mar. Res. 37: 643–656.

Baird, R. H. 1958. On the swimming behaviour of scallops (*Pecten maximus* L.). Proc. Malacological Soc. 33: 67–71.

Barkai, A. 1991. The effect of water movement on the distribution and interaction of three holothurian species on the South African west coast. J. Exp. Mar. Biol. Ecol. 153: 241–254.

Barnes, R. D. 1964. Tube building and feeding in the chaetopterid polychaete, *Spiochaetopterus oculatus*. Biol. Bull. 127: 397–412.

Bertness, M. D., and E. Grosholz. 1985. Population dynamics of the ribbed mussel, *Geukensia demissa*: the costs and benefits of an aggregated distribution. Oecologia (Berlin) 67: 192–204.

Borowsky, B. 1983. Reproductive behaviour of three tube-building peracarid crustaceans: the amphipods *Jassa falcata* and *Amphithoe valida* and the tanaid *Tanais cavolinii*. Mar. Biol. 77: 257–263.

Bosence, D. W. J. 1979. The factors leading to aggregation and reef formation in *Serpula vermicularis* L., p. 299–318. *In* G. Larwood and B. R. Rosen (eds.) Biology and systematics of colonial organisms. Academic Press, London.

Bowers, R. L. 1968. Observations on the orientation and feeding behaviour of barnacles associated with lobsters. J. Exp. Mar. Biol. Ecol. 2: 105–112.

Buhr, K.-J. and J. E. Winter. 1977. Distribution and maintenance of a *Lanice conchilega* association in the Weser estuary (FRG) with special inference to the suspension-feeding behaviour of *Lanice conchilega*, p. 101–113. *In* B. F. Keegan, P. Ó'Ceidigh, and P. J. S. Boaden (eds.) Biology of benthic organisms. Pergamon Press, Oxford.

Caddy, J. F. 1968. Underwater observations on scallop (*Placopecten magellanicus*) behaviour and drag efficiency. J. Fish. Res. Board Can. 25: 2123–2141.

Carey, D. A. 1983. Particle resuspension in the benthic boundary layer by a tube-building polychaete. Can. J. Fish. Aquat. Sci. 40 (Suppl.): 301–308.

Carthy, J. D. 1958. An introduction to the behaviour of invertebrates. George Allen & Unwin, London.
Coté, J., J. H. Himmelman, and M. R. Clareredboudt. 1994. Separating effects of limited food and space on growth of the giant scallop, *Placopecten magellanicus* in suspended culture. Mar. Ecol. Prog. Ser. 106: 85–91.
Creutzberg, F. 1975. Orientation in space: animals. Invertebrates, p. 555–653. *In* O. Kinne (ed.) Marine ecology. Vol. II. Part 2. Wiley, London.
Crisp, D. J. 1953. Changes in the orientation of barnacles of certain species in relation to water currents. J. Anim. Ecol. 22: 331–343.
1960a. Factors influencing growth rate in *Balanus balanoides*. J. Anim. Ecol. 29: 95–116.
1960b. Mobility of barnacles. Nature 188: 1208–1209.
1979. Dispersal and re-aggregation in sessile marine invertebrates, particularly barnacles, p. 319–327. *In* G. Larwood and B. R. Rosen (eds.) Biology and systematics of colonial organisms. Academic Press, London.
Crisp, D. J., and H. G. Stubbings. 1957. The orientation of barnacles to water currents. J. Anim. Ecol. 26: 179–196.
Dadswell, M. J., and D. Weihs. 1990. Size-related hydrodynamic characteristics of the giant scallop, *Placopecten magellanicus* (Bivalvia: Pectinidae). Can. J. Zool. 68: 778–785.
Dame, R. F., R. G. Zingmark, and E. Haskin. 1984. Oyster reefs as processors of estuarine materials. J. Exp. Mar. Biol. Ecol. 83: 239–247.
Demont, M. E. 1990. Tuned oscillations in the swimming scallop, *Pecten maximus*. Can. J. Zool. 68: 786–791.
Denny, M. W. 1982. Forces on intertidal organisms due to breaking ocean waves: design and application of a telemetry system. Limnol. Oceanogr. 27: 178–183.
1985. Wave forces on intertidal organisms: a case study. Limnol. Oceanogr. 30: 1171–1187.
Dolmer, P., M. Karlsson, and I. Svane. 1994. A test of rheotactic behavior of the blue mussel *Mytilus edulis* L. Phuket. Mar. Biol. Cent. Spec. Publ. 13: 177–184.
Eckman, J. E., and A. R. M. Nowell. 1984. Boundary skin friction and sediment transport about an animal-tube mimic. Sedimentology 31: 851–862.
Eckman, J. E., A. R. M. Nowell, and P. A. Jumars. 1981. Sediment destabilization by animal tubes. J. Mar. Res. 39: 361–374.
Eckman, J. E., W. B. Savidge, and T. F. Gross. 1990. Relationship between duration of cyprid attachment and drag forces associated with detachment of *Balanus* amphitrite cyprids. Mar. Biol. 107: 111–118.
Edington, J. M. 1968. Habitat preferences in net-spinning caddis larvae with special reference to the influence of water velocity. J. Anim. Ecol. 37: 675–692.
Elliot, J. M. 1977. Some methods for the statistical analysis of samples of benthic invertebrates. Freshw. Biol. Assoc. Sci. Publ. 25.
Emig, C. C., and F. Becherini. 1970. Influence des courants sur l'ethologie alimentaire des phoronidians. Etude par séries de photographies cycliques. Mar. Biol. 5: 239–244.
Enequist, P. 1949. Studies on the soft-bottom amphipods of the Skagerrak. Zool. Bidr. Upps. 28: 297–492.
Eshleman, W. P., and J. L. Wilkins. 1979. Brachiopod orientation to current direction and substrate position (*Terebratalia transversa*). Can. J. Zool. 57: 2079–2082.

Fager, E. W. 1964. Marine sediments: effects of a tube-building polychaete. Science 143: 356–359.
Fitzsimons, G. 1965. Feeding and tube-building in *Sabellastarte magnifica* (Shaw) (Sabellidae: Polychaeta). Bull. Mar. Sci. 15: 642–671.
Gould, S. J. 1971. Muscular mechanics and the ontogeny of swimming in scallops. Palaeontology 14: 61–94.
Grant, J., C. W. Emerson, and S. E. Shumway. 1993. Orientation, passive transport and sediment erosion features of the sea scallop *Placopecten magellanicus* in the benthic boundary layer. Can. J. Zool. 71: 953–959.
Grosberg, R, K., and J. F. Quinn. 1986. The genetic control and consequences of kin recognition by the larvae of a colonial marine invertebrate. Nature 322: 456–459.
Gruffyd, L. D. 1976. Swimming in *Chlamys islandica* in relation to current speed and an investigation of hydrodynamic lift in this and other scallops. Norw. J. Zool. 24: 365–378.
Hartnoll, R. G. 1967. An investigation of the movement of the scallop. Helg. Wiss. Meeresunters. 15: 523–533.
Helmuth, B., and K. Sebens. 1993. The influence of colony morphology and orientation to flow on particle capture by the scleractinian coral *Agaricia agaricities* (Linnaeus). J. Exp. Mar. Biol. Ecol. 165: 251–278.
Hoerner, S. F. 1965. Fluid-dynamic drag. Midland Park, New Jersey.
Hubbard, J. A. E. B. 1974. Scleractinian coral behaviour in calibrated current environment: an index to their distributional patterns. Proc. Second Int. Coral Reef Symp. Great Barrier Reef Comm. 2: 107–126.
Hughes, R. N. 1979. Coloniality in Vermetidae (Gastropoda), p. 243–253. *In* G. Larwood and B. R. Rosen (eds.) Biology and systematics of colonial organisms. Academic Press, London.
Hutcheson, M. S., and P. L. Stewart. 1994. A possible relict population of *Mesodesma deauratum* (Turton): Bivalvia (Mesodesmatidae) from the southeast shoal, Grand Banks of Newfoundland. Can. J. Fish. Aquat. Sci. 51: 1162–1168.
Johnson, A. S. 1988. Hydrodynamic study of the functional morphology of the benthic suspension feeder *Phoronopsis viridis* (Phoronida). Mar. Biol. 100: 117–126.
Johnson, R. G. 1959. Spatial distribution of *Phoronopsis viridis* Hilton. Science, N.Y. 129: 1221.
Joll, L. M. 1989. Swimming behaviour of the saucer scallop *Amusium balloti* (Mollusca: Pectinidae). Mar. Biol. (Berl.) 102: 299–305.
Kautsky, N. 1982. Growth and size structure in a Baltic *Mytilus edulis* population. Mar. Biol. 68: 117–133.
Koehl, M. A. R. 1977a. Mechanical design of cantilever-like sessile organisms: sea anemones. J. Exp. Biol. 69: 127–142.
1977b. Effects of sea anemones on the flow forces they encounter. J. Exp. Biol. 69: 37–105.
1982. Mechanical design of spicule-reinforced connective tissues: stiffness. J. Exp. Biol. 98: 239–267.
1984. How do benthic organisms withstand moving water? Am. Zool. 24: 57–70.
LaBarbera, M. 1977. Brachiopod orientation to water movement. I. Theory, laboratory behaviour and field observations. Paleobiology 3: 270–287.
1978. Brachiopod orientation to water movement: functional morphology. Lethia 11: 67–79.

Leversee, G. J. 1976. Flow and feeding in fan-shaped colonies of the gorgonian coral *Leptogorgia*. Biol. Bull. 151: 344–356.

Levin, L. A. 1982. Interference interactions among tube dwelling polychaetes in a dense infaunal assemblage. J. Exp. Mar. Biol. Ecol. 65: 107–119.

Luckenbach, M. W. 1986. Sediment stability around animal tubes: the roles of hydrodynamic processes and biotic activity. Limnol. Oceanogr. 31: 779–787.

Manuel, J. L., and M. J. Dadswell. 1991. Swimming behaviour of juvenile giant scallop, *Placopecten magellanicus*, in relation to size and temperature. Can. J. Zool. 69: 2250–2254.

1993. Swimming of juvenile sea scallops, *Placopecten magellanicus* (Gmelin): a minimum size for effective swimming. J. Exp. Mar. Biol. Ecol. 174: 137–175.

Mathers, N. F. 1976. The effects of tidal currents on the rhythms of feeding and digestion in *Pecten maximus* L. J. Exp. Mar. Biol. Ecol. 24: 271–284.

Merrill, R. J., and E. S. Hobson. 1970. Field observations of *Dendraster excentricus*, a sand dollar of western North America. Am. Midl. Nat. 83: 594–624.

Merz, R. A. 1984. Self-generated versus environmentally produced feeding currents: a comparison for the sabellid polychaete *Eudistyla vancouveri*. Biol. Bull. 167: 200–209.

Meyer, D. L., C. A. LaHaye, N. D. Holland, A. C. Arneson, and J. R. Strickland. 1984. Time-lapse cinematography of feather stars (Echinodermata: Crinoidea) on the Great Barrier Reef, Australia: demonstrations of posture changes, locomotion, spawning and possible predation by fish. Mar. Biol. 78: 179–184.

Mills, E. L. 1967. The biology of an ampeliscid amphipod crustacean sibling species pair. J. Fish. Res. Board Can. 24: 305–355.

Millward, A., and M. A. Whyte. 1992. The hydrodynamic characteristics of six scallops of the superfamily Pectinacea, class Bivalvia. J. Zool. Lond. 227: 547–566.

Monismith, S. G., J. R. Koseff, J. K. Thompson, C. A. O'Riordan, and H. M. Nepf. 1990. A study of model bivalve siphonal currents. Limnol. Oceanogr. 35: 680–696.

Moore, J. D., and E. R. Trueman. 1971. Swimming of the scallop *Chlamys opercularis* (L.). J. Exp. Mar. Biol. Ecol. 6: 179–185.

Morton, B. 1980. Swimming in *Amusium pleuronectes* (Bivalvia: Pectinidae). J. Zool. London 190: 375–404.

Muschenheim, D. K. 1987a. The role of hydrodynamic sorting of seston in the nutrition of a benthic suspension feeder, *Spio setosa* (Polychaeta: Spionidae). Biol. Oceanogr. 4: 265–288.

1987b. The dynamics of near-bed seston flux and suspension-feeding benthos. J. Mar. Res. 45: 473–496.

Myers, A. C. 1972. Tube–worm–sediment relationships of *Diopatra cuprea* (Polychaeta: Onnpulidae). Mar. Biol. 17: 350–356.

Nixon, S. W., C. A. Oviatt, C. Rogers, and K. Taylor. 1971. Mass and metabolism of a mussel bed. Oecologia 8: 21–30.

Nowell, A. R. M., and M. Church. 1979. Turbulent flow in a depth limited boundary layer. J. Geophys. Res. 84: 4816–4824.

Okamura, B. 1986. Group living and the effects of spatial position in aggregations of *Mytilus edulis*. Oecologia (Berlin) 69: 341–347.

1990. Behavioural plasticity in the suspension feeding of benthic animals, p. 637–660. *In* R. N. Hughes (ed.) Behavioural mechanisms of food selection. NATO ASI Ser. Vol. G.20. Springer-Verlag, Berlin.
O'Neill, P. L. 1978. Hydrodynamic analysis of feeding in sand dollars. Oecologia 34: 157–174.
O'Riordan, C. A., S. G. Monismith, and J. R. Koseff. 1993. An experimental study of concentration boundary layer formation over a bed of model bivalves. Limnol. Oceanogr. 38: 1712–1729.
O'Riordan, C. A., S. G. Monismith, and J. R. Koseff. 1995. The effect of bivalve excurrent jet dynamics on mass transfer in a benthic boundary layer. Limnol. Oceanogr. 40: 330–344.
Otway, N. M., and A. J. Underwood. 1987. Experiments on orientation of the intertidal barnacle, *Tesseropora rosea* (Krauss). J. Exp. Mar. Biol. Ecol. 105: 85–106.
Peer, D., D. J. Wildish, A. J. Wilson, J. Hines, and M. Dadswell. 1980. Sublittoral macro-infauna of the lower Bay of Fundy. Can. Tech. Rep. Fish. Aquat. Sci. 981: 74.
Richardson, J. R., and J. E. Watson. 1975. Locomotory adaptations in a free-lying brachiopod. Science 189: 381–382.
Riisgård, H. U., and N. M. Ivarsson. 1990. The crown filament pump of the suspension feeding polychaete *Sabella penicillus*: filtration effects of temperature and energy cost. Mar. Ecol. Prog. Ser. 62: 249–257.
Stanley, S. M. 1970. Relation of shell form to life habits of the Bivalvia (Mollusca). Geol. Soc. Am. Inc., Memoire 125. Geological Society of America, Boulder, Colo.
Stephens, P. J. 1978. The sensitivity and control of the scallop mantle edge. J. Exp. Biol. 75: 203–221.
Taghon, G. L., A. R. M. Nowell, and P. A. Jumars. 1980. Induction of suspension feeding spionid polychaetes by high particulate fluxes. Science 210: 562–564.
Telford, M. 1983. An experimental analysis of lunule function in the sand dollar *Mellita quinquiesperforata*. Mar. Biol. 76: 125–134.
Telford, M., and R. Mooi. 1987. The art of standing still. New Scientist 114: 36–39.
Théodor, J. 1963. Contribution à l'etude des Gorgones. III. Trois formes adaptives d'*Eunicella stricta* en fonction de la turbulence et du courant. Vie Milieu 14: 815–818.
Thomas, F. I. M. 1994. Morphology and orientation of tube extensions on aggregations of the polychaete annelid *Phragmatopoma californica*. Mar. Biol. 119: 525–534.
Thompson, R. J. 1984. Production, reproductive effort, reproductive value and reproductive cost in a population of the blue mussel *Mytilus edulis* from a subarctic environment. Mar. Ecol. Prog. Ser. 16: 249–257.
Thorburn, I. W., and L. D. Gruffyd. 1979. Studies of the behaviour of the scallop *Chlamys opercuralis* (L.) and its shell in flowing seawater. J. Mar. Biol. Assoc. U.K. 59: 1003–1023.
Thrush, S. F. 1991. Spatial patterns in soft bottom communities. Trends Ecol. Evol. 6: 75–78.
Trager, G. C., D. Coughlin, A. Genin, Y. Achituv, and A. Gangopodyay. 1992. Foraging to a rhythm of ocean waves: porcelain crabs and barnacles synchronize feeding motions with flow oscillations. J. Exp. Mar. Biol. Ecol. 164: 73–86.

Trager, G. C., J. -S. Hwang, and J. R. Strickler. 1990. Barnacle suspension feeding in variable flow. Mar. Biol. 105: 117–129.
Turner, E. J., and D. C. Miller. 1991a. Behaviour and growth of *Mercenaria mercenaria* during simulated storm events. Mar. Biol. 111: 55–66.
1991b. Behaviour of a passive suspension feeder (*Spiochaetopterus oculatus*) (Webster) under oscillatory flow. J. Exp. Mar. Biol. Ecol. 149: 123–137.
Vincent, R., G. Desrosiers, and Y. Gratton. 1988. Orientation of the infaunal bivalve *Mya arenaria* L. in relation to local current direction on a tidal flat. J. Exp. Mar. Biol. Ecol. 124: 205–214.
Vogel, S. 1981a. Life in moving fluids: the physical biology of flow. Willard Grant Press, Boston.
1981b. Behaviour and the physical world of an animal, p. 179–198. *In* P. P. G. Bateson and P. H. Klopfer (eds.) Perspectives in ethology, Vol. 4. Plenum Press, New York.
1985. Flow-assisted shell reopening in swimming scallops. Biol. Bull. 169: 624–630.
Vosburgh, F. 1977. The response to drag of the reef coral, *Acropora reticulata*, p. 477–482. *In* D. L. Taylor (ed.) Proc. 3rd Intl. Coral Reef Symp. Univ. of Miami, Miami 1.
Wainwright, S. A., and J. R. Dillon. 1969. On the orientation of sea fans (genus *Gorgonia*). Biol. Bull. 136: 130–139.
Waite, J. H. 1983. Adhesion in bysally attached bivalves. Biol. Rev. 58: 209–231.
Waller, T. R. 1975. The behaviour and tentacle morphology of pteriomorphan bivalves: a motion picture study. Bull. Am. Malacol. Union, p. 7–13.
Warner, G. F. 1977. On the shapes of passive suspension feeders, p. 567–576. *In* B. F. Keegan, P. O. Ceidigh, and P. J. S. Boaden (eds.) Biology of benthic organisms. Pergamon Press, Oxford.
1979. Aggregation in echinoderms, p. 375–396. *In* G. Larwood and B. R. Rosen (eds.) Biology and systematics of colonial organisms. Academic Press, London.
Warwick, R. M., and J. R. Davies. 1976. The distribution of sublittoral macrofauna communities in the Bristol Channel in relation to substrate. Est. Coast. Mar. Sci. 5: 111–222.
Wildish, D. J. 1984. Secondary production of four, sublittoral, soft sediment amphipod populations in the Bay of Fundy. Can. J. Zool. 62: 1027–1033.
Wildish, D. J., D. D. Kristmanson, R. L. Hoar, A. M. DeCoste, S. D. McCormick, and A. W. White. 1987. Giant scallop feeding and growth responses to flow. J. Exp. Mar. Biol. Ecol. 113: 207–220.
Wildish, D. J., and U. Lobsiger. 1987. Three-dimensional photography of soft sediment benthos, S.W. Bay of Fundy. Biol. Oceanogr. 4: 227–241.
Wildish, D. J., and D. Peer. 1983. Tidal current speed and production of benthic macrofauna in the lower Bay of Fundy. Can. J. Fish. Aquat. Sci. 40: 309–321.
Wildish, D. J., and A. M. Saulnier. 1992. The effect of velocity and flow direction on the growth of juvenile and adult giant scallops. J. Exp. Mar. Biol. Ecol. 155: 133–143.
Wildish, D. J., and A. M. Saulnier. 1993. Hydrodynamic control of filtration in *Placopecten magellanicus*. J. Exp. Mar. Biol. Ecol. 174: 65–82.
Wildish, D. J., A. J. Wilson, W. Young-Lai, A. M. DeCoste, D. E. Aiken, and J. D. Martin. 1988. Biological and economic feasibility of four grow-out

methods for the culture of giant scallops in the Bay of Fundy. Can. Tech. Rep. Fish. Aquat. Sci. 1658.
Winter, M. A., and P. V. Hamilton. 1985. Factors influencing swimming in bay scallops, *Argopecten irradians* (Lamarck, 1819). J. Exp. Mar. Biol. Ecol. 88: 227–242.
Witman, J. D., and T. H. Suchanek. 1984. Mussels in flow: drag and dislodgement by epizoans. Mar. Ecol. Prog. Ser. 16: 259–268.
Young, G. A. 1985. Byssus-thread formation by the mussel *Mytilus edulis*: effects of environmental factors. Mar. Ecol. Prog. Ser. 24: 261–271.
Yule, A. B., and D. J. Crisp. 1983. Adhesion of cypris larvae of the barnacle, *Balanus balanoides*, to clean and athropodin treated surfaces. J. Mar. Biol. Assoc. U.K. 63: 261–271.
Yule, A. B., and G. Walker. 1987. Adhesion in barnacles, p. 389–402. *In* A. J. Southward (ed.) Barnacle biology. A. A. Balkema, Rotterdam.

7
Benthic populations and flow

The concept of hydrodynamic limitation of macrobenthic animal populations arose from an attempt to explain the importance of flow to benthic populations consisting of two key trophic groups: the deposit and suspension feeders (Wildish 1977). Consistent with the aims outlined in Chapter 1, we are concerned primarily with the latter trophic group, and deposit feeding macrofauna are not dealt with in detail here.

We will first consider the historical background of the benthic limitation by flow concept, as well as a special theory associated with it: trophic group mutual exclusion, and hypotheses derived from the latter. We present turbulent mass transfer models relating seston supply and uptake by populations of suspension feeders with both field and laboratory experiments to test the theory. We also present field evidence that suspension feeding populations can be impoverished by high energy water movement events. For completeness, we briefly consider two possible competing theories of trophic group mutual exclusion which propose that biotic interactions between deposit and suspension feeders explain the exclusion of one of the groups.

Benthic limitation by flow

The concept to which the general theory of benthic limitation by flow belongs is that of limiting factors (Wildish 1977). Limiting factor theory has a long history, beginning with Liebeg's (1840) law of the minimum (see in Odum 1971) and Blackman's (1905) ideas on optima and limiting factors. Both of these authors were dealing with terrestrial plants, and the application of limiting factors in benthic ecology occurred later (e.g. Levinton and Lopez 1977; Wildish 1977). According to the limiting fac-

Table 7.1. *Trophic group classification of benthic macrofauna.*

Letter code	Trophic group
S	Suspension feeding
D	Deposit feeding
C	Carnivores
O	Omnivorous scavengers
A	Algal scrapers

Source: Wildish (1985).

tor concept, multiple interactive physical and biological factors regulate macrofaunal community composition, biomass, and productivity.

Trophic group mutual exclusion (Wildish 1977), a special theory of benthic limitation, depends on key types, or trophic groups, of macrofauna: the deposit and suspension feeders. Mutual exclusion arises because each group responds differently to flow, as is described in detail later in this section. The trophic group concept originated with Hunt (1925), who described the feeding methods of the benthic animals taken in grab samples off the fishing grounds near Plymouth, United Kingdom. Trophic group designations such as those shown in Table 7.1 have subsequently been used to generalize the functional nature of the benthic communities with which ecologists are working (e.g. Sanders 1956, 1960; Rhoads and Young 1970; Pearson 1971; Wildish 1977; Neyman 1979; Pearson and Rosenberg 1987).

The trophic group concept has not been without modern critics; cogent criticisms raised by Dauer, Maybury, and Ewing (1981) are based on the following points:

- Designations of trophic groups are not available for some species.
- Some benthic animals are capable of facultative switching from/to suspension and deposit feeding (Chap. 4).
- The limiting physical factors in suspension feeding production remain unknown in many cases.
- Environmental factors not related to feeding, and hence trophic type, may be critical in controlling local density and productivity of suspension feeders.

Although these criticisms are warranted in some cases, we believe that they can be resolved by further research. Thus, experimental research in laboratory flumes is required to answer the first three points. Facultative suspension–deposit feeders may actually represent only a small fraction of total benthic animals. Such species usually occur in hydrodynamic conditions where the erosional/depositional nature of the sediments and associated water flow patterns is transitional. They can be counted as the feeding method commonly used in the local conditions. That non-trophic factors – the last item in the list – may be critical in controlling production has been incorporated in the model of unimodal response presented herein and can be examined as benthic community impoverishment hypotheses (see Table 7.2 and the section, Water movement impoverishment). As benthic biology matures as a discipline, we would expect the focus to switch from the arbitrary generalization of the trophic group to the individual species.

Seed and Suchanek (1992) considered a comprehensive range of physical and biological environmental factors of importance in controlling population ecology of blue mussels. This methodology is probably applicable to other suspension feeders. The major concern of this chapter is to consider three hydrodynamic mechanisms recognized as significant limits and, hence, controls on species composition, biomass, and productivity. They include the hydrodynamics associated with larval dispersal and recruitment (see also Chap. 3), turbulent seston supply to suspension feeders as food (Wildish 1977), and seawater velocities sufficient to cause macrofaunal impoverishment (Wildish and Kristmanson 1979). Such controls depend directly on water movement and result in the types of benthic communities diagrammed in Fig. 7.1. In Fig. 7.1, water movement is the independent variable which, increasing from left to right, causes flow-dependent effects on benthic animals as well as sedimentary deposition and natural selection of the proportions of deposit–suspension feeders and, hence, type of benthic community (deposit or mixed). The highest levels of suspension feeding benthic production only occur where a suitable stable substrate is present.

Mutual exclusivity of the two key trophic groups arises because they respond differently to water movement forces, as depicted in Fig. 7.1. Deposit feeders are directly and indirectly affected by flow in their physiology, behavior, and ecology. Thus, infaunal *Macoma* spp. (Levinton 1991) and surface deposit feeders (Miller, Bock, and Turner 1992) may be inhibited by flow during feeding; locomotion of surface deposit feeders may be inhibited (Miller et al. 1992) and burrowing enhanced (Levinton et al. 1995) by increasing flows. By their burrowing and feeding activities, deposit feeders influence sediment–water relations (Davis 1993). Such activities cause sediment to be ejected into the overlying seawater, decrease silt–clay cohesiveness, increase sediment

Table 7.2. *Hypotheses concerning deposit or mixed benthic macrofaunal communities which are referable to the special theory of trophic group mutual exclusivity.*

No.	Hypothesis		Reference
1	H_0	Sestonic food does not become limiting above a suspension feeding bed	Wildish and Kristmanson (1979)
	H_1	Downstream seston depletion may occur above a suspension feeding bed	
2	H_0	Suspension feeder production is unrelated to U or potential for turbulent and advective transfer	Wildish and Peer (1983) Wildish and Kristmanson (1985)
	H_1	Suspension feeder production is positively related to U or potential for turbulent and advective transfer	
3	H_0	Suspension feeder production is unaffected by energetic water movement	Wildish and Kristmanson (1979) Warwick and Uncles (1980)
	H_1	Suspension feeder production in soft sediments can be impoverished by high energetic movement	
4	H_0	Deposit feeder production is unrelated to flow velocity	Rhoads (1973) Rhoads et al. (1978)
	H_1	Deposit feeder production is inversely related to the erosional power of flow	
5	H_0	Population density does not affect seston uptake or population growth of suspension feeders	Wildish and Kristmanson (1979), Broom (1982) Fréchette and Bourget (1985)
	H_1	Population density determines the seston uptake or population growth rate of suspension feeders	
6	H_0	Suspension feeder reef path length is unrelated to potential for turbulent and advective transfer or tidal excursion length	Wildish and Kristmanson (1979)
	H_1	Suspension feeder reef path length is determined by potential for turbulent and advective transfer and tidal excursion length	

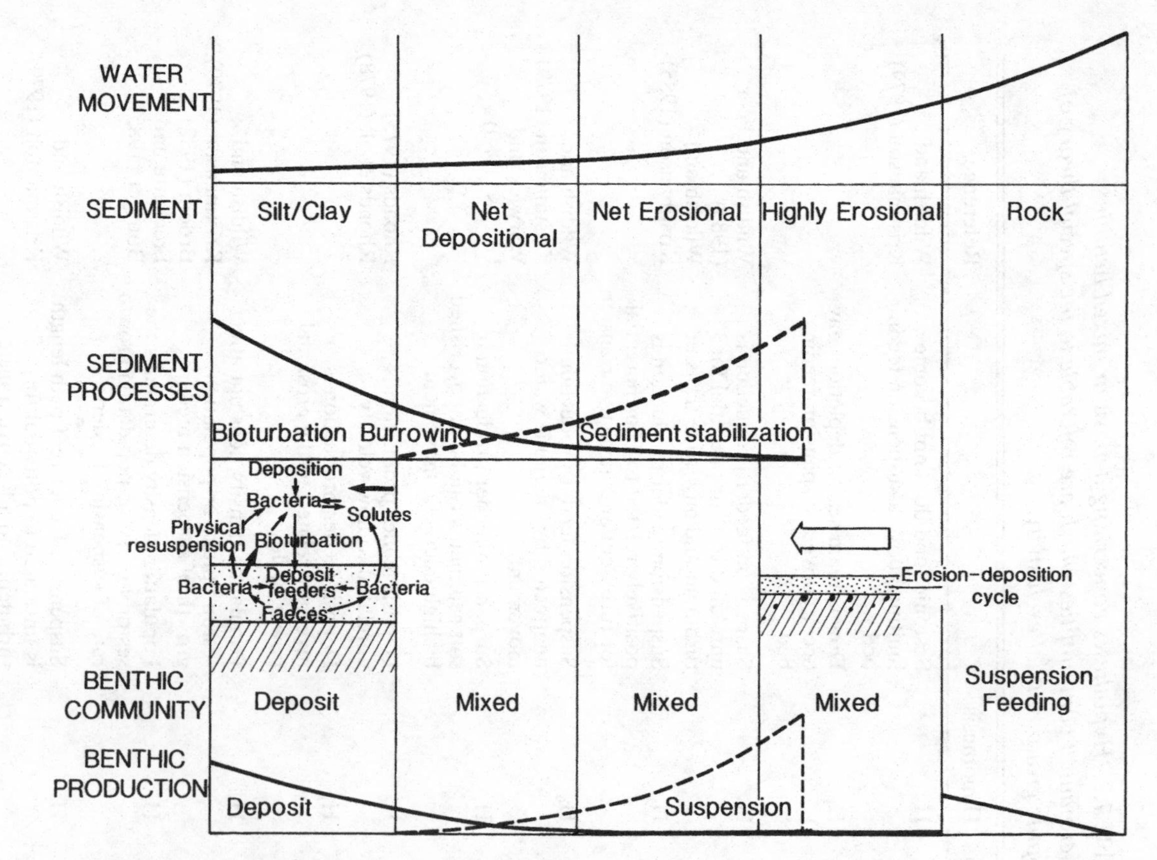

Figure 7.1 Diagrammatic representation of benthic communities of the continental shelf along a gradient of increasing water movement energy.

water content, and thereby increase the possibility of physical resuspension. The degree of bioturbation activity, coupled with benthic shear stress, were considered by Davis (1993) to be most important in determining the type of deposit feeding animals present. Deposit feeders are found on all types of soft sediment, although they reach their highest production levels where water movements are minimal (Fig. 7.1) and where gravity settling mechanisms dominate advective flux. In such conditions, a silt–clay sediment is present and active bioturbation by burrowing deposit feeders is high, resulting in resuspension of the deposits, with subsequent redeposition close to the original bioturbaters. Those deposit feeders associated with net erosional sediments cannot exploit bioturbation because most of the resuspended material would be transported away. Deposit feeders depend trophically on the microbial flora attached to sedimentary particles (Newell 1965; Levinton and Lopez 1977; Jumars and Nowell 1984a). The microbial bacteria involved are aerobic forms limited by substrate and oxygen availability within sedimentary deposits. But on resuspension by biological or physical means the bacteria can grow and multiply (Nickels, King, and White 1979; White et al. 1979; Anderson and Meyer 1986; Muschenheim, Kepkay, and Kranck 1989; Ritzrau and Graf 1992) to increase their contribution to an enhanced food supply for deposit feeders. Because of the presumed energetic cost of bioturbation, it would be wasteful to resuspend sediment in this way, only to have it carried away by flow. On highly erosional sediments (Fig. 7.1), only a few deposit feeders are found; they are specially adapted for deep burrowing to escape the periodic sediment erosion which is characteristic of these sediments.

By contrast, suspension feeders require a higher rate of water movement (Fig. 7.1), and their biomass and productivity depend positively on tidal or wind–wave activity. In soft sediments, suspension feeders tend to bind the sediment to prevent washout and produce at rates related to the turbulent seston supply. The ability of benthic organisms to de-stabilize or stabilize sediments was considered by Jumars and Nowell (1984b), who concluded that no classification was yet possible because of lack of data. Flume experiments with a wide range of benthic organisms in realistic simulated sediment–water interface conditions are required, but not yet available, to achieve a satisfactory empirical classification.

For soft sediments there is an upper limit beyond which seston flux cannot further increase productivity because energetic water movement may result in feeding inhibition and eventual washout and impoverishment. Some suspension feeders, e.g. barnacles with specialized attach-

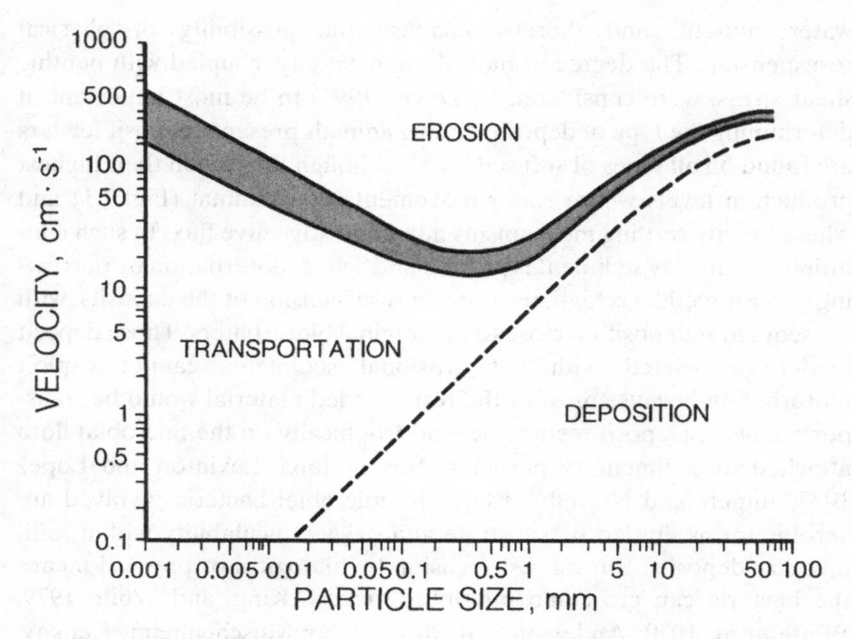

Figure 7.2 Relationship between current velocity and its ability to erode, transport, and deposit sedimentary particles of various sizes (after Perkins 1974).

ment mechanisms, may be able to survive on stable hard substrates (Fig. 7.1) at extremely high flows.

Specific hypotheses referable to the trophic group mutual exclusion theory which concerns suspension feeders are dealt with in this chapter, as shown in Table 7.2, and are considered in the order shown. It is important in developing the mutual exclusion concept to determine what constitutes food and to understand the mechanism of capture. However, it is not necessary to use subcategories in trophic group typing only to determine whether or not the benthic organism is suspension or deposit feeding. Sanders (1960) and Rhoads and Young (1970) were among the first to stress the fundamental importance of suspension and deposit feeding in food web process studies. Inputs of food reach the benthic boundary layer and are transferred by advection and diffusion to suspension feeders or transferred as sedimentary particles after settlement, which are ingested by deposit feeders.

The relationship between benthic communities and substrates in Fig. 7.1 is really more complex than shown because it is medium sand-sized particles (250- to 500-μm median particle diameter) which are susceptible to erosion at the lowest velocities (Fig. 7.2). By contrast, clay-sized particles (4- to 62-μm median particle diameter) do not erode unless the velocities reach nearly an order of magnitude greater than for medium sand-sized particles. Depositional rates, on the other hand, are faster for sand- than for clay-sized particles because the larger, more dense sand particles settle at a greater rate than smaller, lighter clay particles. Because of biological activity involving binding of surficial sediments (Madsen, Nilsson, and Sundback 1993), the figured critical sediment erosion velocities in Fig. 7.2 are approximate.

Hydrodynamic modelling of suspension feeding populations

The earliest attempt to apply a turbulent mass transfer model to the problem of seston supply and uptake by a bivalve reef was made by Wolff et al. (1976). Wildish and Kristmanson (1979) independently developed a turbulent mass transfer model very similar to the earlier one. In this formulation, the rate at which seston is supplied to the reef is assumed to be limited by turbulent convection near the top of the reef. This rate is defined as N_A (*g* seston/$h^{-1}\,m^2$ of reef). The consumption of seston per unit area of reef is given by

$$N_A = PR\alpha C' \tag{7.1}$$

where

- C' is the seston concentration near the inhalant opening, or filtration surface, $g \cdot m^3$
- P is the density of suspension feeders, number $\cdot\, m^{-2}$
- R is the average filtration rate of each individual in the population, $m^3 \cdot h^{-1}$
- α is the average filtration efficiency (0% to 100%) of each individual in the population

The term $PR\alpha$ is equivalent to the more familiar term *clearance rate*, defined as cubic metres cleared per hour per square metre of reef. It has the units of velocity, $m \cdot h^{-1}$. The model shown in (7.1) does not deal with seston deposition or resuspension. The flux is driven only by the suspension feeding of the benthos, which creates a concentration deficit near the reef, and turbulent transport. Assumptions necessary are that the suspension feeder reef is of semi-infinite extent; that the benthic boundary layer is well developed, that is, that the velocity and concentration profiles do not change in the downstream direction; and that there is no mass transfer resistance attributable to a laminar sublayer.

The limiting turbulent transport rate can be estimated by assuming that seston is transported by the same mechanisms as is momentum. Bottom shear stress data as in Sternberg (1968) can be used to develop a flux equation:

$$N_A = \gamma U(C_0 - C') \tag{7.2}$$

where

C_0 is the upstream seston concentration, $g \cdot m^3$
γ is a dimensionless hydrodynamic term approximately equal to 0.003 and dependent on roughness
U is the velocity at the top of the benthic boundary layer, $m \cdot h^{-1}$

Combining (7.1) and (7.2) we obtain

$$\frac{(C_0 - C')}{C'} = \frac{PR\alpha}{\gamma U} \tag{7.3}$$

where the left-hand term is a measure of seston depletion. The right-hand term in (7.3), which we will call the *seston depletion index* (SDI), is a useful indicator of whether depletion is an important factor in a particular locality. The SDI is dimensionless and represents the ratio of the total clearance rate of all suspension feeders occupying a unit area of bottom to the rate at which turbulence can deliver seston to them.

Equation (7.3) can also be written as

$$\frac{C_0}{C'} = 1 + \text{SDI} \tag{7.4}$$

SDI can be calculated for a particular site if the total clearance rate can be estimated and typical velocities of the bulk flow are known. If "significant" seston depletion is defined as $C' \leq 0.9C_0$, then it can be expected to be observed if SDI > 0.11.

Data for sublittoral mussel reefs suggest that depletion effects can be expected in the field. Jørgensen (1990) provides clearance rate data for natural blue mussel reefs of the highest density found in the North Sea and Canada as from 6–12 $m \cdot h^{-1}$. Dare (1976) reported that a blue mussel reef in Morecambe Bay, United Kingdom, of 1.4 kg of soft tissue per m^2 of reef had a total clearance rate of 12 $m \cdot h^{-1}$. Using this value for $PR\alpha$ and $\gamma = 0.003$, SDI can be calculated for a range of velocities (Table 7.3). These results show that the highest density mussel reefs will deplete the benthic boundary layer of seston at typical tidal velocities. Other

Table 7.3. *Dependence of SDI on velocity for mussel reefs at high density (1.4 kg dry body weight/m²).*[a]

U, cm·s^{-1}	5	10	20	50	100	500
SDI	22	11	5.5	2.2	1.1	0.2

[a] For the sublittoral of Morecambe Bay, UK, at maximum clearance rate of 12 m·h^{-1}.
Source: Data reported by Dare (1976).

sublittoral mussel reefs show similar total clearance rates (Jørgensen 1990). Values found by Smaal et al. (1986) in the Oosterschelde, Netherlands, were lower: a maximum total clearance rate per square meter for cultured reefs of blue mussels of 0.58 m·h^{-1} and for cockles *Cardium edule* of 0.2 m·h^{-1}. Under the Oosterschelde conditions, water samples showed depletion of seston over the mussel reef, but not the cockle reef.

The preceding analysis is based on the idea that a balance has been achieved between the two processes of turbulent mass transfer of seston and its consumption at the boundary. This condition is presumably satisfied if the population of filter feeders is evenly distributed over a large area and the analysis is applied at a point well downstream of the leading edge of the reef. Upstream of this hypothetical point, the two processes may not be in balance and the concentrations of seston near the bottom will vary with depth and distance from the leading edge of the reef. The distribution of seston (the "concentration boundary layer") will be as shown in Fig. 7.3. According to Fig. 7.3, upstream mussels will receive more seston, and, therefore, growth rates will be higher than for downstream ones where seston depletion is well developed. At locations where tidal flows regularly reverse, e.g. in estuaries, growth of individuals will be highest at the edges rather than the middle of the reef as a result of the seston depletion effect. O'Riordan, Monismith, and Koseff (1993) have presented a concentration boundary layer study based on a reef of model bivalves (625 animals·m^{-2} simulating *Tapes japonica*) in a recirculating flume at realistic pumping rates to characterize the seston depletion effect at different points in the reef (see also the section, Infauna, Chap. 6).

To predict under what circumstances seston depletion will become important in controlling population filtration, and hence growth, a solution of the mass conservation equation is required:

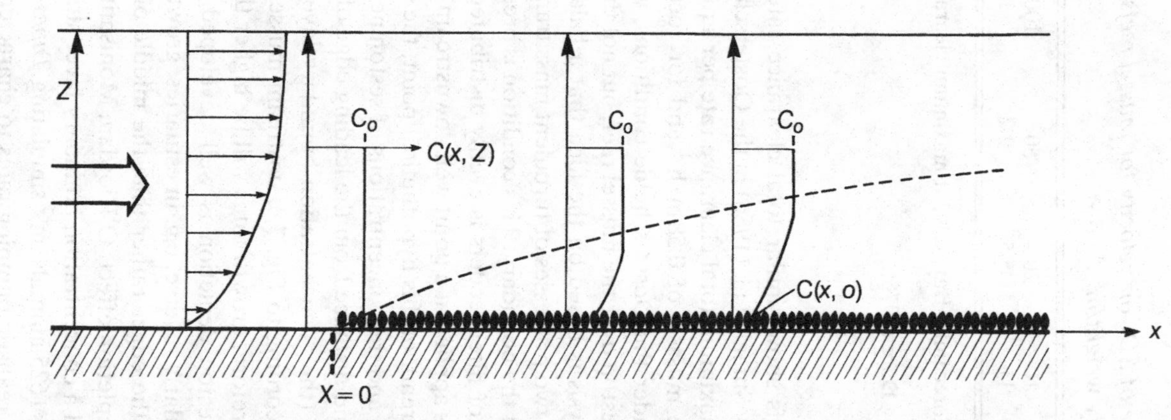

Figure 7.3 The concentration boundary layer which develops over a blue mussel reef in a unidirectional flow: Z, water depth; X, mussel bed length; Co, initial seston concentration.

$$\frac{dc}{dt} + U\frac{dc}{dx} - \frac{d}{dz}\left(E_m \frac{dc}{dz}\right) - \text{sinks} + \text{sources} = 0 \tag{7.5}$$

where

c is the concentration of seston
U is the velocity at (x, z) and
E_m is the vertical mixing coefficient

Equation (7.5), called the convective-diffusion equation, is of time averaged form and does not consider changes in the cross-stream direction. The first term represents temporal change in concentration of seston at the point (x, z). The second term represents the change of advection of seston by the average flow in the downstream direction. The third term describes the vertical diffusion of seston due to turbulence. Several authors, including Wildish and Kristmanson (1984), Smaal et al. (1986), and Fréchette, Butman, and Rockwell Geyer (1989), have proposed solutions for the convective-diffusion equation.

Wildish and Kristmanson (1984) suggested an analysis of a developing flume boundary layer in which the velocity boundary layer thickened in the downstream direction so that velocity profiles could not be assumed to be fully developed. The concentration boundary layer was assumed to be well mixed in the vertical direction and to have the same thickness as the velocity defect layer. The mean concentration in the boundary layer was calculated by integrating the differential mass balance. This analysis was extended by Verhagen (1986) and Smaal et al. (1986), who provided an analytical solution and introduced a more realistic treatment of the seston within the boundary layer. For the case of seawater >3 m deep where steady unidirectional flows prevailed, using realistic profiles of mean velocity and vertical eddy diffusivity, Verhagen (1986) also predicted the seston depletion to be expected at various downstream distances along the reef (Fig. 7.4). The abscissa can be considered as the ratio of the velocity generated by the mussels (the number of mussels per unit reef area times the volumetric flow per mussel) to the mean stream velocity. It is therefore similar to the ratio SDI earlier proposed as a measure of the tendency to depletion. An analytical solution is also provided:

$$\frac{C_0 - C(x,0)}{C_0} = \frac{N_A \alpha}{U}\left(\frac{3Ux}{2D_1}\right)^{0.5} \tag{7.6}$$

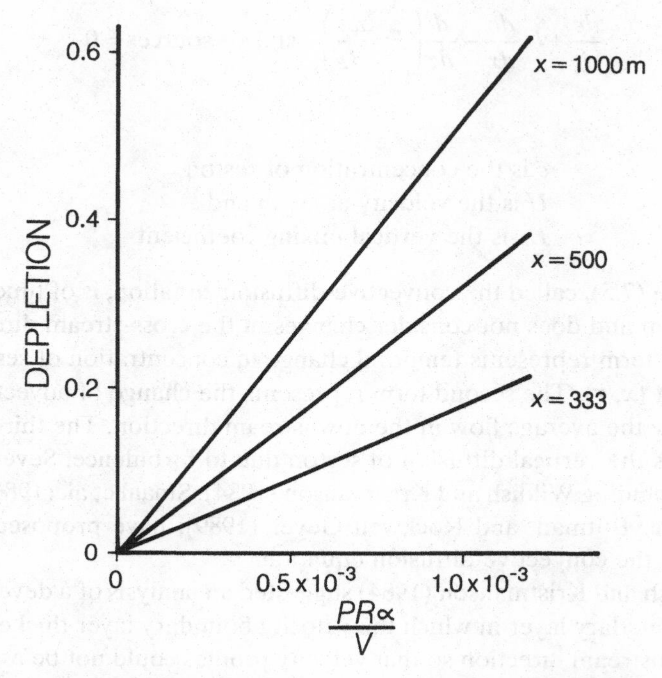

Figure 7.4 Seston depletion as a function of $PR\alpha/V$ (Verhagen 1986), where X is the mussel reef path length in meters.

where

x is the distance from the upstream boundary of the reef
$C(x, 0)$ is the seston concentration at the inhalants
D_1 is the vertical eddy diffusivity near the bottom

The units of the turbulent flux term are as in (7.1) and (7.2). This solution is limited to small values of $NR\alpha\delta/D_1$ where δ = seston boundary layer thickness. Note that the depletion is directly proportional to the product of two dimensionless groups, the abscissa of Fig. 7.4, which depends on the population density and the current speed, and the square root term, which is of the same form as the Peclet number defined in Chapter 5 to express the ratio of advective transport to diffusional transport. Depletion occurs when seston is moved rapidly from the boundary by pumping, and smearing by turbulent diffusion cannot smooth out the gradients thus created. The Peclet number term is determined by the mechanics of the flow and will become smaller if the

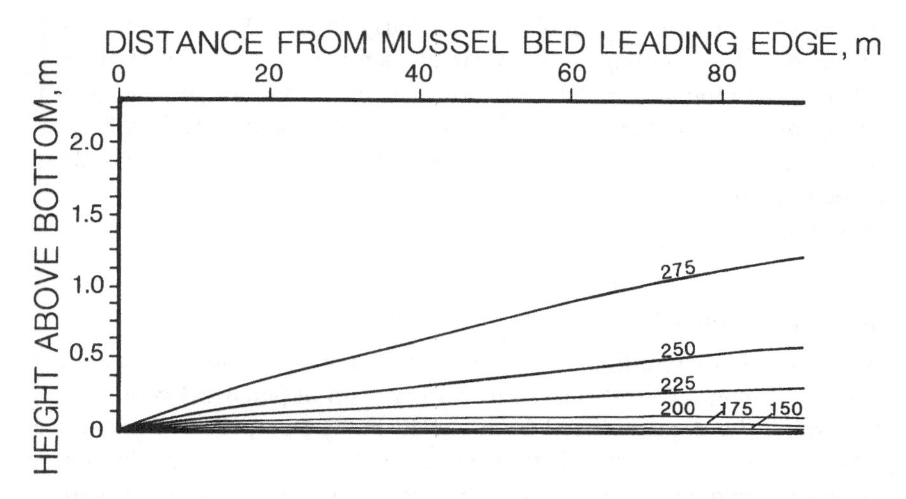

Figure 7.5 Model generated contours of seston concentration in a unidirectional flume flow in arbitrary fluorescence units: conditions: $U = 15\,\mathrm{cm \cdot s^{-1}}$ at $Z = 2.25\,\mathrm{m}$; $Z_0 = 0.1\,\mathrm{cm}$, $U_* = 0.78\,\mathrm{cm \cdot s^{-1}}$; mussel density $= 430 \cdot \mathrm{m^{-2}}$, dry weight biomass $= 497\,\mathrm{g \cdot m^{-2}}$, bed length $= 100\,\mathrm{m}$; mussel inhalant height above the bottom is 3 cm (from Fréchette et al., 1989).

boundary is made rougher. Verhagen (1986) also provided a solution of the shallow water (<3 m deep) case in which it can be assumed that the seston is well mixed in the vertical direction and the z-dimension can be neglected.

Fréchette et al. (1989) solved the convective-diffusion equation numerically to analyze the vertical and horizontal distribution of natural seston over a blue mussel reef. The mean velocity profile was developed by using field measurements, and the vertical eddy diffusivity was assumed equal to the eddy viscosity. Their treatment thus differs from that of Verhagen, who assumed that depletion near the bottom was dominated by the mixing produced by the exhalant streams from the mussels. The sinking rate of phytoplankton was assumed to be small enough to be ignored. The rate of consumption of phytoplankton by the mussels was estimated from physiological chamber tests and from measurements of the size–frequency distribution and total biomass of mussels in the reef. The unit area clearance rate found in this way was $6\,\mathrm{m \cdot h^{-1}}$. Their results for a model reef similar to their field site are shown in Fig. 7.5. In this simulation, the phytoplankton is represented by equivalent fluorescence units, with the upstream value 300. The velocity at 2.25 m depth is

$0.15\,m \cdot s^{-1}$. The effect of depletion is clearly shown. Time series measurements at a sampling station near the centre of a mussel reef about 100 m long in the axis of the flow showed seston depletion at a depth of 0.05 m of approximately 50% at similar current speeds. Although the model represents a simplification of the field conditions, the extent of the agreement between the model and the fluorometric measurements is impressive.

Seston depletion effect

An early demonstration of seston depletion was provided by Glynn (1973a, b) over a Caribbean coral reef made up of patches 3–50 m wide and where the depths ranged from 1 to 1.5 m. The purpose of this study was to test quantitatively whether the reef removed substantial amounts of seston to determine whether the reef relied mainly on symbiotic photosynthesis or suspension feeding of plankton as the major contributor to its energy needs. Wind–wave energy moved seawater across the *Porites* reef flat and resulted in effective vertical mixing. Surface samples were taken at the upstream edge, on the reef, and at a downstream position landward. Glynn (1973b) was able to show, by weekly sampling at the same three locations across the reef for a 2-year period, that numerically diatoms could be reduced by 91% and zooplankton by 60% in passing across it.

Buss and Jackson (1981) designed a field experiment to measure seston depletion of seawater passed through a 2.4-m-long, flowthrough box by aligning the open ends in the mean flow direction in Discovery Bay, Jamaica, at 40 m depth. Each box contained panels on which various densities of bryozoans and sponges were allowed to grow during the 26-mo-long experiment. Upstream and downstream seawater samples were taken simultaneously through corked holes in the upper wall of the box by SCUBA divers. Samples taken with a 25-ml syringe were analyzed for densities of bacteria and phytoplankton (mostly flagellates). Measurements of seston depletion were made at different times during the experiment. Because different percentages of cover had developed on the test panels at these times, suspension feeder density was varied. Results suggested that percentage cover by sponges and Bryozoa was the primary factor determining seston depletion, and if these two variables were plotted on logarithmic coordinates, they were linearly related. During this experiment, equal numbers of flow boxes were set up so that one end was blocked off in one series of boxes as controls, and in the treatment boxes both ends were left open to test the effect of flow.

In the first study to use adequate hydrodynamic field measurements, Fréchette and Bourget (1985) showed that an intertidal blue mussel reef in the St. Lawrence estuary could produce seston depletion effects within the benthic boundary layer. By direct sampling at 3 to 4 cm height above the mussel reef, fluorometric measurements showed that seston depletion during the first week of observation was negatively correlated with current velocity. During the next week, seston depletion was only observed for half of the sampling time and was negatively correlated with wave energy. Fréchette and Bourget (1985) thus observed a temporal change in the presence/absence of a seston depletion effect, depending on the seston concentration as well as the type of water movement energy available. During the first week, seston depletion was associated with tidal flows; during the second week, this association was partially destroyed by wind–wave-caused oscillating flows which resulted in resuspension and better mixing.

In a field study on an intertidal sand flat in Shark Bay, Western Australia, the ability of infaunal suspension feeding bivalves to deplete the flooding tidal seawater was tested by Peterson and Black (1991). Seawater samples were pumped, approximately isokinetically, from 5 to 10 cm above the bottom over a 205-min period in the same water mass (indicated by rhodamine B dye marker). Chlorophyll a had declined by 25% during this period.

A few laboratory experiments have also been designed to demonstrate that *populations* of suspension feeding animals can deplete the seston as it passes over the reef, as is required by all hydrodynamic models discussed in the section, Hydrodynamic modelling of suspension-feeding populations. Wildish and Kristmanson (1984) used an outdoor flow-through flume of 5 m length which was supplied with unfiltered seawater from a submersible pump. Seasonal seston concentrations and temperatures in the Bay of Fundy were not modified in the flume during the experiment except by the feeding activities of mussels. Sampling was from 1 cm above the flume bottom at the inlet end and 3.6 m away at the downstream end of the mussel reef. Sampling was approximately isokinetic since the pumping rate used was adjusted to be isokinetic at $4\,cm \cdot s^{-1}$. Results suggest that significant ATP–seston depletion occurred in six of eight experiments (Table 7.4), whereas for bacterial numbers only three of eight experiments showed significant depletion.

Experiments undertaken by Butman et al. (1994) in a large recirculating flume containing a 5-m-long reef of blue mussels showed that seston depletion could be detected in one experimental run although not in two others. For the experimental run showing seston

Table 7.4. *Depletion of seston from natural seawater by adult horse mussels* Modiolus modiolus *(Expts. 1–3) and blue mussels* Mytilus edulis *(Expts. 4–8) over reef of 3.6 m path length.*[a]

Expt.	Bulk flow $cm \cdot s^{-1}$	Temperature °C	Mussel density number $\cdot m^{-2}$	ATP–seston % depletion[b]	Bacterial nos. % depletion
1	3.92	7.5	164	–36[c]	–39[c]
2	3.78	7.7	164	–80[c]	–39[c]
3	3.35	7.8	164	+3	–53[c]
4	4.10	13.6	2123	–58[c]	+7
5	4.47	12.8	2123	–46[c]	–36
6	3.20	12.7	2123	–81[c]	–23
7	4.18	12.2	1062	–16	+11
8	3.84	10.9	1062	–84[c]	—

[a] Wet biomass $3.672 kg \cdot m^{-2}$ (Expts. 1–3), $4.848 kg \cdot m^{-2}$ (Expts. 4–6), $2.424 kg \cdot m^{-2}$ (Expts. 7–8).
[b] ATP, adenosine triphosphate.
[c] $P<0.5$.
Source: Wildish and Kristmanson (1984).

depletion, the vertical distribution of seston, as chlorophyll a, was as predicted by the model of Fréchette et al. (1989). Unfiltered seawater and natural seston were used in these experiments with high mussel densities and biomasses. Low concentrations of natural seston may be the explanation for the low filtration activity of the mussel reefs in the experiments in which seston depletion was absent.

Even though the experimental flume results discussed are for a developing boundary layer, the SDI measure appears to be robust enough to give reasonable values (Table 7.5), which are all >1, indicating that seston depletion is feasible. This is in agreement with independent concentration boundary layer measurements, which were made in both studies shown in Table 7.5.

Flow and suspension feeder growth and production

The study of energy exchanges which occur between communities or populations of suspension feeders at the sediment–water interface and the water column can involve a detailed understanding of consumption, absorption, excretion, assimilation, production, respiration, and gonadal output (Crisp 1984). In this account, we have avoided the detail and used

Table 7.5. *Free-stream velocities and calculated SDI values for some flume experiments with mussels.*[a]

Species	Biomass, kg AFDW·m^{-2}	U, cm·s^{-1}	SDI	Reference
Mytilus edulis	0.245	8	3.3	Butman et al. (1994)
Mytilus edulis	0.245	19	1.3	Butman et al. (1994)
Mytilus edulis	0.243–0.485	3.2–4.5	3.3–8.7	Wildish and Kristmanson (1984)
Modiolus modiolus	0.367	3.4–3.9	2.1–2.5	Wildish and Kristmanson (1984)

[a]The following were used in the calculations: R=2.83 L·h^{-1} for *M. edulis* and R=10.98 L·h^{-1} for *M. modiolus* (Wildish and Kristmanson 1984, based on Jørgensen 1990); γ=0.006 (Butman et al. 1994); wet weight converted to ash-free dry weight (AFDW) by multiplying by 0.1.

simple measures in describing the energetics of suspension feeders as follows:

- *Biomass*, or amount of living substance measured as wet or dry weight per unit area (g·m^{-2}). The wet measure uses crude units inclusive of non-living tissues such as valves. *Biomass* is here used synonymously with "standing crop" or "standing stock."
- *Production*, the rate of change of biomass from that part of the assimilated food that is retained and incorporated in the biomass of the population (g·m^{-2}·y^{-1}). Our measure does not separate gonad from somatic growth. We use short-term direct individual *growth rate* estimates (g·d^{-1}) rather than actual observations of population growth. The latter involves integration of survivorship–biomass curves for each year-class (Crisp 1984).

Model predictions from the previous section suggest that seston depletion above a suspension feeder bed will be affected by turbulent transfer processes; the most important variables are velocity, seston concentration, and bed roughness. Other variables which determine whether seston depletion occurs are reef path length, bed density, and average body size of suspension feeders. Diagrammatic representation of the effect of different velocities over a mussel reef shows how population growth may be increased if the ambient flow rate is increased (Fig. 7.6), assuming that depth is shallow and that the benthic boundary layer intersects the seawater surface.

In Chapter 4, we saw how some active suspension feeders filter optimally in particular environmental conditions, e.g. velocity and seston concentration. If these suspension feeders are filtering at reduced rates in

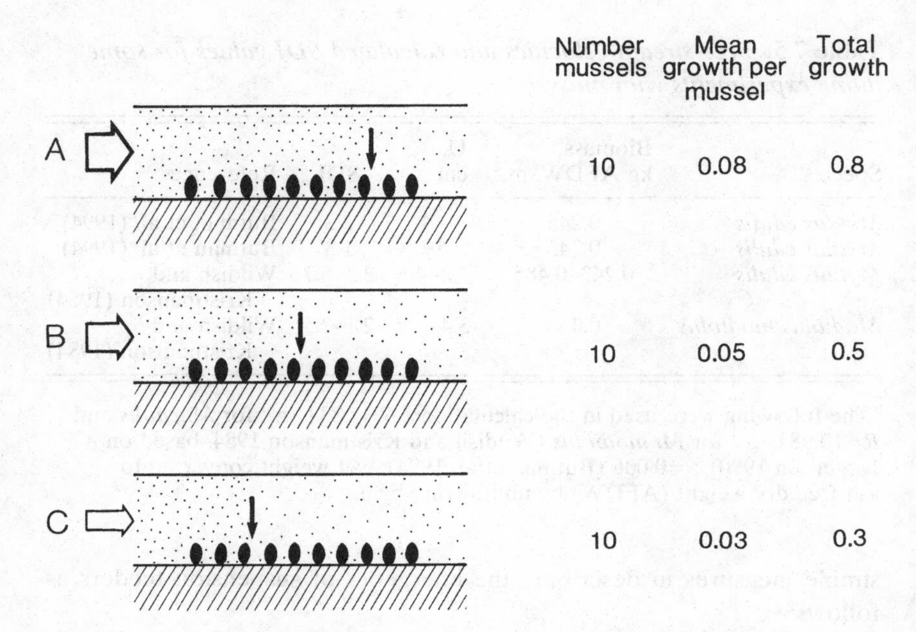

Figure 7.6 Effect of unidirectional flow at different velocities on mussel growth (arbitrary units): A, high; B, medium; C, low; mussel density and seston concentration at the upstream end are constant. The arrow indicates the point at which N_A becomes limiting.

environmental conditions below their optimum, e.g. in seston concentration, then further increases in seston concentration have the potential to increase downstream growth rate, beyond the point indicated by the arrow in Fig. 7.6, where N_A is limiting. If there is a permanent increase in seston flux, then the point where N_A becomes limiting also moves downstream in the reef. Considering a longer temporal scale, it may also allow an increased reef length after immigration or settlement of new recruits.

Wildish and Kristmanson (1985) investigated the effect of current velocity on growth rates of blue mussels in a Kirby-Smith growth tube apparatus. Although this apparatus does not simulate the benthic boundary layer as it occurs in the sea, mussels are common fouling organisms in pipes carrying seawater. Natural seston was supplied concurrently to eight tubes, all with seawater flowing at different velocities, with each tube containing 10 mussels. The results (Fig. 7.7) show that growth was asymptotic with respect to velocity and that up to ~2 $cm \cdot s^{-1}$ growth increases with velocity. Examination of individual growth showed that

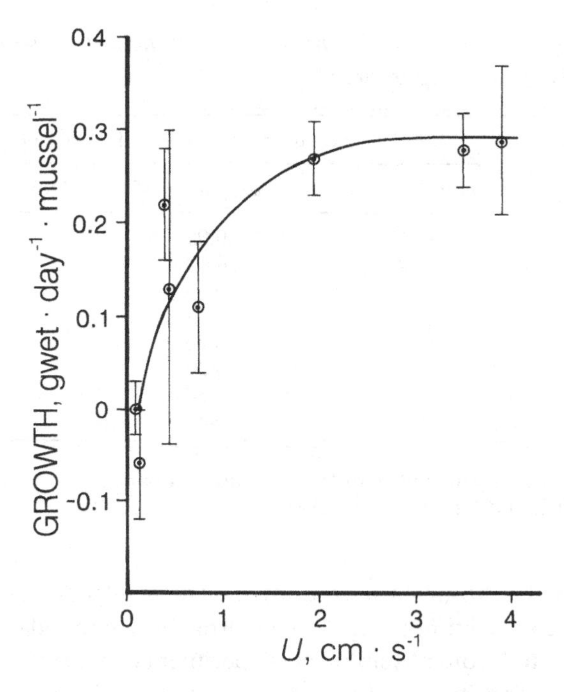

Figure 7.7 Blue mussel growth rates (mean and range) as a function of velocity (U, cm·s^{-1}) in 31-d-long concurrent experiments in Kirby-Smith growth tubes; 10 mussels per tube ≈ 364 mussels·m^{-2} with each mussel approximately 2.54 g wet weight (Wildish and Kristmanson 1985); natural seawater from the Bay of Fundy, Canada.

upstream (the first five mussels in the tube), mussels grew better than downstream ones (Table 7.6) at flows $<2\,cm \cdot s^{-1}$, but at $>2\,cm \cdot s^{-1}$, there was no significant difference between upstream and downstream individual mussel growth. The theory predicts that not only velocity but also seston concentration in relation to the composition of the suspension feeder bed (mean size, density, and reef path length) affected average growth, so this result is not universally valid. Thus, if the seston concentration were increased or mussel density reduced, quite different growth results could be expected.

To understand fully the ecological effects of flow on suspension feeding populations, a wide flow range must be tested. Flow velocities from ~1–24 cm·s^{-1} were provided in a multiple-channel flume (Chap. 2) with four separate channels in giant scallop growth experiments (Wildish et al. 1987). Each channel had the same seston concentration but a

Table 7.6. *Growth rates of blue mussels as a function of velocity in a concurrent 31-d-long experiment.*[a]

Mean velocity $cm \cdot s^{-1}$	Mean growth, g wet $\cdot d^{-1} \cdot mussel^{-1}$		
	Overall, 1–10	Upstream, 1–5	Downstream, 6–10
0.10	0	0.05	−0.05
0.13	−0.06	0.02	−0.15
0.40	0.22	0.33	0.15
0.45	0.13	0.44	−0.13
0.75	0.12	0.26	−0.01
1.95	0.27	0.30	0.31
3.50	0.28	0.30	0.30
3.89	0.29	0.34	0.27

[a]Natural seston at a constant density of 10 mussels per tube ≡ 364 mussels $\cdot m^{-2}$.
Source: Wildish and Kristmanson (1985).

different velocity during each experiment. The results of six experiments at different seasonal times (Fig. 7.8) confirmed the unimodal response to velocity expected from physiological experiments on filtration (Chap. 4). At the scallop density tested in each channel of ~27 adult scallops $\cdot m^{-2}$, there were no differences in growth between the inlet and outlet ends of the flume, so a seston depletion effect was absent. This is because low values of $PR\alpha$ were associated with this density of scallops.

Earlier growth results with bay scallops obtained by Kirby-Smith (1972) can also be explained as a unimodal response to velocity, although because of the complexity of the flow in the tube and because volumetric flow estimates were used to calculate velocities, they are not directly comparable. One way in which growth experiments under field conditions can be misleading is if for a substantial part of the time, strong tidal flows are directly growth inhibiting. This possibility was examined with giant scallops in a multiple-channel flume, but with a variable flow so that various periods of inhibitory flows (>12 $cm \cdot s^{-1}$) and optimum flows (~4 $cm \cdot s^{-1}$) were present (Wildish and Kristmanson 1988). Results (Table 7.7) suggest that if between 33% and 66% of the time the flows are optimal, then the scallops can adjust and produce growth rates equivalent to those of controls which 100% of the time are at optimum flows.

Grizzle and Morin (1989) describe a field growth experiment of the hard clam *Mercenaria mercenaria* in a tidal lagoon in New Jersey, United States. Their 3 × 3 factorial experiment was designed to test, over a

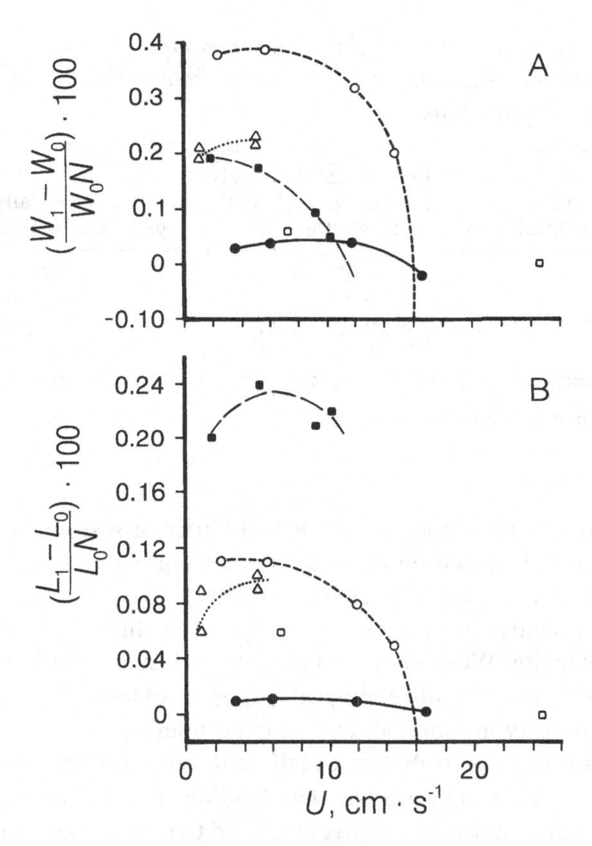

Figure 7.8 Giant scallop *Placopecten magellanicus* specific daily growth rates of adults (60 to 75 g wet weight) determined in 25- to 33-d-long experiments at different times of the year, as a function of velocity (U, $\mathrm{cm \cdot s^{-1}}$). Four channels each with 16–24 scallops $\simeq 18$–$27 \cdot \mathrm{m^{-2}}$ were run concurrently in a multiple-channel flume using natural seawater from the Bay of Fundy, Canada: A, W = total wet weight, g; B, L = valve height, cm; N = number of days in experiment (Wildish et al. 1987).

15-wk summer growth period, the effect of sediments (mud, sand, and undisturbed control) at three different sites in the local area. Current speed was measured at each site for such a small portion of the experimental period that it is possible that the data are not representative of conditions during the experiment. Grizzle and Morin (1989) concluded that, whereas sediments had no effect on clam growth, site differences did. Because environmental variables such as tidal velocity and seston concentration were not controlled during the experiment, we consider

Table 7.7. *Effect of inhibitory flows as percentage of total experimental time available on mean growth rates of giant scallops* Placopecten magellanicus.

Treatment: percentage of time flow is inhibitory	Percentage daily valve growth: valve height (cm) $\cdot$ scallop^{-1} $\cdot$ d^{-1}	Percentage daily biomass growth g wet $\cdot$ scallop^{-1} $\cdot$ d^{-1}
100	0.053 ± 0.01	0.37 ± 0.02
60	0.056 ± 0.01	0.36 ± 0.03
33	0.104 ± 0.01	0.41 ± 0.02
0	0.095 ± 0.01	0.41 ± 0.02

Source: Wildish and Kristmanson (1988).

that the positive relationship established between seston flux rate and hard clam growth, based on three data points, may be fortuitous. Other environmental variables such as seston quality, dissolved oxygen levels, or inhibitory wind–wave effects could explain the higher growth rates at one particular site. We concur with Grizzle and Lutz (1989) that there is a need to determine the physiological responses of hard clams to velocity and seston density in more controlled experiments.

Other authors (e.g. Cahalan, Sidall, and Luckenbach 1989; Wildish and Saulnier 1993) have shown that feeding is not simply related to seston flux, but rather is interactively related to seston concentration and velocity. In passive suspension feeders, too, Leonard (1989, and see Chap. 4) found that in the crinoid he studied, the feeding responses were to prey concentration and velocity separately.

In a 3-mo-long field experiment in the northern Gulf of Mexico, Judge, Coen, and Heck (1992) tested the effect of seawater flow on hard clam growth rates. The water depth was 0.6 m and tidal range small (<15 cm). Wind–wave effects were an important driving force for flows at their study locality. An open-ended four-channel flume, without a bottom, was built in the field. The flume was 7 m long with 1.2-m-high walls, and the flows within each channel were controlled by the width of the entrance. Comparative flows in each channel were assessed on one day so that three treatments of 2% and 40% reduction, or 65% increase in flows, were provided. Velocity was not measured continuously throughout this experiment, although spot-check determinations suggested a range of 0–27 cm $\cdot$ s^{-1}. Six *Mercenaria mercenaria* were placed in the middle of each channel in a plot of 0.25 m^{-2} (=24 clams · m^{-2}), but whether other suspension feeders were present in the natural sand of the rest of the flume was not clear. Since growth rates between all treatments and the control were not significantly different, Judge et al. (1992) concluded that velocity, independent of seston depletion, does not influence hard clam growth. We believe that this conclusion is not justified by the methods used, particularly because we know nothing of the temporal distribution of flow throughout the experiment. This is important because of the stochastic nature of wind–wave effects and the demonstrated ability of bivalves to adjust and compensate for the varying flow periods (Wildish and Kristmanson 1988).

Comparative growth studies between siphonate (*M. mercenaria*) and non-siphonate (*Crassostrea virginica*) bivalves were conducted by Grizzle, Langan, and Howell (1992) in a multiple-channel flume. Unfortunately, only a limited range of free-stream velocities (<8 cm · s^{-1}) were tested so the results must be regarded as preliminary. For 1991 results, oyster growth (valve increments) was unimodally related to velocity with a peak ~1 cm · s^{-1}, while hard clam growth over the same flow range showed only a and b of the unimodal response, with a broad b peak between 2 and 4.5 cm · s^{-1}.

Laboratory and field studies in Narragansett Bay, on the eastern coast of the United States, were designed to determine the effect of flow, seston concentration, and temperature on the feeding and population biology of the acorn barnacle, *Semibalanus balanoides* (Sandford et al. 1994). In simple flow tank studies, the acorn barnacles were exposed to unidirectional flows in the range 0–21 cm · s^{-1} and at temperatures of 10°–25°C, which were typical of local conditions. Barnacle feeding was in all cases passive filtering, except at zero flow, where slow cirral beating was observed. The percentage of barnacles passively filtering was shown to increase up to a flow of 21 cm · s^{-1} and to peak at a temperature ~15°C for both adults and juveniles. Locally available seawater with its naturally occurring seston resulted in variations in the percentage of barnacles feeding so that the locality with the highest chlorophyll a values had most barnacles feeding. These results were generally confirmed by growth experiments in various local conditions. The study by Sandford et al. (1994) might have benefitted from the use of an integrated environmental/physiological model for feeding and growth as suggested for the giant scallop (Chap. 4, the section, Bivalves). The limited range of unidirectional flows tested with respect to barnacle feeding and growth and the absence of simulated wind–wave action render laboratory results obtained by Sandford et al. (1994) in respect to water movement preliminary.

So far, the measures which have been used to discuss the effect of flow are short-term growth rates. It should be realized that we cannot simply multiply mean growth rates by population density to arrive at a satisfactory secondary benthic production estimate. Not only are these changes in growth rate seasonal in boreal climates, but positive and negative changes to population numbers must be accounted for if an accurate idea of annual production is to be obtained. This involves increases in density due to recruitment (larval settlement, migration) or decreases caused by mortality or emigration. It is these population changes which allow benthic suspension feeder production to adjust to the several orders of magnitude difference in carrying capacity between local habitats. There is some evidence that larval settlement occurs preferentially where reefs of suspension feeders are already established and that migration by juveniles to adult reefs may occur (Chap. 3).

Emigration of adult *Sanquinolaria nuttalli*, a deep-burrowing suspension feeding bivalve of the California coast, was noted after transplant experiments involving high densities of two other deep-burrowing species (Peterson and Andre 1980). Mortality rates of the blue mussel *Mytilus edulis* were reported by Incze et al. (1980) from a Maine, United States, estuary during a period of high ambient temperatures. The deaths were thought to result from a sudden crash in the phytoplankton population and hence reduction in ration coincident with metabolic stress due to high temperature.

We have another way of demonstrating that velocity is a key variable in controlling suspension feeder production: the use of field-observed standing crops or crude secondary production data calculated from them. The results of benthic sampling by grab or corer can be expressed as the species number, as biomass (g wet $\cdot$ m^{-2}), or, if an estimate of the $P:B$ ratio for each species is available, as crude secondary production (g wet $\cdot$ m$^{-2} \cdot$ y^{-1}) by methods described in Wildish and Peer (1983). For five estuaries and at seven stations within the Bay of Fundy, Wildish and Kristmanson (1979) partitioned benthic biomass into deposit and suspension feeding (Fig. 7.9). Deposit feeding biomass was inversely related to mean tidal flows, whereas suspension feeding biomass was a unimodal function of velocity. This field result is in general agreement with the model proposed in Fig. 7.1.

Purdy (1964) reanalyzed H. L. Sanders's benthic data from Buzzards Bay and Long Island Sound showing that the proportion of deposit

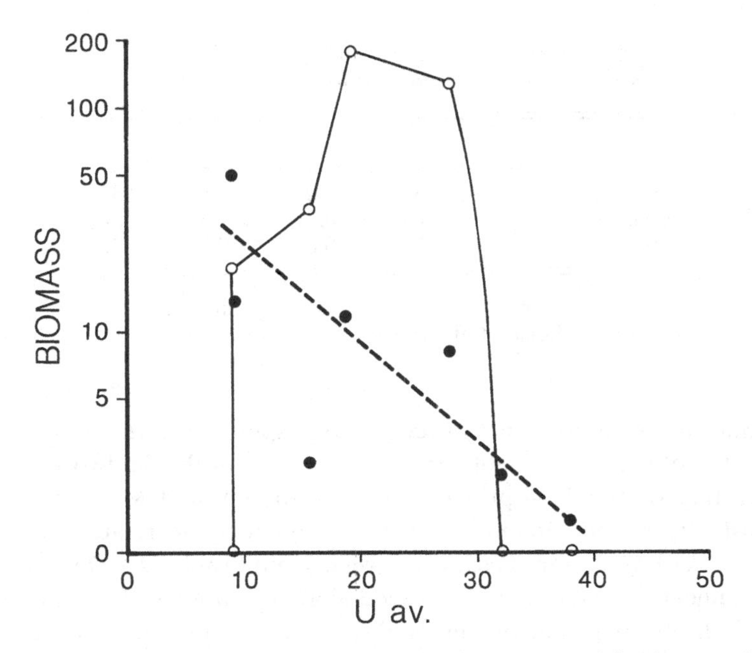

Figure 7.9 Deposit feeder (•) and suspension feeder (○) biomass (g wet weight · m^{-2}) at seven estuarine or nearshore locations within the southwestern Bay of Fundy, Canada (Wildish and Kristmanson 1979): *x*-axis, mean current velocity as $cm \cdot s^{-1}$ over one tidal period at 1–2 m above the sediment–water interface.

feeder biomass was positively related to percentage of clay in the sediment samples, while the proportion of suspension feeders was inversely related to the percentage of silt and clay in the sediment. These field results are also generally supportive of the concept shown in Fig. 7.1.

The suspension feeder production in the lower Bay of Fundy – mostly horse mussel, *Modiolus modiolus* – was positively correlated with velocity (Wildish and Peer 1983). The mean tidal velocity at each station was determined from a numerical model of circulation in the Bay (see Wildish and Peer 1983). From a subset of these data, reanalysis by Wildish and Kristmanson (1993) showed that the important variable controlling the greater than three orders of magnitude in production was velocity. Velocity, U in $m \cdot s^{-1}$, predicted production in the lower Bay of Fundy as

$$\text{Log}_{10} \text{ benthic production, } g \cdot m^{-2} \cdot y^{-1} = -1.155 + 5.862\,U, \quad n = 29, \quad r = 0.85, \quad p > 99\% \tag{7.7}$$

Table 7.8. *Bay of Fundy suspension feeder production for the equation* log_{10} x=a+by.[a]

Location	a	b	r	N
Lower Bay	−1.38	0.06	0.84	33
Minas Basin	1.42	0.01	0.36	34
Chignecto Bay	−0.71	0.02	0.25	35

[a] x = g wet wt · m^{-2} · yr^{-1}; y = mean velocity, cm · s^{-1}.
Source: Wildish et al. (1986).

by omitting 6 stations with reduced, or no, suspension feeder production from the total of 35 available. In a continuation of the fieldwork in the upper Bay of Fundy – primarily Chignecto Bay and Minas Basin – Wildish, Peer, and Greenberg (1986) found that the relationship for these upper Bay stations became non-significant (Table 7.8). Despite this clear indication that the current speed alone was not a sufficiently complete indicator of production in the Bay of Fundy for suspension feeders, this finding does circumstantially support the views of many earlier biologists (e.g. Kerswill 1949) that production was influenced by water movements.

Wind–wave energy sufficient to cause sediment resuspension could be one way in which particles of nutritive value to suspension feeders could replenish depleted layers within the benthic boundary layer. Fréchette and Grant (1991) conducted a field experiment with blue mussels *Mytilus edulis* to test the relative importance of seston depletion and resuspension events on growth rates at two intertidal locations, 0 and 1 m off bottom, in the St. Lawrence estuary. Growth rates were estimated in two ways: by tissue weight changes over time and by scope for growth predictions based on filtration rates and calculations performed on calm, windless days. The former measure should integrate growth due to seston depletion and resuspension, while the latter should yield predicted growth rates free of resuspension events. Growth responses at both sites showed that summer growth rates were not augmented by resuspension phenomena.

Bock and Miller (1994) showed in a field study on an intertidal sand flat in Delaware, United States, where water movement was dominated by wind–wave effects, that growth of *Mercenaria mercenaria* was correlated with the percentage of organic matter in wave-resuspended seston

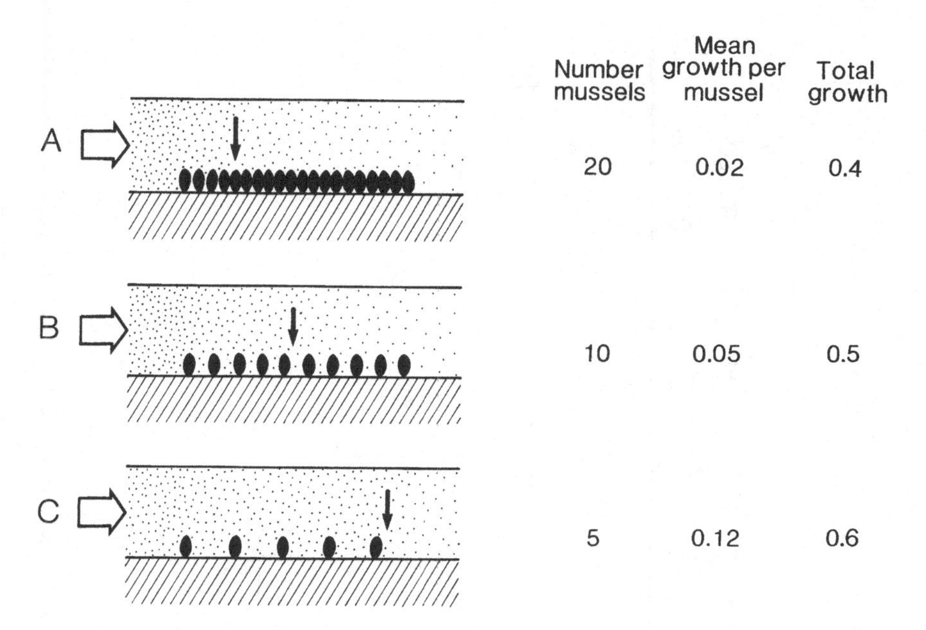

Number mussels
Mean growth per mussel
Total growth
A
20
0.02
0.4
B
10
0.05
0.5
C
5
0.12
0.6

Figure 7.10 Effect of mussel density on growth (arbitrary units): flow velocity and seston concentration are equal at the upstream ends of A, B, and C; arrow, point where N_A becomes limiting.

rather than with total seston concentration. The disparity between this and the previous results suggests a species-specific difference due to resuspended sediment and its effect on bivalve growth.

Effects of population density

The effect of density on suspension feeder growth predicted by hydrodynamic models of suspension feeding is shown diagrammatically in Fig. 7.10. The model applies where the benthic boundary layer intersects the surface and turbulent transfer rates near the suspension feeder become limiting to growth along a mussel reef, as occurs in pipe or flume flows. Mean growth rate per animal is inversely related to density.

Experimental results in a 7-cm-diameter Kirby-Smith growth tube (Wildish and Kristmanson 1985) at various densities of blue mussels *Mytilus edulis* confirm this prediction at least at one low flow speed, $0.09\,cm \cdot s^{-1}$ (Fig. 7.11B). At a higher velocity – $1.44\,cm \cdot s^{-1}$ – at all densities tested and at fixed path length of 1.5 m for this experiment, all

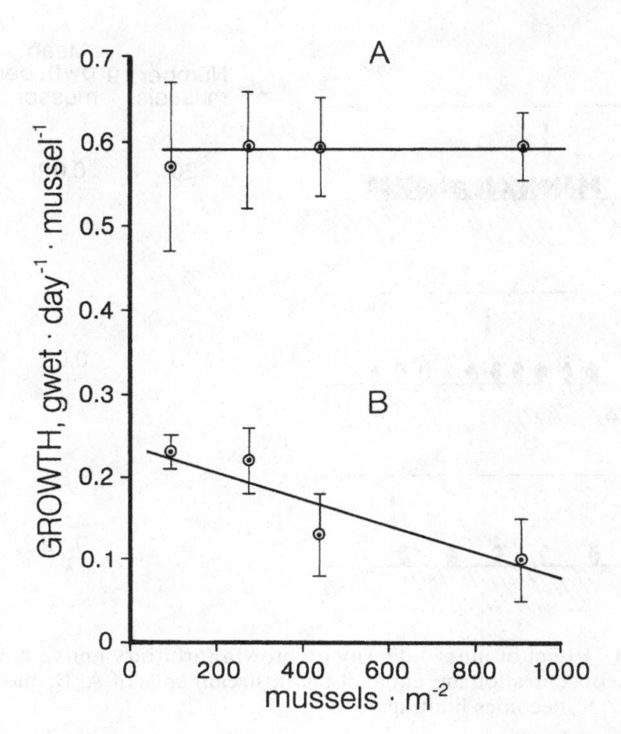

Figure 7.11 Blue mussel growth rates, mean and range, as a function of density in 34-d-long concurrent experiments in Kirby-Smith growth tubes. Natural seawater from the Bay of Fundy, Canada: A, $1.44\,cm \cdot s^{-1}$; B, $0.09\,cm \cdot s^{-1}$ (Wildish and Kristmanson 1985).

mussels grew at the maximum possible rate, suggesting that no seston depletion occurred, given the seston concentration available (Fig. 7.11A). The laminar flow speeds chosen for this experiment, however, render the results strictly outside the theory, which refers to turbulent flows.

Broom (1982), in field studies of a locally exploited cockle, *Anadara granosa* from Malaysia, showed that growth rates, as mean wet weight per square metre, are inversely related to the log of density. Field experiments in which *Anadara* of different year-classes were stocked at different densities confirmed that density did influence growth rates. A number of bivalve density manipulation experiments all showed that the percentage of growth relative to a low density control decreased in inverse proportion to the stocking density (Table 7.9). Study localities include the U.S. West Coast (Peterson 1982) and East Coast (Eldridge,

Table 7.9. *Field experimental studies of bivalves in which densities were manipulated, showing individual valve growth rates inversely proportional to density.*

Species	Density range, number · m^{-2}	Bed path length, m	Maximum growth decrease, %	Reference
Andara granosa	125–2500	10.0	?–70	Broom (1982)
Protothaca staminea	5–96	1.0	38–49	Peterson (1982)
Chione undatella	0.5–48	1.0	38–49	
Circe lenticularis	5–30	1.0	40–90	
Anomalocardia squamosa	5–30	1.0	–70	Peterson and Black (1987)
Callista impar	10–60	1.0	10–30	
Katelysia scalarina	20–160	1.0	35	Peterson and Black (1988)
K. rhytiphora	20–160	1.0	40	
Mercenaria mercenaria	290–1159	0.6–1.2	50	Eldridge et al. (1979)
M. mercenaria	10–80	1.0	18	Peterson and Beal (1989)

Eversole, and Whetstone 1979; Peterson and Beal 1989) and Australia (Peterson and Black 1987, 1988).

Field experiments involving intertidal blue mussels *Mytilus edulis* have shown that for patches of mussels, the individual "edge" mussels grew faster than the control ones (Okamura 1986; Newell 1990). Here, the oscillating flows of the shore enable "edge" mussels to be upstream for at least some of the time, while the central mussels experience the seston depletion effect and thus grow at slower rates.

Water movement impoverishment

In a conventional benthic grab sampling study of Bay of Fundy estuaries, limited to soft sediment and depths of 8–35 m, referred to earlier, Wildish and Kristmanson (1979) described three distinct communities: mixed, deposit, and impoverished. The basic distinction is between mixed and deposit feeding; thus the mixed community consists of both suspension and deposit feeders, whereas both other types contain no suspension feeders at all. The characteristic feature of an impoverished community is that it has high water velocities and high turbulent fluxes of seston, yet no suspension feeders (see Fig. 7.9). Subsequently, Wildish and Peer (1983) described specialized species of deposit-feeding benthic macrofauna – usually deep burrowers – capable of resisting sediment erosion by tidal or wind–wave action. In the Bay of Fundy, deep-burrowing polychaetes such as *Spiophanes bombyx, Scoloplos armiger*, and *Scolelepis squamatus*; a rapid-burrowing amphipod, *Psammonyx nobilis*; and a massively built sand dollar able to burrow, *Echinarachnius parma*, are examples of species associated with water movement conditions which cause impoverishment of suspension feeders. Specialized water movement–impoverished macrofaunal communities of soft sediments are generally of low biomass and production, e.g. in the Bay of Fundy (Wildish and Peer 1983). The few species involved are deposit feeders specialized for high energy niches and also do not conform to the exclusion principle outlined in Fig. 7.1. Some field sampling evidence supporting this characterization is shown in Fig. 7.9.

In their field study of benthic macrofauna in the Bristol Channel, Warwick and Uncles (1980) related classically defined communities to sediment bed stress. Bed shear stress, τ_0, for each location was determined from

$$\tau_0 = \frac{1}{2}\gamma\rho V_{max}^2 \tag{7.8}$$

where γ is a dimensionless sediment friction coefficient and ρ is the density of seawater, usually expressed in units of dynes per square centimetre. The maximum tidal current velocities of stations within the Bristol Channel were obtained from a numerical model. Results suggested that each community was associated with a particular set of bed stress values so that a map of bed stress values closely followed that of the independently defined communities. Associated with the highest shear stresses of 22 $dynes \cdot cm^{-2}$, Warwick and Uncles (1980) describe a reduced soft bottom community similar to impoverished soft bottom communities of the Bay of Fundy. Use of τ_0 as the independent variable in this study is similar to the use of mean velocity by Wildish et al. (1986), as described earlier, with the exception that this study made no allowance for bottom roughness implicit in the term γ.

One type of water movement which has not yet been discussed is direct wind–wave activity on benthic communities. Wind effects are generally discrete, stochastic events which may have direct influence on the bottom at depths down to ~100 m (Emerson 1989). As for tidal currents, we can expect both positive effects of wind–wave activity on benthic production as well as negative ones, e.g. storm damage to benthic communities (Yeo and Risk 1979; Wildish 1980; Dobbs and Vozarik 1983; Shanks and Wright 1986; Peterson et al. 1989; Nehls and Thiel 1993). Emerson (1989) presented a statistical study of published secondary benthic production data which included biotic (phytoplankton primary production, benthic primary production, meiofaunal, macrofaunal, and total benthic secondary production) as well as physical variables (temperature, salinity, depth, tidal height, sediment type, wind stress, and two wind shelter indices). From 201 published data sets, a Pearson correlation matrix determined that wind stress was weakly inversely correlated with total (macrofauna, meiofauna, microflora) and macrofaunal secondary production ($r^2 = 0.32$ and 0.12, respectively). Benthic production was negatively correlated with wind stress, suggesting that the overriding effect of this variable in the benthic environment is negative, i.e. that increasing wind stress at a locality results in decreased production. Further multiple regression analysis on the log-transformed benthic secondary production data (Emerson 1989) showed that wind stress, tidal height, wind shelter indices, and seawater temperature explained ~90% of the variance in total secondary benthic production. This a posteriori analysis suggests that water movement variables are major controls on benthic energy flow, as is postulated in the trophic group mutual exclusion theory.

A spatial survey of the benthic macrofauna (biomass, numbers of species) of high energy sediments on the coast of South Africa, inclusive of intertidal and nearshore sublittoral regions, could be interpreted as benthic limitation by water movement (McLachlan, Cockcroft, and Malan 1984). In the wave breaking region of the high energy surf zone transect, the extreme turbulence prevents any macrofauna from becoming established, and it is only when more stable sediment–water interfaces occur that macrofaunal biomass and species numbers increase.

Emerson and Grant (1991) have observed one way in which wind–wave activity could affect production on intertidal soft shell clam (*Mya arenaria*) flats. Juvenile clams were eroded and transported across the flats coincident with peaks of bedload transport associated with wind–wave activity.

Biological interactions in mixed benthic communities

The trophic group amensalism hypothesis of Rhoads and Young (1970) suggested that deposit feeding infauna exclude other species, particularly suspension feeders, by their feeding and burrowing activities, which also result in resuspension of sedimentary particles. High sedimentary particulate loads cause clogging of the filtration mechanism of suspension feeders and hence results in reduced growth and/or increased mortality rate. Posey (1990) has reviewed the scant evidence pertaining to trophic group amensalism. The deposit feeding ghost shrimp *Callianassa* sp. in tropical lagoons caused reduced growth rates of suspension feeders, such as corals or polychaetes (Aller and Dodge 1974), by its bioturbating activities. Another species of *Callianassa* present in a California lagoon apparently restricted the introduced suspension-feeding hard clam *Mercenaria mercenaria* by its bioturbating activity (Murphy 1985). Field and laboratory experiments undertaken by Murphy (1985) confirmed that hard clam growth was reduced by sediment resuspension produced by the ghost shrimp.

Rhoads and Young (1970) realized the limited applicability of their theory because it did not adequately describe the factors that actually separate deposit feeding and suspension feeding communities. Examples of field studies of macrobenthic distribution which are at odds with trophic group amensalism include those of Posey (1986), who found that dense populations of the ghost shrimp *C. californiensis* were positively correlated with a suspension feeder on the Oregon, United States, coast and Dittmann (1990), who studied a blue mussel (*Mytilus edulis*) reef on the island of Sylt in the North Sea and showed that this species facilitates the establishment of a rich deposit feeding infauna associated with hydrodynamic protection within the reef and which fed on the faecal wastes produced by the mussels. A similar rejection of trophic group amensalism is presented in the review of Snelgrove and Butman (1994).

The adult–larval interaction theory of Woodin (1976) was designed to explain the occurrence of deposit feeders, suspension feeding clams, and tube builders on the U.S. West Coast. The mechanism proposed was that burrowing deposit feeders disturbed the sediment to the extent that epifaunal tube bases and siphon holes of other species were disrupted, thus inhibiting their growth and development. In addition, recently settled larvae were consumed by the deposit feeders, thus helping to ensure that clams and tube builders could be excluded from these highly reworked sediments. The suspension feeding of clams served to remove settling larvae, as did the predatory and defaecatory activities of tube builders, thus tending to exclude deposit feeders.

Posey (1990) reviewed the evidence for the adult–larval interaction hypothesis, listing 12 studies which provide some evidence in favor of the alternate hypothesis that larval settlement is inhibited by feeding and locomotion of adult infauna. The work of Elmgren et al. (1986) in the soft sediments of the Baltic Sea with two deposit feeding amphipods (*Pontoporeia affinis* and *P. femorata*) and the facultative suspension feeding bivalve *Macoma balthica* may be cited as an example. In the Baltic, *Macoma* is absent or rare in those areas where dense populations of *Pontoporeia* sp. occurred. Experimental tests in flowthrough aquaria containing sediments (2-L glass jars) in which adult *P. affinis* and newly settled spat of *M. balthica* were placed, showed that the latter decreased with increasing densities of *Pontoporeia*. Apparently, *P. affinis* caused direct physical injury to the spat, presumably ingesting the soft tissue, although in nature the spat are not an important food source for *Pontoporeia*.

Ólafsson (1989) studied two contrasting localities in the Baltic containing *Macoma balthica* on the coast of southern Sweden. The first was in a sheltered bay with limited water movement in a *Zostera* meadow where the *Macoma* were deposit feeders. The second locality was exposed and had a sandy sediment and more energetic water movement – mainly by wind–wave action – where adult *Macoma* were primarily suspension feeding. Adult density manipulation experiments in the field showed that juvenile *Macoma* growth and density levels were negatively affected at the first, but not the second, location. Juvenile *Macoma* in both habitats are deposit feeders, and the field result was explained by dietary resource overlap, which is absent where the water movements are more energetic and where adult *Macoma* are suspension feeders. Experimental tests suggested that *Hydrobia* sp., another deposit feeder, was negatively affected by high densities of deposit feeding *Macoma*, whereas high densities of suspension feeding *Macoma* had no effect on deposit-feeding oligochaetes of the second, water movement–energetic, sand habitat.

André and Rosenberg (1991) in manipulative field experiments involving adjusting the densities of *Cerastoderma edule* and *Mya arenaria* in the Baltic showed that adults reduced the settlement of bivalve larvae and hence reduced the recruitment of new individuals proportional to adult density. In a further laboratory flume study (André et al. 1993), these findings were confirmed. Survival of settling larvae was proportional to adult density of *C. edule* and increases in flume velocity caused only a slight rise in predation risk.

There is as much evidence against the adult–larval interaction hypothesis as there is for it (Posey 1990). Thus, deposit feeders, tube builders, and suspension feeding clams do not always occur in discrete communities and may occur together in many natural situations (Posey 1990). A direct field test involving manipulated densities of deposit- and suspension-feeding clams and their effect on co-existing tube dwelling and infaunal deposit-feeding polychaetes suggested no exclusion effects (Hines, Posey, and Haddon 1989). Posey (1990) concluded that for a number of communities examined, the predictions implied in the adult–larval interaction hypothesis were not met and other factors must be involved.

Summary

The importance of benthic production limitation by flow, inclusive of trophic group mutual exclusion, has been emphasized in recent reviews (e.g. Olafsson et al. 1994; Hall 1994). However, flow influences food supply to suspension feeders not only by controlling the turbulent supply of seston, but by direct water movement impoverishment of both suspension and deposit feeders. The mechanisms involved have, by comparison, received relatively little study and could include larval incompetency, and adult washout during storm events.

Using the well-developed theory of hydrodynamics to help formulate specific hypotheses about suspension feeders at all hierarchial levels has proved to be, and we believe will continue to be, most rewarding. At the level of the population the seston depletion effect (number 1, Table 7.2) has been validated by both downstream reductions in seston concentration and downstream reductions in growth. How flow and seston concentration interact to produce increased production has begun to be documented for a few species (number 2, Table 7.2). The references cited for number 2, Table 7.2, are observational confirmation of H_1, although experimental results (e.g. Wildish and Kristmanson 1985) which also support it are available. More experimental results to cover a wider range of variables (velocity, seston concentration) are needed. Another ecological mechanism which may be important in the population response to increases in flow involves the addition of suspension feeders, by immigration and larval recruitment, and their subtraction, by emigration and death. All of these mechanisms allow three to four orders of magnitude variation in suspension feeder production in the natural marine environment, dependent largely on velocity and seston concentration.

Of the four other hypotheses outlined in Table 7.2, there is field evidence that the growth and production of suspension feeders can be limited by energetic water movement (number 3), although the mechanisms remain to be investigated in detail. The effect of suspension feeder density on seston uptake or growth rates of populations (number 5) has been demonstrated, although this is not the case with the effect of variable reef path lengths on these rates (number 6). Because it is outside the scope of this book, we have not considered the relationship between deposit feeder production and the erosional power of flow (number 4), although pertinent studies have been undertaken (e.g. by Rhoads 1973; Rhoads, Yingst, and Ullman 1978; Wainwright 1987, 1990; Davis 1993).

The trophic group amensalism hypothesis of Rhoads and Young (1970) has received little support despite some efforts to test it and should be discarded as a general explanation (see also Snelgrove and Butman 1994). Similarly, the larval interaction hypothesis of Woodin (1976) has received some support as well as contradictory evidence from field studies. By comparison with the hydrodynamic hypotheses presented in Table 7.2, the two biotic interaction hypotheses are limited in application and are not central to the exclusion principle operative among suspension and deposit feeders in mixed benthic communities.

Wildish and Kristmanson (1993) recognized an apparent paradox when comparing physiological responses of filtration at variable flows (Chap. 4) to the ecological or population responses to flow presented in this chapter. The blue mussel data of Fig. 7.12 are for the filtration rate of an individual, while the horse mussel data are for the annual production of mussels from many different localities in the Bay of Fundy. Figure 7.12 is a depiction of this paradox where, although the two y-axis values are incommensurate for the reasons just mentioned, the x-axis is the same for both y variables. The latter was achieved by measuring flows near blue mussel inhalants and by recalculating the horse mussel ambient velocities at the inhalant level from free-stream averages (Wildish and Kristmanson 1993). If the filtration rate of an individual suspension feeder is considered to indicate feeding and thus growth rate, one might expect more concordance between the physiological and ecological data than is shown in Fig. 7.12. We believe that the paradox can be described better as an emergent property of the population (reef or aggregation) of suspension feeders. Salt (1979) provided the following operational definitions:

- "An emergent property of an ecological unit is one which is wholly unpredictable from observation of the components of that unit."
- "An emergent property of an ecological unit is only discernible by observation of that unit itself."

By these considerations the average filtration and feeding rate of a population as a function of velocity cannot be determined by studying

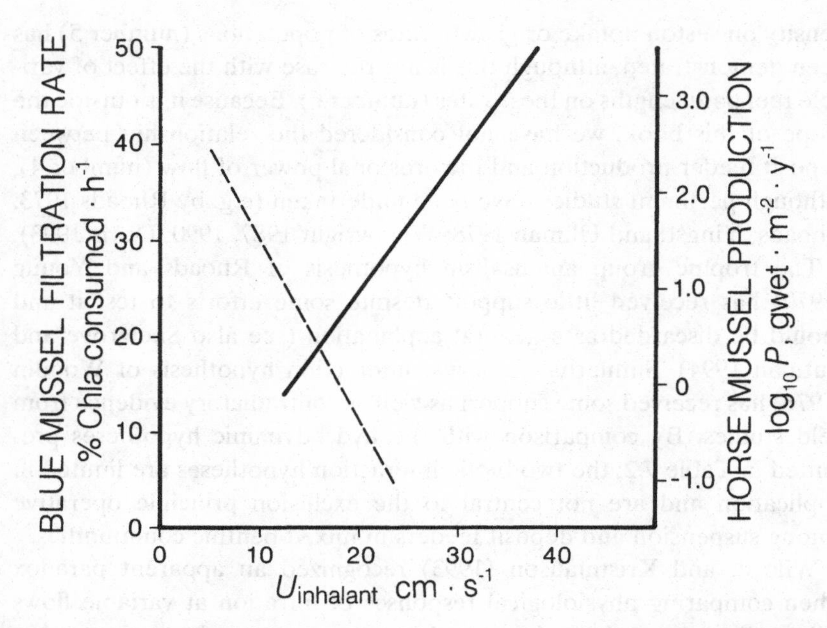

Figure 7.12 Blue mussel *Mytilus edulis* filtration rates, as percentage chlorophyll a consumed per hour. Dashed line (based on Wildish and Miyares 1990); solid line, horse mussel *Modiolus modiolus* production (Wildish and Peer 1983). The *x* axis shows ambient velocities at inhalant level for both species. Chla, chlorophyll a.

the physiological feeding rate of one or a few individuals even though the flow conditions used are the same. The change in hierarchical level from the individual to the population is, of course, crucial in appreciating the nature of this phenomenon. If feeding rates are an emergent property of a population of suspension feeders, then methods presently used to determine population feeding rates and hence growth, e.g. in bivalves (Fréchette et al. 1989; Fréchette and Grant 1991), must be misleading. This is because population level phenomena, such as growth, determined in this way exclude recruitment mortality and the seston depletion effect, and so cannot reliably predict population growth.

References

Aller, R. C., and R. E. Dodge. 1974. Animal–sediment relations in a tropical lagoon, Discovery Bay, Jamaica. J. Mar. Res. 32: 209–232.

Anderson, F. E., and L. M. Meyer. 1986. The interaction of tidal currents on a disturbed intertidal bottom with a resulting change in particulate matter quantity, texture and food quality. Estuar. Coastal Shelf Sci. 22: 19–30.

André, C., P. R. Jonsson, and M. Lindegarth. 1993. Predation on settling bivalve larvae by benthic suspension feeders: the role of hydrodynamics and behaviour. Mar. Ecol. Prog. Ser. 97: 183–192.

André, C., and R. Rosenberg. 1991. Adult–larval interactions in the suspension-feeding bivalves *Cerastoderma edule* and *Mya arenaria*. Mar. Ecol. Prog. Ser. 71: 227–234.

Blackman, F. F. 1905. Optima and limiting factors. Ann. Bot. 19: 281–295.

Bock, M. J., and D. C. Miller. 1994. Seston variability and daily growth in *Mercenaria mercenaria* on an intertidal sandflat. Mar. Ecol. Prog. Ser. 114: 117–127.

Broom, M. J. 1982. Analysis of the growth of *Anadara granosa* (Bivalvia: Arcidae) in natural, artificially seeded and experimental populations. Mar. Ecol. Prog. Ser. 9: 69–79.

Buss, L. W., and J. B. C. Jackson. 1981. Planktonic food availability and suspension-feeder abundance: evidence of in situ depletion. J. Exp. Mar. Biol. Ecol. 49: 151–161.

Butman, C. A., M. Fréchette, W. Rockwell Geyer, and V. R. Starczak. 1994. Flume experiments on food supply to the blue mussel *Mytilus edulis* L. as a function of boundary layer flow. Limnol. Oceanogr. 39: 1755–1768.

Cahalan, J. A., S. E. Sidall, and M. W. Luckenbach. 1989. Effects of flow velocity, food concentration and particle flux on growth rates of juvenile bay scallops *Argopecten irradians*. J. Exp. Mar. Biol. Ecol. 129: 45–60.

Crisp, D. J. 1984. Energy flow measurements, p. 284–372. *In* N. A. Holme and A. D. McIntyre (eds.) Methods for the study of marine benthos. 2nd ed. Blackwell, Oxford.

Dare, P. J. 1976. Settlement, growth, and production of the mussel, *Mytilus edulis* l., in Morecambe Bay, England. Fish. Invest. Ser. II, 28: 1–25.

Dauer, D. M., C. A. Maybury, and R. M. Ewing. 1981. Feeding behaviour and general ecology of several spionid polychaetes from the Chesapeake Bay. J. Exp. Mar. Biol. Ecol. 54: 21–38.

Davis, W. R. 1993. The role of bioturbation in sediment resuspension and its interaction with physical shearing. J. Exp. Mar. Biol. Ecol. 17: 187–200.

Dittmann, S. 1990. Mussel beds: amensalism or amelioration for intertidal fauna? Helgolander. Meeresunters. 44: 335–352.

Dobbs, F. C., and J. M. Vozarik. 1983. Immediate effects of a storm on coastal infauna. Mar. Ecol. Prog. Ser. 11: 273–279.

Eldridge, P. J., A. G. Eversole, and J. M. Whetstone. 1979. Comparative survival and growth rates of hard clams, *Mercenaria mercenaria*, planted in trays subtidally and intertidally at varying densities in a South Carolina estuary. Proc. Nat. Shellfish. Assoc. 69: 30–39.

Elmgren, R., S. Ankar, B. Marteler, and G. Edjung. 1986. Adult interference with postlarvae in soft sediment – the *Pontoporeia–Macoma* example. Ecology 67: 827–836.

Emerson, C. W. 1989. Wind stress limitation of benthic secondary production in shallow, soft-sediment communities. Mar. Ecol. Prog. Ser. 53: 65–77.

Emerson, C. W., and J. Grant. 1991. The control of soft-shell clam (*Mya arenaria*) recruitment on intertidal sand flats by beadload sediment transport. Limnol. Oceanogr. 36: 1288–1300.

Fréchette, M., and E. Bourget. 1985. Energy flow between the pelagic and

benthic zones: factors controlling particulate organic matter available to an intertidal mussel bed. Can. J. Fish. Aquat. Sci. 42: 1166–1170.
Fréchette, M., C. A. Butman, and W. Rockwell Geyer. 1989. The importance of boundary layer flows in supplying phytoplankton to the benthic suspension feeder, *Mytilus edulis* L. Limnol. Oceanogr. 34: 19–36.
Fréchette, M., and J. Grant. 1991. An in situ estimation of the effect of wind driven resuspension on the growth of the mussel *Mytilus edulis* L. J. Exp. Mar. Biol. Ecol. 148: 201–213.
Glynn, P. W. 1973a. Ecology of a Caribbean coral reef, the *Porites* reef–flat biotope. Part I. Methodology and hydrography. Mar. Biol. 20: 297–318.
1973b. Ecology of a Caribbean coral reef, the *Porites* reef–flat biotope. Part II. Plankton community with evidence for depletion. Mar. Biol. 22: 1–21.
Grizzle, R. E., R. Langan, and W. H. Howell. 1992. Growth responses of suspension-feeding bivalve molluscs to changes in water flow: differences between siphonate and non-siphonate taxa. J. Exp. Mar. Biol. Ecol. 162: 213–228.
Grizzle, R. E., and R. A. Lutz. 1989. A statistical model relating horizontal seston fluxes and bottom sediment characteristics to growth of *Mercenaria mercenaria*. Mar. Biol. 102: 95–106.
Grizzle, R. E., and P. J. Morin. 1989. Effect of tidal currents, seston, and bottom sediments on growth of *Mercenaria mercenaria*: results of a field experiment. Mar. Biol. 102: 85–94.
Hall, S. J. 1994. Physical disturbance and marine benthic communities: life in unconsolidated sediments. Ocean. Mar. Biol. Annu Rev. 32: 179–239.
Hines, A. H., M. H. Posey, and P. J. Haddon. 1989. Effects of adult suspension- and deposit-feeding bivalves on recruitment of estuarine infauna. Veliger 32: 109–119.
Hunt, O. D. 1925. The food of the bottom fauna of the Plymouth fishing grounds. J. Mar. Biol. Assoc. U.K. 13: 560–599.
Incze, L. S., R. A. Lutz, and L. Watling. 1980. Relationship between effects of environmental temperature and seston on growth and mortality of *Mytilus edulis* in a temperate northern estuary. Mar. Biol. 57: 147–156.
Jørgensen, C. B. 1990. Bivalve filter feeding: hydrodynamics, bioenergetics and ecology. Olsen and Olsen, Fredensborg, Denmark.
Judge, M. L., L. D. Coen, and K. L. Heck. 1992. The effect of long-term alteration of *in situ* water currents on the growth of *Mercenaria mercenaria* in the northern Gulf of Mexico. Limnol. Oceanogr. 37: 1550–1559.
Jumars, P. A., and A. R. M. Nowell. 1984a. Fluid and sediment dynamic effects on marine benthic community structure. Am. Zool. 24: 45–55.
1984b. Effects of benthos on sediment transport: difficulties with functional grouping. Cont. Shelf Res. 3: 115–130.
Kerswill, C. J. 1949. Effects of water circulation on the growth of quahaugs and oysters. J. Fish. Res. Board Can. 7: 545–551.
Kirby-Smith, W. W. 1972. Growth of the bay scallop: the influence of experimental currents. J. Exp. Mar. Biol. Ecol. 8: 7–19.
Leonard, A. B. 1989. Functional responses in *Antedon mediterranea* (Lamarck) (Echinodermata: Crinoidea): the interaction of prey concentration and current velocity on a passive suspension feeder. J. Exp. Mar. Biol. Ecol. 127: 81–103.
Levinton, J. S. 1991. Variable feeding behaviour in three species of *Macoma*

(Bivalvia: Tellinacea) as a response to water flow and sediment transport. Mar. Biol. 110: 375–383.
Levinton, J., and G. R. Lopez. 1977. A model of renewable resources and limitation of deposit feeding benthic populations. Oecologia 31: 177–190.
Levinton, J. S., D. E. Martinez, M. M. McCartney, and M. L. Judge. 1995. The effect of water flow on movement, burrowing, and distributions of the gastropod *Ilyanassa obsoleta* in tidal creeks. Mar. Biol. 122: 417–424.
Madsen, K. N., P. Nilsson, and K. Sundback. 1993. The influence of benthic microalgae on the stability of a subtidal sediment. J. Exp. Mar. Biol. Ecol. 170: 159–177.
McLachlan, A., A. C. Cockcroft, and D. E. Malan. 1984. Benthic faunal response to a high energy gradient. Mar. Ecol. Prog. Ser. 16: 51–63.
Miller, D. C., M. J. Bock, and E. J. Turner. 1992. Deposit and suspension feeding in oscillatory flows and sediment fluxes. J. Mar. Res. 50: 489–520.
Murphy, R. C. 1985. Factors affecting the distribution of the introduced bivalve, *Mercenaria mercenaria*, in a California lagoon: the importance of bioturbation. J. Mar. Res. 43: 673–692.
Muschenheim, D. K., P. E. Kepkay, and K. Kranck. 1989. Microbial growth in turbulent suspension and its relation to marine aggregate formation. Neth. J. Sea Res. 23: 283–292.
Nehls, G., and M. Thiel. 1993. Large-scale distribution patterns of the mussel *Mytilus edulis* in the Wadden Sea of Schleswig-Holstein: do storms structure the ecosystem? Neth. J. Sea Res. 31: 181–187.
Newell, R. 1965. The role of detritus in the nutrition of two marine deposit-feeders, the prosobranch *Hydrobia ulvae* and the bivalve *Macoma balthica*. Proc. Zool. Soc. London 144: 25–45.
Newell, C. R. 1990. The effects of mussel (*Mytilus edulis*, Linnaeus, 1758) position in seeded bottom patches on growth at subtidal lease sites in Maine. J. Shellfish Res. 9: 113–118.
Neyman, A. A. 1979. Soviet investigations of the benthos of the shelves of the marginal seas, p. 269–284. *In* M. V. Dunbar (ed.) Marine production mechanisms. Cambridge University Press, Cambridge.
Nickels, J. S., J. D. King, and D. C. White. 1979. Poly-β-hydroxybutyrate accumulation as a measure of unbalanced growth of the estuarine detrital microbiota. Appl. Environ. Microbiol. 37: 459–465.
Odum, E. P. 1971. Fundamentals of ecology. W. B. Saunders, Philadelphia.
Ólafsson, E. B. 1989. Contrasting influences of suspension-feeding and deposit-feeding populations. Mar. Ecol. Prog. Ser. 55: 171–179.
Ólafsson, E. B., O. H. Peterson, and W. G. Ambrose. 1994. Does recruitment limitation structure populations and communities of macroinvertebrates in marine soft sediments: The relative significance of pre- and post-settlement processes. Ocean. Mar. Biol. Annu. Rev. 32: 65–109.
O'Riordan, G. A., S. G. Monismith, and J. R. Koseff. 1993. A study of concentration boundary layer formation over a bed of model bivalves. Limnol. Oceanogr. 38: 1712–1729.
Pearson, T. H. 1971. Studies on the ecology of the macrobenthic fauna by comparison of feeding groups. Vie Milieu 22 (Suppl.): 53–91.
Pearson, T. H., and R. Rosenberg. 1987. Feast and famine: structuring factors in marine benthic communities, p. 373–395. *In* J. H. R. Gee and P. S. Giller (eds.) Organization of communities past and present. Blackwell, Oxford.

Perkins, E. J. 1974. The biology of estuaries and coastal waters. Academic Press, London.

Peterson, C. H. 1982. The importance of predation and intra- and interspecific competition in the population biology of two infaunal suspension-feeding bivalves, *Protothaca staminea* and *Chione undatella*. Ecol. Monogr. 52: 437–475.

Peterson, C. H., and S. V. André. 1980. An experimental analysis of interspecific competition among marine filter feeders in a soft-sediment environment. Ecology 61: 129–139.

Peterson, C. H., and B. F. Beal. 1989. Bivalve growth and higher order interactions: importance of density, site and time. Ecology 70: 1390–1404.

Peterson, C. H., and R. Black. 1987. Resource depletion by active suspension feeders on tidal flats: influence of local density and tidal elevation. Limnol. Oceanogr. 32: 143–166.

Peterson, C. H., and R. Black. 1988. Density-dependent mortality caused by physical stress interacting with biotic history. Am. Nat. 131: 257–270.

1991. Preliminary evidence for progressive sestonic food depletion in incoming tide over a broad tidal sand flat. Estuar. Coast. Shelf Sci. 32: 405–414.

Peterson, C. H., H. C. Summerson, S. R. Fegley, and R. C. Prescott. 1989. Timing, intensity and sources of autumn mortality of adult bay scallops *Argopecten irradians concentricus* Say. J. Exp. Mar. Biol. Ecol. 127: 121–140.

Posey, M. H. 1986. Changes in the benthic community associated with dense beds of a burrowing deposit feeder, *Callianessa californiensis*. Mar. Ecol. Prog. Ser. 31: 15–22.

1990. Functional approaches to soft-substrate communities: how useful are they? Rev. Aquat. Sci. 2: 343–356.

Purdy, E. G. 1964. Sediments as substrates, p. 238–271. *In* J. Imbre and N. Newell (ed.) Approaches to paleoecology. Wiley, New York.

Rhoads, D. C. 1973. The influence of deposit-feeding benthos on water turbidity and nutrient recycling. Am. J. Sci. 273: 1–22.

Rhoads, D. C., J. Y. Yingst, and W. J. Ullman. 1978. Seafloor stability in Central Long Island Sound. Part 1. Temporal changes in erodibility of fine grained sediments, p. 221–244. *In* M. L. Wiley (ed.) Estuarine interactions. Academic Press, New York.

Rhoads, D. C., and D. K. Young. 1970. The influence of deposit feeding organisms on sediment stability and community trophic structure. J. Mar. Res. 28: 150–178.

Ritzrau, W., and G. Graf. 1992. Increase of microbial biomass in the benthic turbidity zone of Kiel Bight after resuspension by a storm event. Limnol. Oceanogr. 37: 1081–1086.

Salt, G. W. 1979. A comment on the use of the term emergent properties. Am. Nat. 113: 145–148.

Sanders, H. L. 1956. Oceanography of Long Island Sound 1952–1954. X. The biology of marine bottom communities. Bull. Bingham Oceanogr. Collection, Yale Univ. 15: 345–414.

1960. Benthic studies in Buzzards Bay. III. The structure of the soft-bottom community. Limnol. Oceanogr. 5: 138–153.

Sandford, E., D. Bermudez, M. D. Bertness, and S. D. Gaines. 1994. Flow, food supply and acorn barnacle population dynamics. Mar. Ecol. Prog. Ser. 104: 49–62.

Seed, R., and T. H. Suchanek. 1992. Population and community ecology of *Mytilus*, p. 87–169. *In* E. Gosling (ed.) The mussel *Mytilus*: ecology, physiology, genetics and culture. Elsevier, Amsterdam.
Shanks, A. L., and W. G. Wright. 1986. Adding teeth to wave action: the destructive effects of wave borne rocks on intertidal organisms. Oecologia 69: 420–428.
Smaal, A. C., J. H. G. Verhagen, J. Coosen, and H. A. Haas. 1986. Interaction between seston quantity and quality and benthic suspension feeders in the Oosterschelde, The Netherlands. Ophelia 26: 385–400.
Snelgrove, P. V. R., and C. A. Butman. 1994. Animal–sediment relationships revisited: cause versus effect. Ocean. Mar. Biol. Annu. Rev. 32: 111–177.
Sternberg, R. W. 1968. Friction factors in tidal channels with differing bed roughness. Mar. Geol. 6: 246–260.
Verhagen, J. H. G. 1986. Tidal motion, and the seston supply to the benthic macrofauna in the Oosterschelde. DHL Report R1310–14.
Wainright, S. C. 1987. Stimulation of heterotrophic microplankton production by resuspended marine sediments. Science 238: 1710–1712.
1990. Sediment-to-water fluxes of particulate material and microbes by resuspension and their contribution to the planktonic food web. Mar. Ecol. Prog. Ser. 62: 271–281.
Warwick, R. M., and R. J. Uncles. 1980. Distribution of benthic macrofauna associations in the Bristol Channel in relation to tidal stress. Mar. Ecol. Prog. Ser. 3: 97–103.
White, D. C., W. M. Davis, J. S. Nickels, J. D. King, and R. J. Bobbie. 1979. Determination of the sedimentary microbial biomass by extractible lipid phosphate. Oecologia 40: 51–62.
Wildish, D. J. 1977. Factors controlling marine and estuarine sublittoral macrofauna. Helgolander. Wiss. Meeresunters. 30: 445–454.
1980. Reproductive bionomics of two sublittoral amphipods in a Bay of Fundy estuary. Int. J. Invertebr. Rep. 2: 311–320.
1985. Geographical distribution of macrofauna on sublittoral sediments of continental shelves: a modified trophic ratio concept. J. Mar. Biol. Assoc. U. K. 65: 335–345.
Wildish, D. J., and D. D. Kristmanson. 1979. Tidal energy and sublittoral macrobenthic animals in estuaries. J. Fish. Res. Board Can. 36: 1197–1206.
1984. Importance to mussels of the benthic boundary layer. Can. J. Fish. Aquat. Sci. 41: 1618–1625.
1985. Control of suspension-feeding bivalve production by current speed. Helgolander. Wiss. Meeresunters. 39: 237–243.
1988. Growth response of giant scallops to periodicity of flow. Mar. Ecol. Prog. Ser. 42: 163–169.
1993. Hydrodynamic control of bivalve filter feeders: a conceptual view, p. 299–324. *In* R. F. Dame (ed.) Bivalve filter feeders in estuarine and coastal ecosystem processes. NATO ASI Ser. Vol. G33. Springer-Verlag, Berlin.
Wildish, D. J., and M. P. Miyares. 1990. Filtration rate of blue mussels as a function of flow velocity: preliminary experiments. J. Exp. Mar. Biol. Ecol. 142: 213–219.
Wildish, D. J., and D. Peer. 1983. Tidal current speed and production of benthic macrofauna in the lower Bay of Fundy. Can. J. Fish. Aquat. Sci. 40(Suppl. 1): 309–321.
Wildish, D. J., D. L. Peer, and D. A. Greenberg. 1986. Benthic macrofaunal

production in the Bay of Fundy and possible effects of a tidal power barrage at Economy Point-Cape Tenny. Can. J. Fish. Aquat. Sci. 43: 2410–2417.

Wildish, D. J., D. D. Kristmanson, R. L. Hoar, A. M. DeCoste, S. D. McCormick, and A. W. White. 1987. Giant scallop feeding and growth responses to flow. J. Exp. Mar. Biol. Ecol. 113: 207–220.

Wolff, W. J., F. Vegter, H. G. Mulder, and T. Meijs. 1976. The production of benthic animals in relation to the phytoplankton production: observations in the saline Lake Grevelingen, The Netherlands. Proc. 10th Eur. Symp. Mar. Biol. 2: 653–672.

Woodin, S. A. 1976. Adult-larval interactions in dense infaunal assemblages: patterns of abundance. J. Mar. Res. 34: 25–41.

Yeo, R., and M. J. Risk. 1979. Intertidal catastrophes: effects of storms and hurricanes on intertidal benthos of the Minas Basin, Bay of Fundy. Can. J. Fish. Aquat. Sci. 36: 667–669.

8
Ecosystems and flow

Our aim in this chapter is to examine whether water movement is fundamentally important in aquatic ecosystems to productivity and materials cycling in the ways briefly outlined in the next section. The testing of a suitable ecosystem level null hypothesis in which water movement is excluded is probably an unrealistic task. Yet, the idea of such a test should be kept in mind as we examine the real-world ecosystems which follow.

We note that the thrust of this chapter, ecosystems interacting with flow, is contradictory to the book's theme. This is because, in most of the book, we concentrate on suspension feeders and largely ignore the rest of the ecosystem of which they are a part. Yet, for ecosystem analysis we must be concerned with all of the biological and physical variables that are important in energy flow. Thus, for this chapter only, we have relaxed our definition to mean complete ecosystems including hydrodynamic effects on macroflora and macrofauna, inclusive of plant canopies, which are clearly not suspension feeders, but important primary producers. Our review of the contemporary benthic literature suggests that plant canopies and bivalve and coral reef ecosystems are best studied from the hydrodynamic perspective.

The examples we have chosen to include in this chapter are plant canopies, e.g. seaweed and seagrass ecosystems, and marine/estuarine environments with suspension feeding bivalve reefs. We have excluded coral reef ecosystems because at least two recent reviews (Sorokin 1993; Bakus 1994) have been published, the literature on this subject is vast, and the authors have no firsthand experience with corals.

Background

The aquatic ecosystem concept of materials cycling, measured as energy, carbon, or nutrient fluxes, was initiated by freshwater ecologists. Juday

(1940) described the annual energy budget of an inland lake in which the phytoplankton and attached aquatic plants utilized only about 1% of the available subsurface solar energy during photosynthesis. Such energy balances depend directly on the first law of thermodynamics – that energy cannot be created or destroyed. Lindeman (1942), also working in a lacustrine environment, showed how the food web complexity of an ecosystem could usefully be simplified by categorizing the constituent organisms into trophic–dynamic groups, hence, primary producers, herbivorous consumers, predators, and decomposers. Lindeman (1942) produced a diagram of a food cycle (web) for a lake in which the loss in respiratory energy at each trophic level increased with distance from the original solar energy supply. This conforms with the second law of thermodynamics because the transformation of energy at each trophic level, e.g. light to chemical energy in plant tissue, involves an energy loss.

Both Juday and Lindeman tacitly assumed that solar radiation was of most importance in freshwater ecosystems. Munk and Riley (1952) presented a quantitative theory that nutrient uptake by marine pelagic and attached plants, e.g. seaweeds and seagrasses, was influenced by flow in removing mass transfer limitation. Absorption of nutrients was considered to be dependent on the ambient concentration gradients and velocities. Thus, Munk and Riley (1952) showed how water movement, up to a threshold level, was an integral part of materials cycling. The physiological cast of their quantitative model, applicable to plant material exchanges and water movement, precludes its central use in this chapter, which is concerned with ecosystem level phenomena.

Odum (1971) and Odum, Finn, and Franz (1979) proposed that water movement was unimodally related to attached plant biomass or productivity (Fig. 8.1) in an ecosystem level extension of part of Munk and Riley's mass transfer limitation theory. Odum (1980) considered that increases in attached marine plant production were caused by increases in water movement which allowed the following:

- Increased removal of waste gases and greater supply of bicarbonate for photosynthesis
- Increased nutrient flux availability to plant absorptive surfaces
- Increased water column recycling of plant nutrients
- Increased sediment–water interface fluxes leading to greater plant nutrient cycling

The reference to this as a "tidal energy subsidy" by Odum (1980) is clearly in error since no energy which subsidizes or enhances the production process is directly supplied. Rather, the tidal energy supplied is

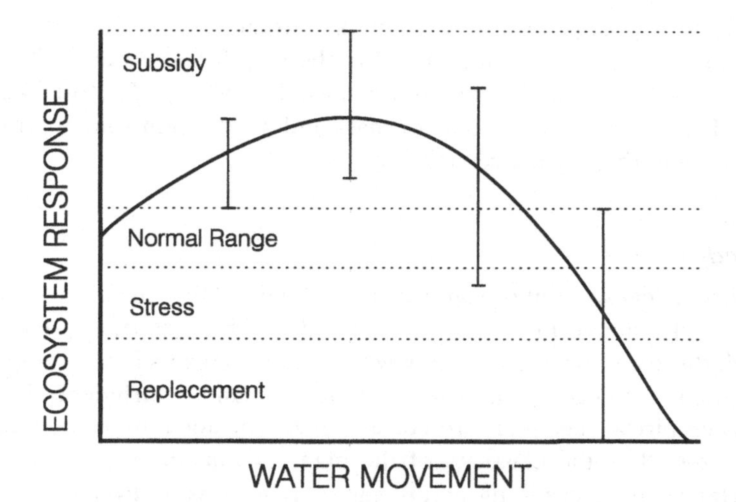

Figure 8.1 Theoretical ecosystem response to water movement perturbation (adapted from Odum et al. 1979).

involved in removal of mass transfer limitation, as proposed by Munk and Riley (1952), and it is this which allows production to proceed at a higher, up to an optimum, rate.

Odum et al. (1979) recognized in the perturbation theory that ecosystem level effects were a unimodal response of important environmental variables such as water motion. Ecosystem level stresses resulted if water movement reached certain levels and could result in physiological stress of the biota present, replacement of sensitive by hardier species, or death of the resident biota if the stresses were sufficiently large or persistent.

Marine plant canopies: kelp forests

Kelp forests form the characteristic nearshore vegetation of the hard substrates of most temperate and boreal latitudes. The dominant kelps of the world (Mann 1973) include *Laminaria* of the North Atlantic and eastern North Pacific and Okhotsk Sea; *Macrocystis* of the Pacific Ocean along South and North American shores, which may reach a lamina length of 40 m (Jackson and Winant 1983); and *Ecklonia* of the southern tip of Africa and southeast Australia. Species of *Macrocystis* are also present at the two latter localities, as well as on the rocky coasts of New Zealand.

Understorey kelps, where the lamina is usually only a few metres in length, may also be present and add to the kelp forest productivity. As pointed out by Eckman, Duggins, and Sewell (1989), understorey kelps may affect horizontal transport of mass and momentum somewhat differently than the larger overstorey kelps.

Hydrodynamics

The theoretical account of Raupach and Thom (1981) concerning turbulence in and above plant canopies provides a good starting point, although these authors were dealing with terrestrial plants in air. Raupach and Thom (1981) conclude that the subject of canopy turbulence is still largely empirical. The plant structures which protrude into the boundary layer, as well as the elasticity of the plants within the canopy, are of importance in determining aerodynamic as well as hydrodynamic responses associated with plant canopies.

The fluid dynamics of seawater flow through a bushy red alga, *Gelidium nudifrons* from Southern California, was studied by Anderson and Charters (1982). *Gelidium* is found on semi-exposed shores, and field measurements of orbital wave action of 2–10 $cm \cdot s^{-1}$ which passed through the branches of the plant emerged as turbulent flow. Laboratory studies in a low velocity water tunnel specially designed for this purpose used dyes and a hot film anemometer to characterise flow. *Gelidium* plants of 35 cm height and with branches of 0.05 cm thickness, when tested with unidirectional flows in the water tunnel, damped turbulent flows $<$6–12 $cm \cdot s^{-1}$, producing a laminar exit flow. At flows $>$6–12 $cm \cdot s^{-1}$, the flow leaving the plant was always turbulent.

A large kelp forest of 7- × 1-km area near San Diego, California, and a nearby reference site which was kelp-free were compared by current meter observations by Jackson and Winant (1983). The dominant kelp was *Macrocystis pyrifera*, with some specimens reaching 40 m in stipe length. These authors tested the null hypothesis that kelp plants had no effect on flow patterns within the kelp forest (number 1, Table 8.1). Vector measuring current meters were deployed at three depths in a central location of the kelp bed and similarly at the reference site. Records were obtained at both sites for a 7-d period and showed that the average longshore currents within the kelp forest were ~1 $cm \cdot s^{-1}$. This velocity was 43%–54% less than the root mean square velocity at the reference site, thus supporting H_1 in number 1, Table 8.1. Calculations from these data suggest that seawater residence times within the kelp

Table 8.1. *Kelp forest ecosystem hypotheses.*

No.	H_0/H_1	Description	Reference
1	H_0	Current velocity is independent of kelp presence/absence	Jackson and Winant (1983)
	H_1	Current velocity is inhibited by the presence of kelp	
2	H_0	Sedimentation rates within are similar to those outside the kelp ecosystem	Eckman et al. (1989)
	H_1	Sedimentation rates within are greater than those outside the kelp ecosystem	
3	H_0	Key predators have no effect in structuring kelp forest ecosystems	Mann (1973) Estes and Palmisano (1974)
	H_1	Key predators structure kelp forest ecosystems by consuming key herbivores such as algal grazing sea urchins	
4	H_0	Reduced velocities and increased sedimentation rates in kelp forests have no effect on suspension feeder production	Duggins et al. (1989) Eckman and Duggins (1991)
	H_1	Reduced velocities and increased sedimentation in kelp forests enhance suspension feeder production or growth	
5	H_0	Most POM and DOM derived from kelp forests are exported	Duggins and Eckman (1994)
	H_1	Most POM and DOM from kelp forests are utilized within the kelp forest by pelagic and benthic suspension feeders	

forest were >7 d. Measured velocities were always strongest at the kelp forest edge, but slowed to average values within 100 m into the kelp forest. In a further study of the same area, Jackson (1984) placed current meters with temperature recorders along a transect across the kelp forest. As a result of the high drag of the kelp plants, present at densities of 0.02–0.14 plants $\cdot$ m^{-2}, internal waves which passed through the forest were considerably dampened (Fig. 8.2), as evidenced by a smoothing of high frequency fluctuations.

Understorey kelps dominated by *Agarum fimbriatum* were studied on

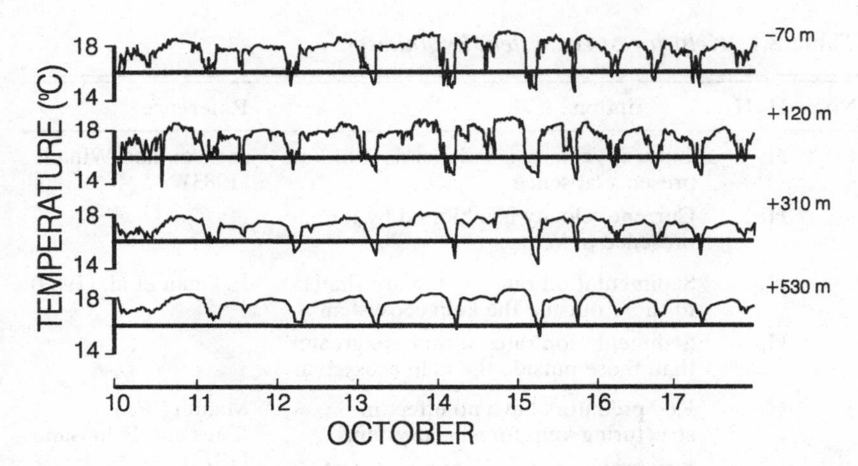

Figure 8.2 Continuous temperature records over a four site deployment at Point Loma kelp forest, near San Diego, California. The first site (−70 m) is 70 m outside, the next (+120 m) is 120 m into the kelp, and so on (Jackson 1984).

the San Juan Archipelago, at 48°N, 123°E, by Eckman et al. (1989). The relative degree of water movement activity, assessed by the gypsum flux method, showed that the kelp plants impeded flow and dampened variability in currents relative to that of a reference location nearby. As with the California data outlined previously, the effect was accompanied by a reduction in mass transport and shear within the kelp bed. Because of the reduced flow within the kelp, Eckman et al. (1989) reasoned that increased sedimentation should occur (Table 8.1). Sedimentation rates were measured by collection of the particles on acrylic plastic (Plexiglas) plates, followed by pumping into deflated plastic bags, filtering, and then weighing. Because this method did not properly distinguish sedimentation and resuspension events, a particle impaction experiment was also employed. It involved glass beads of 22 µm diameter and with a settling velocity of $0.0276\,\text{cm}\cdot\text{s}^{-1}$, which were slowly released upstream of microscope slides coated with vacuum grease and on which the beads became trapped. Results suggested to Eckman et al. (1989) that particle trapping was reduced in the presence of kelp plants, consistent with a reduction in the intensity of turbulent mixing within the canopy. The increased sedimentation within kelp canopies is explained by the greater retention times that particles must spend within the canopy and hence a greater proportion of particles which must settle. During flood–ebb tidal changes, settled particles may be resuspended and the resuspension–sedimentation cycle repeated many times.

These results suggest that flow may pass through and around individual or groups of macroalgal plants, depending on the ambient velocity. In passing through the bed, flow is impeded by the elastic lamina and stipe. A consequent reduction in mass transport and shear stress results in increased sedimentation within the macroalgal bed.

Materials cycling

A functional account of the *Laminaria*-dominated kelp forests of Atlantic Canada is given by Mann (1973). From this and related studies, a partial carbon budget can be constructed (Fig. 8.3). Such a budget represents the following:

- An estimate of the amount of energy passing through the ecosystem
- A simplification of a more complex cycle than is shown in Fig. 8.3
- Hypotheses about the pathways of energy, carbon, or minerals in the kelp ecosystem

Salient features of this budget are the very high primary production levels achieved by *Laminaria longicornis*, *L. digitata*, and *Agarum cribosum*, which are based on energetic water movement, principally wind–wave action, which enhances photosynthesis by supplying and removing gases, bicarbonate, and nutrients. The proportion of dissolved organic matter (DOM) utilized in the pelagic microbial loop was not estimated, although it was probably small. Both DOM and particulate organic matter (POM) are exported from the kelp forest. Estimates of net export of POM suggest that it is a large proportion of the total because the principal herbivores, the sea urchin *Strongylocentrotus droebachiensis* and periwinkle *Littorina littorea*, utilized <10% of the available kelp carbon. High density aggregations of urchins sometimes occurred and could be very destructive of the kelp plants by biting through the stipe bases, thus forming cleared patches referred to as "sea urchin barrens" (Mann 1973). Whether the feeding aggregations of *S. droebachiensis* are controlled by decapod key predators, as suggested by Mann (1973), or by some other mechanism, inclusive of innate urchin behavior (Hagen and Mann 1994), is still a matter of debate.

In field studies conducted on the Aleutian Archipelago, Alaska, Estes and Palmisano (1974) chose two study sites: Rat Island at 52°N, 178°E, and Near Island at 52°N, 174°E. The two sites were 400 km apart but

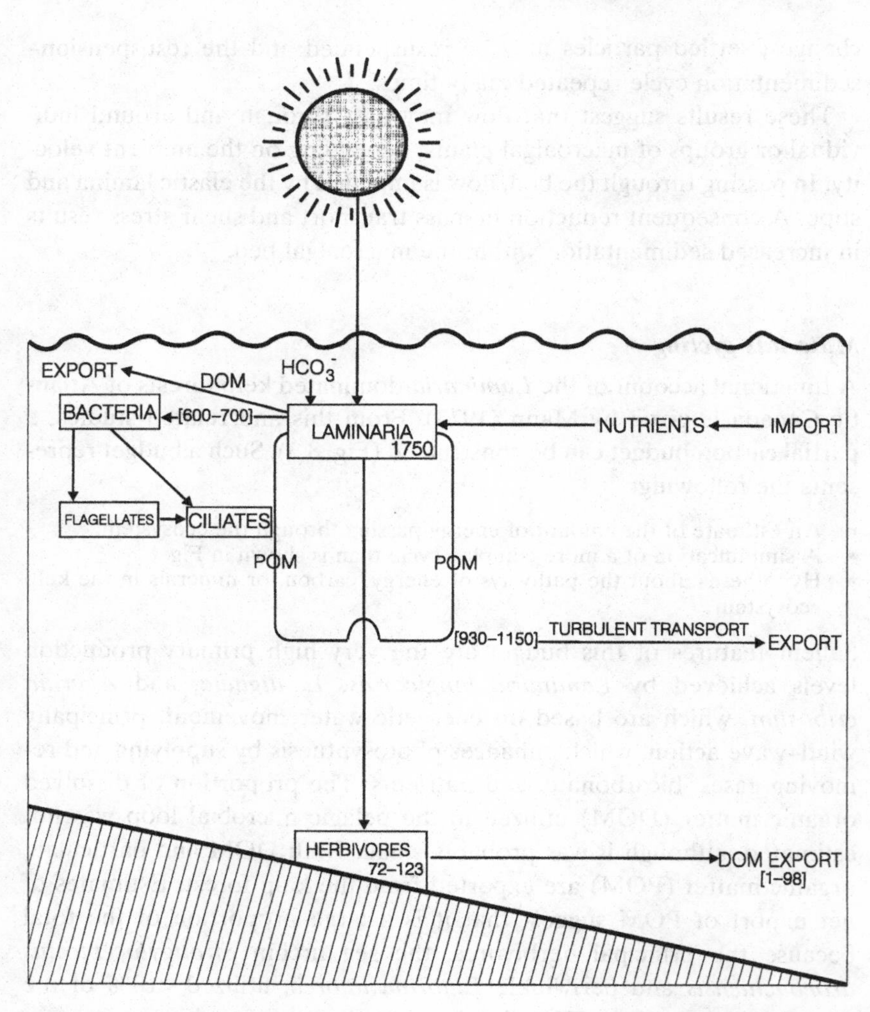

Figure 8.3 Partial carbon budget, $g\,C \cdot m^{-2} \cdot yr^{-1}$, for a *Laminaria*-dominated kelp forest in St. Margaret's Bay, Nova Scotia (based on Mann 1972, 1973; Miller and Mann 1973).

otherwise were similar in physical characteristics, differing only in that Rat Island had a rich and diverse littoral and sublittoral kelp forest dominated by *Laminaria* sp., whereas at Near Island kelps were virtually absent sublittorally and replaced by a sea urchin barrens dominated by *Strongylocentrotus* sp. Estes and Palmisano (1974) suggested that the marked difference between the kelp-free site at Near and kelp-rich site at Rat Island was due to the presence of a key predator, the sea otter *Enhydra lutris*, at the latter (Table 8.1, number 3). High densities – 20–30 sea otters $\cdot$ km^{-2} – removed the larger sea urchins, thereby preventing destructive damage by them to the kelp. Circumstantial field evidence gathered in support of this hypothesis was limited; it included the following: sea urchin densities were much higher at the kelp-free sites; at those sites where sea otters were present at high density, only small urchins, less than 35-mm test diameter, were found; and some differences in higher trophic level species between sites, e.g. the presence of harbour seals and bald eagles, were related to the higher primary production available at kelp-rich Rat Island.

Further study at these sites was undertaken by Duggins, Simenstad, and Estes (1989) to determine the importance of detrital pathways within coastal ecosystems (Table 8.1, number 4). Comparative field measurements of growth for two suspension feeders, *Balanus glandula* attached to test plates and *Mytilus edulis*, were made by locating marked individuals sublittorally within kelp beds at Rat Island and at the same depth in the kelp-free environment at Near Island. After 1 yr of growth, the rates were compared (Fig. 8.4); barnacle growth was five times and mussel growth two to four times greater at the kelp-rich site. Because the carbon ingested during feeding by these suspension feeders could have originated from sources other than kelp, e.g. from phytoplankton, it was necessary to use a method capable of tracing the feeding habits of suspension feeders. Because the $\delta\ ^{12}C:\ ^{13}C$ signatures of phytoplankton and kelp were so different, it was possible to indicate the source of the carbon on which they had been feeding. Of 11 species examined by this technique from the Alaskan sites, Duggins et al. (1989) found that 10 species had enriched levels of $\delta\ ^{13}C$ from Rat Island, suggesting that they had been feeding indirectly on kelp (on average 58% C from kelp). Only *Mytilus edulis* did not show enrichment in this way, suggesting that it may prefer a phytoplankton diet. By contrast, $\delta\ ^{13}C$ enrichment at the Near Island site was much lower (on average 32% C from kelp).

Both field growth experiments and measurements of carbon isotope ratios in naturally co-occurring benthic macrofauna of kelp beds led Duggins et al. (1989) to conclude that the kelp plants themselves are a significant carbon source for them. Because of the nature of the measurements, the results give no indication about how important this pathway is to suspension feeders. Nor do they allow an estimation of the relative importance of kelp carbon export from the kelp forest.

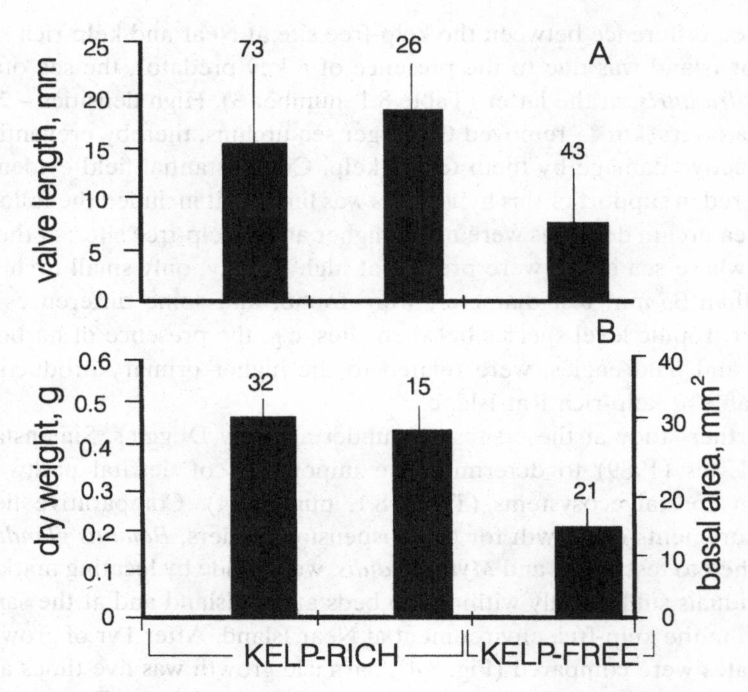

Figure 8.4 Growth experiments on the Aleutian archipelago, Alaska: A, *Mytilus edulis*; B, *Balanus glandula* (from Duggins et al. 1989). Two kelp-rich sites from Rat Island, one kelp-free site from Near Island. Number above each bar, sample size.

For the understorey kelp ecosystem on the Washington coast described from the physical point of view earlier Eckman and Duggins (1991) conducted a field manipulative experiment to test hypothesis number 4 (Table 8.1) further. Four sites, which differed in mean free-stream velocity from 16 to 24 $cm \cdot s^{-1}$ and, therefore, in sedimentation rate, were chosen. Kelp and kelp-free treatments were created and maintained by translocating or removing kelp plants in plots of 10–20 m^2. Field growth experiments with a range of typical kelp suspension feeders, attached to either upward- or downward-oriented test plates bolted to the substrate, established that sedimentation within the kelp bed was inhibiting for growth. Despite this, upwardly oriented plates with suspension feeders such as *Mytilus* sp., *Balanus glandula*, and a serpulid polychaete, *Pseudochitmopoma occidentalis* within the kelp plots all grew significantly faster than those in adjacent kelp-free plots. For two additional species, a bryozoan and a sponge, no differences in growth in kelp and kelp-free plots were evident because other limiting factors were involved. The results are at least partly consistent with H_1 in hypothesis 4, Table 8.1.

In a further field experiment on the Washington coast, Duggins and Eckman (1994) attempted to de-couple velocity and sedimentation effects (hypothesis 4, Table 8.1) from those due to in situ production of DOM and POM from kelps. These experiments involved creating real and plastic model kelp understorey in 10-m^2 plots and testing whether suspension feeders benefitted trophically from the autochthonously produced POM and DOM (hypothesis 5, Table 8.1). Of the six species of suspension feeders tested – *Mytilus trossulus*, *Chlamys hastata*, *Balanus glandula*, *Pseudochitmopoma occidentalis*, *Chelyosoma productum*, and spirorbid polychaetes – only one, *B. glandula*, grew significantly faster in the artificial kelp plot. These results support the null hypothesis (number 5, Table 8.1) and are clearly different from the results obtained by Duggins et al. (1989) in the Aleutian Islands, suggesting that responses are governed by multiple limiting factors and rendering field experimentation difficult to interpret.

By comparison with the closed steady-state model of a *Laminaria*-dominated kelp forest in a small bay on the northwest Atlantic coast (Fig. 8.3), an ecosystem simulation model of a kelp forest dominated by *Ecklonia maxima* and *Laminaria pallida* on the west coast of South Africa has been described (Wulff and Field 1983; Wickens and Field 1986). The model simulations suggested that local water movements of the Benguela current were integrally involved in materials cycling of the kelp ecosystem. Thus, down-welling was brought about by relaxation of offshore prevailing southeast winds or by strong onshore winds, and up-welling caused by strong prevailing southeast winds. During the southern summer (September–April), the wind patterns favor up-welling; during winter (May–August), down-welling (Wulff and Field 1983). Under up-welling conditions, the major consumers' trophic requirements, including the mussel *Aulacomya ater*, are not met, as kelp POM and faeces are also being exported from the kelp forest. During down-welling, offshore phytoplankton is continuously being advected through the kelp and provides a rich source of food for suspension feeders which are part of the kelp ecosystem. If a closed system model is used for the South African kelp ecosystem, then a large portion of the nitrogen flow is depicted to be present in bacteria associated with kelp POM and faeces (Wickens and Field 1986). Modelling this system as an open one in which local water movements are included shows that bacteria have insufficient time to accumulate before being advected from the kelp forest.

Kelp productivity

Overstorey kelps are present on rocky substrates of the nearshore coastal environment to a maximum depth of ~40 m. Hence, they form a narrow zone close to the shore and are light-limited at greater depths. Primary production estimates made here suggest very high levels, e.g. 1750 g $C \cdot m^{-2} \cdot yr^{-1}$, for a *Laminaria*-dominated forest on the Atlantic coast of Canada (Mann 1973). A West Coast kelp forest dominated by the sea palm *Postelsia palmaeformis* is reported (Leigh et al. 1987) to have an even higher productivity, 3000 g $C \cdot m^{-2} \cdot yr^{-1}$, on the most wave-exposed parts of the coast of Washington.

A cursory review of the data concerning kelp productivity in relation to wind–wave exposure suggests apparently conflicting results. Thus, *Laminaria* kelp forests were less productive at the most exposed sites on the Atlantic coast of Canada studied by Gerard and Mann (1979), and *Macrocystis* sp. beds grew best at intermediate levels of wave exposure on the southern coast of Chile (Dayton 1985). Yet, on the coast of Washington, *Postelsia* sp., referred to earlier, as well as the shrubby kelps, *Lessoniopsis littoralis*, are found only on the most wave-exposed locations (Leigh et al. 1987). Unfortunately, although Leigh et al. (1987) determined kelp productivity at a number of different sites on the Washington coast, the ranges of productivity were lumped and not related individually to an estimate of wave power, thus rendering the results unconvincing.

An obvious difficulty in comparing the work of different authors on kelp productivity is the lack of a commonly used absolute measure of wave exposure. This problem was considered in Chapter 2 and the gypsum dissolution method proposed is an inexpensive way of comparatively measuring wind–wave activity.

Despite the lack of a wave exposure measure for the data cited, we believe that the gross differences in exposure suggested are real and result in the responses by kelp forests suggested. If this is so, the key to understanding the conflicting results lies in the probability that kelps are adapted to different amounts of wave exposure, both inter- and intraspecifically. Indeed, the form of a sheltered versus an exposed frond of *Laminaria longicornis* (Fig. 8.5) bears this out. Further examples of intraspecific lamina form changes, e.g. within *Macrocystis* sp., which are

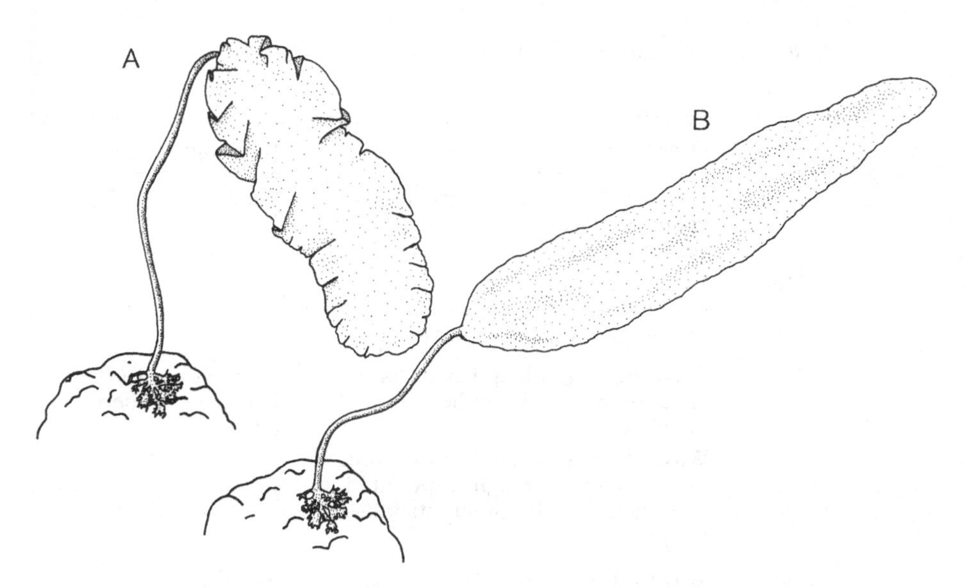

Figure 8.5 Growth forms of *Laminaria longicornis* from the Canadian Atlantic coast: A, sheltered site; B, exposed site (Gerard and Mann 1979).

dictated by the hydrodynamic conditions experienced, are provided by Wheeler (1980), and within *Nerocystis* by Koehl and Alberte (1988). It is reasonable also to expect that different kelp species will have different wave exposure optima.

Finally, we consider possible ways that wave energy may influence kelp productivity. Two major kinds of mechanisms can be identified: *physical* effects and interacting *physical–biological* effects. They are formulated as hypotheses in Table 8.2 with numbers 1–3 representing physical and numbers 4–5 representing physical–biological effects.

As has already been pointed out, increased diffusion of materials across plant surfaces results when water movement is increased up to an asymptotic point. This idea was introduced in marine ecology by Munk and Riley (1952). Although there is available evidence in both the field and the laboratory that undirectional free-stream velocities increase this flux and enhance photosynthesis, we know of no studies where simulated wave action has been tested for the same effect. An example of a good laboratory study with unidirectional flows and the kelp *Macrocystis pyrifera* is provided by Wheeler (1980), who showed that velocities up to $4\,cm \cdot s^{-1}$ increase the photosynthetic rate of the lamina.

Table 8.2. *Hypotheses regarding wind–wave effects on kelp productivity.*

No.	H_0/H_1	Description	Reference
1	H_0	Diffusion of gases or nutrients across lamina surfaces is unrelated to wave activity	Leigh et al. (1987)
	H_1	Enhanced diffusion of gases or nutrients across lamina surfaces results from increased wave activity	
2	H_0	Wave stirring of kelp fronds has no effect on net photosynthetic rate of the plant	Leigh et al. (1987) Koehl and Alberte (1988)
	H_1	Wave stirring of kelp fronds causes more fronds to be light exposed, thus increasing net photosynthetic rate of the plant	
3	H_0	Wave exposure does not produce shearing stresses to remove all, or part, of the kelp plant	Koehl (1986)
	H_1	Wave exposure produces shearing stresses which can remove all, or part, of the kelp plant, resulting in its death	
4	H_0	Strong wave action has no effect on the kelp–herbivore interaction	Connell (1972) Menge and Sutherland (1976)
	H_1	Strong wave action protects kelps from herbivory	
5	H_0	Strong wave action has no effect on kelps and their competitors	Dayton (1971) Dayton (1975) Paine (1979)
	H_1	Strong wave action allows kelps to outcompete other species present	

For the other hypotheses shown in Table 8.2 (numbers 2–5), there is some evidence for the alternative hypotheses, as described in the listed references.

Marine plant canopies: seagrass meadows

Seagrass meadows represent the common attached macrophytic ecosystem of nearshore soft sediments at temperate and boreal latitudes (Den

Hartog 1970). Over 50 species of seagrass are known; the following genera are common: *Zostera*, *Spartina*, *Halodule*, *Thalassia*, *Halophilia*, *Posidonia*, and *Enhalus*. Seagrasses may be present intertidally and/or subtidally, and each plant consists of a shoot and leaves with an underground root and rhizomes which serve to stabilize both plant and surrounding soft sediment.

Primary production by seagrass ecosystems is generally high, although somewhat less than that of overstorey kelps. Thus, for *Zostera marina* ecosystems near the low tide mark on Atlantic shores of North America, the maximum production was ~1500 $g\,C \cdot m^{-2} \cdot yr^{-1}$, and for a *Spartina* salt marsh in the same area near high water mark, the maximum production found was ~897 $g\,C \cdot m^{-2} \cdot yr^{-1}$ (Mann 1973).

Hydrodynamics

Laboratory studies in a recirculating flume 6 m long by 0.15 m wide by Fonesca et al. (1982) were designed to determine the effect of *Zostera marina* shoot density on flows within the meadow. This work was criticized by Gambi, Nowell, and Jumars (1990) because the flume did not simulate dynamically similar conditions to those present in the field, in particular, the flume was too narrow, resulting in flow blockages when seagrass shoots were present.

A larger annular flume with an 8-m-long working section, 2 m wide, was used by Gambi et al. (1990). This flume had a water depth of 25 cm and a 4-cm-deep sand layer on the flume floor into which young *Z. marina* plants were inserted. The eelgrass plants used were ~16 cm in height so that flow could form above the bed. The experimental eelgrass patch (15 by 100 cm long) was placed where the flume boundary layer was fully developed and so that the plants were 30 cm from the walls. Flow was deflected around and over the patch so that there was a sharp interface near the shoot tips where high shear and turbulence intensity occurred. Within the eelgrass shoots, velocity was reduced as a function of shoot density and distance into the meadow (Table 8.3). In these experiments, the penetration of flow effects into the eelgrass bed is more marked than the density effect.

Field studies by Ackerman (1983), Fonesca et al. (1983), and Eckman (1987) also suggest that reduced velocities occur within natural *Zostera* meadows. Calculations by Eckman (1987) to determine the flux rates through a high density (604–1134 $plants \cdot m^{-2}$) and a low density (183–271 $plants \cdot m^{-2}$) eelgrass meadow showed that at slower free-stream veloci-

Table 8.3. *Percentages of water flux reduction at different freestream velocities, shoot densities, and sampling positions in flume experiments with* Zostera marina.

Density (shoots · m^{-2})	Downstream sampling positions			
	25 cm	50 cm	75 cm	100 cm
Freestream velocity 5 cm · s^{-1}				
1200	20.7	38.3	43.1	35.2
1000	11.0	31.4	26.7	40.6
800	7.55	19.9	21.4	28.9
600	3.58	12.3	13.7	20.3
400	9.6	18.7	29.7	21.8
Freestream velocity 10 cm · s^{-1}				
1200	11.5	30.0	30.7	31.5
1000	5.6	20.6	30.9	31.5
800	12.3	23.0	31.4	32.6
600	0.9	9.8	19.5	20.5
400	8.0	19.4	27.6	22.1
Freestream velocity 20 cm · s^{-1}				
1200	8.9	16.6	19.9	34.2
1000	10.4	19.6	23.16	21.0
800	2.2	13.4	21.0	18.3
600	5.2	13.9	21.7	19.8
400	2.0	8.3	15.2	14.7

Source: Gambi et al. (1990).

ties, there was a 16% reduction in flux due to the higher density. At a higher velocity, difference was increased to 52%. The field results do not agree very well with the flume results shown in Table 8.3, perhaps because the latter are artifacts of the small bed size tested.

Field observations with a dual axis electromagnetic current meter suggested to Ackerman and Okubo (1993) that the plants themselves were generating mechanical turbulence. The frequency generated by the eelgrass plants of 6.4–8.0 s^{-1} results because they are hydroelastically translating fluid energy.

The ecosystem implications of seawater flows to and in seagrass meadows include larval supply (Chap. 3), effects on seagrass production, nutrient exchange between sediment and seawater inclusive of resuspension and deposition, direct shear stress on seagrass plant stability, and flux reduction within seagrass which results in increased sedimentation and limits suspension feeder production. Some of these are considered in the following sections, although additional flow-related phenomena – effects on diffusion of gases across leaf surfaces, seagrass pollination, and increased seawater turbidity, which results in reduced light and therefore less photosynthesis – are not because they are considered more appropriate to discussion of lower hierarchical levels of biology than ecosystem level.

Seagrass productivity

The effect of tidal flows on seagrass productivity (Table 8.4, number 2) was investigated, apparently independently, by Conover (1968) on eelgrass meadows of the U.S. East Coast and by Odum (1961, 1971) and Steever, Warren, and Niering (1976) on cordgrass meadows on the same coast.

In Conover's study, eelgrass standing crops of various locations are presented as a function of the mean hourly tidal velocity averaged over a complete springs to neaps period (Conover 1968). Where the current meter was deployed in this study was not stated; presumably free-stream velocities seaward of the eelgrass meadow are meant. The data for Charlestown Pond, Rhode Island, are shown in Fig. 8.6. Standing crop data in Fig. 8.6 are probably a good indication of eelgrass primary productivity because herbivores are unimportant in this ecosystem. A characteristic unimodal response to velocity is seen. Conover (1968) suggested that the positive relationship could be explained by an increased photosynthetic activity as flows increased, and that the inverse relationship was due to an inhibition of metabolism at high flows and/or the shearing effect of high flow, which prevented the plants from becoming established. Further results from Texas and from Massachusetts and Rhode Island along the U.S. East Coast for both seagrasses and seaweeds (Fig. 8.7) show a similar trend. The data confirm that individual species of canopy-forming macroalgae have distinctively different responses to tidal flows, confirming the suggestion made previously for kelp.

Steever et al. (1976) compared the standing crops of *Spartina alterniflora* at various marshes along the Connecticut, United States, coast. For 10 such marshes in 1971, the standing crop dry weight of cordgrass was linearly related to mean tidal height (how this was estimated was not stated) with a calculated slope equal to 544 g dry

Table 8.4. *Ecosystem hypotheses referable to seagrass meadows.*

No.	H_0/H_1	Description	Reference
1	H_0	Velocity in seagrass meadows is unaffected by the presence of seagrass shoots	Fonesca et al. (1982) Eckman (1987) Gambi et al. (1990)
	H_1	Velocity within a seagrass meadow is dependent on the seagrass shoot density	
2	H_0	Primary production within seagrass meadows is unaffected by tidal factors	Odum (1961, 1971) Conover (1968)
	H_1	Primary production of seagrass meadows is a unimodal function of tidal factors	
3	H_0	Intermediate pools of sedimentary nitrogen are unaffected by seagrass cover	Kenworthy (1982)
	H_1	Intermediate pools of sedimentary nitrogen are enhanced in seagrass by sediment trapping	
4	H_0	Productivity of suspension feeders within seagrass meadows is independent of shoot density related fluxes	Kerswill (1949) Peterson et al. (1984) Eckman (1987)
	H_1	Productivity of suspension feeders within seagrass meadows is controlled by shoot density related fluxes	
5	H_0	POM and DOM arising from the seagrass meadow have no effect on resident suspension feeders	Lin (1989)
	H_1	POM and DOM arising from the seagrass meadow provide additional food, which enhances suspension feeder productivity	
6	H_0	Energetic cost of maintaining suspension feeder shore position is unrelated to local velocities	Myers (1977) Peterson and Beal (1989)
	H_1	Energetic cost of maintaining suspension feeder shore position is directly related to local velocities	Irlandi and Peterson (1991)

weight $\cdot$ m^{-2} for every increase of 1 m in mean tidal range and $r = 0.96$. Similar data for other years and sites suggest a similar trend but with increased scatter, consistent with involvement of other factors besides tidal height. Although not observed in the data reported by Steever et al.

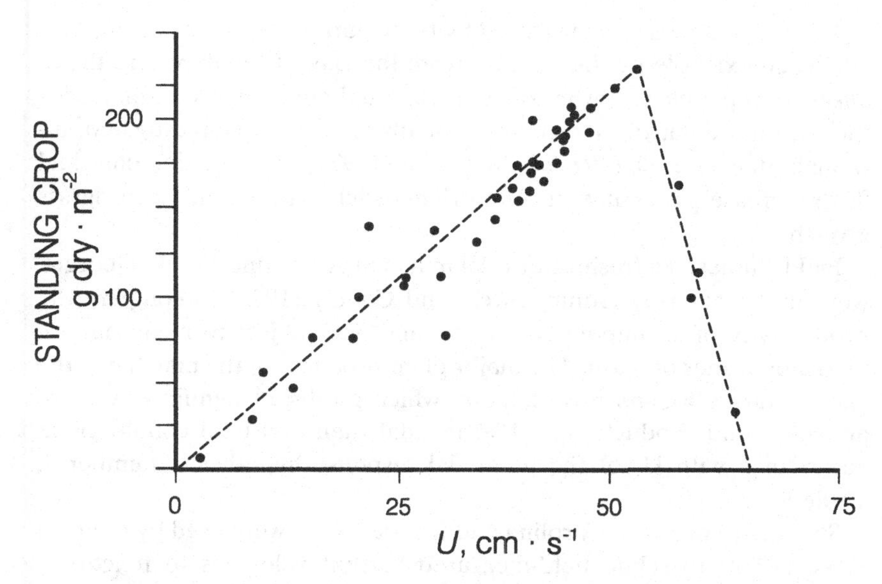

Figure 8.6 Standing crop of *Zostera marina* in Charlestown Pond, Rhode Island, 1962–63, as a function of velocity (average lunar period velocity expressed, cm · s^{-1}) at different sites (Conover 1968).

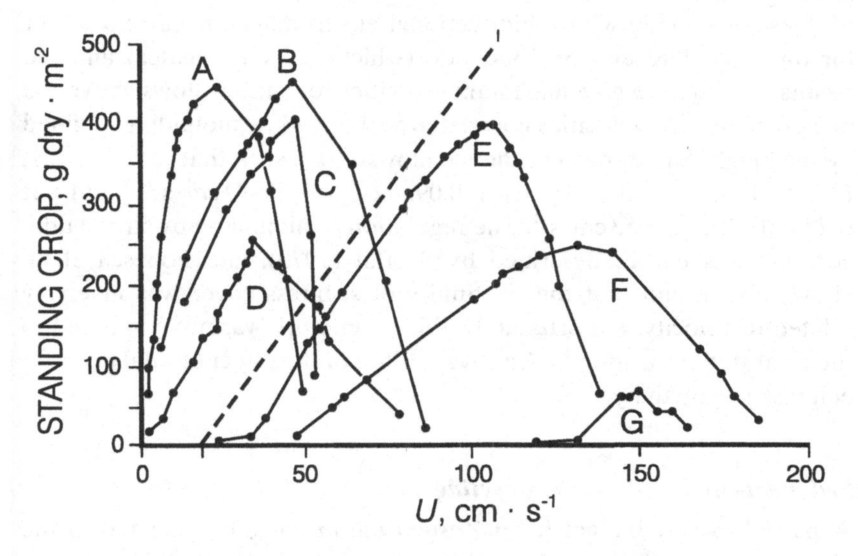

Figure 8.7 Standing crop data for seaweeds and seagrasses for various locations on the U.S. East Coast where summer temperatures are ~23°–28°C and salinities are ~29–31 o/oo; dashed line divides lagoon (to left) from channel and coastal ecosystems: A, *Ruppia maritima*; B, *Thalassia testudinum*; C, *Stylophora rhizoides*; D, *Zostera marina*; E, *Chondrus crispus*; F, *Laminaria agardhii*; G, *Bryopsis plumosa* (Conover 1968).

(1976), the response is considered by these authors to be unimodal; they cite as an example published data from the Bay of Fundy where the *S. alterniflora* plants are dwarfed and mean tidal ranges exceed 8 m. As for the eelgrass data, the mechanisms involved were not investigated, although Steever et al. (1976) follow Odum (1971) in suggesting that tidal flows increase the availability of nutrients, such as nitrogen, for cordgrass growth.

Field studies in a freshwater tidal marsh in Augustine Creek, Georgia, were undertaken by Odum, Birch, and Cooley (1983) to compare the productivity of an impounded marsh and one subject to mean natural tidal amplitudes of 0.5 m. The major plant producer in the marsh was the giant cutgrass *Zizaniopsis miliacea*, which produced significantly more above-ground production, ~31%, in tidal than non-tidal conditions in accordance with H_1 of the unimodal response hypothesis (number 2, Table 8.4).

Study sites in North Carolina and Rhode Island were used by Fonesca et al. (1982) to relate field-measured current velocities to functional aspects of eelgrass ecosystems. One of the study locations was at Charlestown Pond, worked earlier by Conover (1968), although this reference is absent from the reference list of Fonesca et al. Current profiles were made with a bidirectional electromagnetic current meter for one complete ebb or flood tide (whichever was greater) and the results corrected to give maximum velocities for surface flows above the meadow. Surface velocities correlated with meadow mounding, defined by the height/length ratio of the meadows, $h \cdot l^{-1}$, such that at $h \cdot l^{-1} > 0.1$, $U_{max} \geq 94\,cm \cdot s^{-1}$, at $h \cdot l^{-1} = 0.1$–0.09, $U_{max} \simeq 53$–$94\,cm \cdot s^{-1}$, and at $h \cdot l^{-1} < 0.09$, $U_{max} < 53\,cm \cdot s^{-1}$. The regulation of mounding by strong tidal currents was earlier described by Scoffin (1970), and Fonesca et al. (1982) also found that the continuity of seagrass cover was inversely related to velocity. Unfortunately, this latter work was not compared to the clear demonstration by Conover (1968) of the effect of tidal flows on eelgrass productivity.

Sedimentation and nutrient cycling

A partial energy budget for a *Zostera marina* meadow located in the Newport River estuary, North Carolina, was presented by Thayer, Adams, and LaCroix (1975). We have adapted this, using carbon equivalents, in constructing Fig. 8.8. Three sources of primary production are available in this particular eelgrass meadow – from eelgrass itself 62%

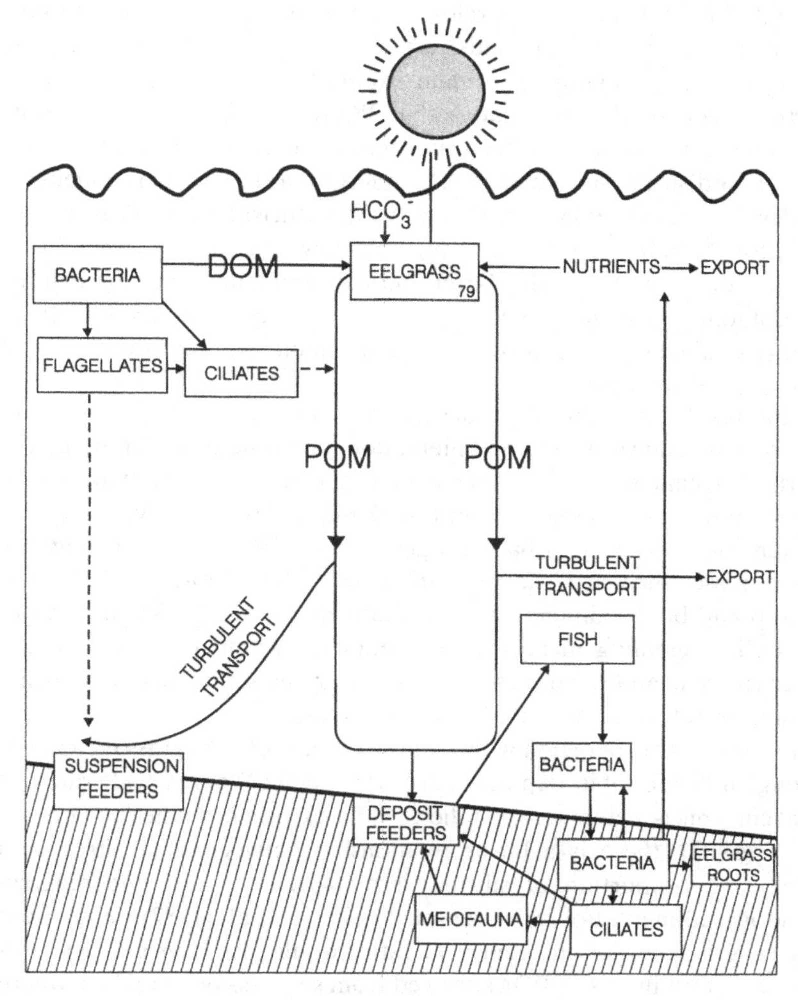

Figure 8.8 Partial carbon budget, $gC \cdot m^{-2} \cdot yr^{-1}$, for a North Carolina estuary *Zostera marina* seagrass meadow (Thayer et al. 1975).

of total primary production, equivalent to $1545 kCal \cdot m^{-2} \cdot yr^{-1}$; from phytoplankton ~30%; and from benthic microalgae ~8%. The two latter minor primary producers are omitted from Fig. 8.8 in order to simplify the diagram.

In contrast to the kelp forest budget of Fig. 8.3, the carbon of the eelgrass meadow passes through a primarily sedimentary detritus cycle and there are no important herbivores of the seagrass plants. Microbial degradation of dead seagrass releases nutrients, e.g. nitrogen and phosphorous containing compounds, which are available for recycling by growing seagrass plants. The main export from the meadow may be partially degraded plant material in POM. In Fig. 8.8, the benthic epifauna and infauna have been shown as their functional groups, hence deposit and suspension feeders. The question as to whether suspension feeders consume POM and its attached microflora, and/or phytoplankton largely advected into the meadow, was not determined by Thayer et al. (1975). The latter authors estimate that the benthic macrofauna – suspension and deposit feeders – consume an amount of energy equivalent to 55% of the total available primary production of the eelgrass meadow.

The result of reduced velocities and turbulence intensity within seagrass meadows is that sedimentation rates increase. Thus, in the North Carolina *Zostera marina* meadow studied by Thayer et al. (1975), sedimentation rates ranging from 1 to 57 $mm \cdot yr^{-1}$ occurred with a carbon content on a dry weight basis ranging from 0.7% to 5.6%. The ability of seagrasses such as *Thalassia testudinum* in Bimini Lagoon, Bahamas, to trap and bind sediments was qualitatively studied by Scoffin (1970). As well as creating increased sedimentation within the meadow, the seagrass roots and rhizome serve to stabilize the sediments by reducing erosion events due particularly to wave action.

The silt–clay proportion within the sediments of a seagrass meadow is a rough indicator of its depositional environment. Where a wide range of tidal currents is available, as on the U.S. East Coast in studies by Fonesca et al. (1982), there is a linear relationship between shear stress and decreasing proportion of silt–clay size particles within the seagrass meadow sediment. For these reasons, Fonesca et al. (1982) pointed out that seagrass beds of different $h \cdot l^{-1}$ ratios would also have characteristically different inputs of POM derived from seagrass plants. Thus, where $h \cdot l^{-1} > 0.1$ and significant mounding was present, the discrete patches of seagrass would be sources of POM, whereas where $h \cdot l^{-1} < 0.09$ and a greater cover of seagrass was present, such areas would be greater sinks of seagrass-derived POM.

Nitrogen cycling at Back Sound, North Carolina, was studied by Kenworthy, Zeiman, and Thayer (1982), who conclusively showed that the seagrass meadow was significantly enriched by comparison to a reference location where seagrasses were absent (Table 8.4, number 3). Both *Zostera marina* and *Halodule wrightii* were present and resulted in elevated levels of silt–clay, organic matter, exchangeable ammonium, ammonium dissolved in pore water, and total nitrogen within the meadow. The enrichment effect was strongest at mid-meadow, intermediate at the edges or in isolated, mounded patches, and small or absent in unvegetated reference locations. Although the major mechanism involved in this enrichment of the seagrass meadow involves the hydrodynamic–seagrass meadow interaction mentioned earlier, another mechanism was recognized by Jordan and Valiela (1982) and Bertness (1984). Ribbed mussels *Geukensia demissus* present within a *Spartina alterniflora* meadow in New England caused a similar enrichment effect. During feeding, the mussels capture seston advected into the bed, as well as seagrass-derived POM, and produce faeces and pseudofaeces which are deposited within the meadow. After aerobic microbial breakdown, such materials release plant nutrients, e.g. nitrogen and phosphorus compounds, which can be used by the seagrass plants during photosynthesis. It has been confirmed experimentally that blue mussels *Mytilus edulis* also increase *Zostera marina* growth in this way in the western Baltic (Reusch, Chapman, and Gröger 1994).

Nitrogen cycling within a tropical seagrass meadow in Jamaica dominated by *Halodule beaudetti* was studied by Blackburn, Nedwell, and Wiebe (1994), who found active nitrogen and carbon cycling within the root–sediment environment. Nitrogen-fixing bacteria associated with the roots, as confirmed by Shieh, Simidu, and Maruyama (1989) for *Z. marina* roots, may be a major pathway, although direct uptake from seawater and after POM decay within sediments may be an alternate pathway. Blackburn et al. (1994) found an ammonia accumulation within the sediment which occurred overnight and which disappeared during the day as a result of active plant uptake. Measured rates of nitrogen fixation in the tropical sediment were lower than denitrification rates, rendering a conclusion about the source of the ammonia pool as problematic.

Nutrient cycling studies clearly require a detailed knowledge of the mechanisms of transport of nutrients as both imports and exports of the seagrass ecosystem. Such studies include that by Valiela et al. (1978) in Great Sippewissett marsh, Massachusetts. This marsh was dominated by *Spartina alterniflora* and *S. patens*. Dissolved plant nutrients could enter the seagrass ecosystem via ground water or rain or during tidal flooding. During tidal ebbing, the seawater may transport nutrients away from the meadow. Seasonal observations and calculations suggest that overall net

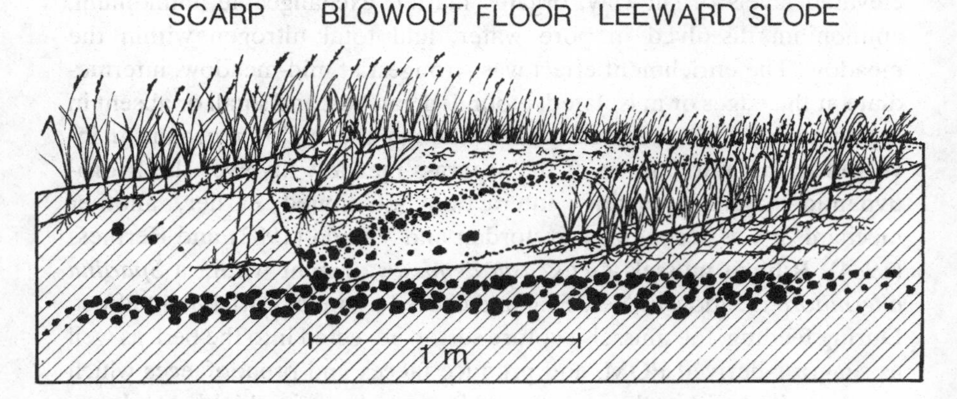

Figure 8.9 Diagrammatic cross section of a crescent-shaped "blowout" in a *Thalassia testudinum* meadow at Bath, Barbados (from Patriquin 1975).

flux into or out of the seagrass meadow is complex, but generally is a net importer during the summer when the *Spartina* plants are growing and a net exporter during winter. Wolaver et al. (1983), working on another *S. alterniflora* meadow in Virginia, United States, also found pronounced seasonal differences in fluxes to and from this ecosystem. These studies were aided by deploying a field flume (see Chap. 2) to aid in sampling tidal fluxes to and from the seagrass meadow.

The extensive body of work on nutrient fluxes to and from seagrass meadows has been reviewed by Hemminga, Harrison, and van Leut (1991). The inputs include nitrogen fixation, sedimentation, and direct uptake from seawater advected to the marsh. The losses include diffusion and tidal transport away from the meadow, denitrification, export of POM, loss due to herbivory, and possible transport away from the meadow, as well as direct exudation of nutrients from seagrass roots and leaves.

Seagrass stability

Qualitative observations regarding the stability of seagrass meadows have been made by Scoffin (1970), Patriquin (1975), Orth (1977), Fonesca and Kenworthy (1987), and Fonesca et al. (1983). In general, the "blowouts" – crescent-shaped areas of erosion (Fig. 8.9) where water movement has torn away the seagrass plants – are associated with isolated patches or mounds of seagrass on the seaward face of the meadow.

As classified by Fonesca et al. (1983), these areas are also associated with the highest water movement energies.

The blowouts are usually associated with extreme wave energy, as may occur during storms and may result in uprooting and formation of large drifting mats of seagrasses in the Gulf of Mexico and the Caribbean (Patriquin 1975). Scoffin (1970) described blowouts in Bimini Lagoon which were caused by locally high tidal velocities. Preliminary observations made by Scoffin (1970) in an underwater flume (Chap. 2) suggest that the shear forces associated with unidirectional flows result in erosion of sediment around the roots of *Thalassia testudinum* in the free-stream velocity range of 50–150 $cm \cdot s^{-1}$.

Suspension feeder production

Hydrodynamic baffling by the shoots within a seagrass meadow could result in limits to suspension feeding and hence growth and survival of suspension feeders present there. This results because the flux of seston passing through the seagrass meadow is reduced by comparison with unvegetated adjacent areas and hence reduces the availability of seston as food (Table 8.4, number 4). Some field tests of hypothesis 4 are presented in Table 8.5. Of three independent tests involving the hard clam, one (Kerswill 1949) supports H_1 and two (Peterson, Summerson, and Duncan 1984; Irlandi and Peterson 1991) support H_0. In *Argopecten irradians*, and for 1 year of the two studied, H_1 is supported, while the null hypothesis is supported in *Anomia simplex*. Thus, of five independent field experiments, only two support the alternative hypothesis that a reduced seston flux in the seagrass meadow results in reduced feeding and production.

Clearly the results of Table 8.5 indicate a more complex situation than suggested by the simple flux hypothesis tested. The low bivalve densities and short path lengths used (Table 8.5) preclude a population level seston depletion effect (Chap. 7) causing a downstream reduction in feeding/growth except, perhaps, in the experiment of Kerswill (1949). Thus, if feeding effects *are* involved in the results on bivalve growth rate, it is individual physiological effects which can be expected, and thus velocity, seston concentration, or seston quality (Chap. 4) should be the controlling variable. Unfortunately, little is known about the feeding and growth response of *M. mercenaria* to these variables.

Other possibilities which might explain field experimental results to test the intensity of turbulent mixing on bivalve growth, both within and

Table 8.5. *Field experiments to compare the effect of lower turbulent mixing intensities on suspension-feeding bivalve growth within seagrass meadows compared to an unvegetated nearby reference location.*

Seagrass meadow	Bivalve	Tidal flow ($cm \cdot s^{-1}$)		Bivalve density ($no. \cdot m^{-2}$)	Bivalve path length (m)	Bivalve growth rate		References
		U_{max}	U_{av}			Seagrass	Reference	
"Eelgrass"	*Mercenaria mercenaria*	>20	8	69–101	0.6–1.2	Low	High	Kerswill (1949)
Zostera marina and *Halodule wrightii*	*M. mercenaria*	7	0.7–1.3	9	—	High	Low	Peterson et al. (1984)
Z. marina and *H. wrightii*	*M. mercenaria*	27	0.3	49	1.0	High	Zero	Irlandi and Peterson (1991)
Z. marina	*Argopecten irradians*	13–98	?	22	—	Low	High	Eckman (1987)
Z. marina	*Anomia simplex*	13–98	?	35	—	Same	Same	Eckman (1987)

outside a seagrass meadow (Table 8.5), include the food subsidy hypothesis of Table 8.4 (number 5) within seagrass meadows and the physiological cost for bivalve feeders to maintain their position on the shore, which is proportional to local velocities (number 6). Field experiments undertaken by Lin (1989) in a North Carolina *Spartina alterniflora* meadow involved the ribbed mussel *Geukensia demissa*, which occurred in a 1- to 2-m wide strip on the seaward side. Manipulative experiments involving singleton or small groups of 10 or 30 mussels were undertaken at different tidal levels and positions within the seagrass meadow. Seston depletion effects could be excluded as an explanation for differences in growth because treatments at the same tidal level gave identical results. At one of the two sites studied, ribbed mussels grew much faster in the middle of the seagrass meadow than at the edge, consistent with presence of a food subsidy (number 5, Table 8.4). Judge, Coen, and Heck (1993) found that seston concentrations were enriched at $\leq$5 cm above the sediment–water interface. The chlorophyll-containing seston in these samples was made up of 90% pennate diatoms, although an estimate of non-living POM was not made. For both of these studies the link showing that the living or non-living POM enhanced bivalve growth was not made.

That the energetic cost of maintaining shore position is higher where water movement energies are higher and sediments less stable (Table 8.4, number 6) has been suggested repeatedly. Irlandi and Peterson (1991), for example, were able to show by seagrass-canopy trimming experiments that this reduced hard clam growth, consistent with increased velocities in these plots and circumstantial evidence to support H_1 in number 6. They also observed that hard clams grew significantly more slowly at the seagrass meadow edge where higher velocities were also present.

Because multiple limiting factors are probably involved in interpreting the Table 8.5 results, one should also consider biotic factors. The effect of seagrass beds on predator–prey relations was reviewed by Orth, Heck, and van MontFrans (1984). Effects could be separated into those involving the seagrass root rhizome and others involving the shoots. Peterson (1982) showed that seagrass meadows provide sufficient below-ground structural complexity to hinder the efficiency of whelk predators, *Busycon* sp., capturing *Mercenaria mercenaria*. Siphon nipping of the suspension-feeding bivalve *Protothaca staminea* by predaceous fish was found to be 9–11 times greater outside the seagrass meadow than within it (Peterson and Quammen 1982). Simulated siphon nipping of *M.*

Table 8.6. *Dry weight of* Zostera marina *and density of* Mercenaria mercenaria *in six replicate samples of 1225 m² each, within the seagrass and at nearby sandflat reference locations.*

Sample replicate no.	Average g dry weight of seagrass · m^{-2}		Average *Mercenaria* no. · m^{-2}	
	Seagrass	Sandflat	Seagrass	Sandflat
1	41.6	0.0	9.68	1.44
2	113.6	0.0	10.24	2.00
3	164.8	0.0	7.32	1.88
4	78.4	0.0	9.12	1.00
5	57.6	0.0	9.12	1.32
6	64.0	0.0	8.76	1.88
Average	86.8	0.0	9.04	1.60

Source: Peterson et al. (1984).

mercenaria was shown to have different impacts on growth, dependent on whether the clams were present within a seagrass meadow (Coen and Heck 1991). Nipped clams within the meadow grew more slowly than nipped clams on sand. The siphon nipping effects on growth included reduced feeding opportunities and costs of regenerating the siphon in addition to the actual tissue removal by the fish.

Although it is not possible to conclude exactly what mechanisms are at work in the experiments outlined in Table 8.5, the following are some possibilities. Seston quality and quantity are greater within the meadow and result in higher hard clam growth there (Table 8.6). Alternatively, higher flows at the reference location may inhibit growth directly or involve higher energetic costs in maintaining position in the less stable sediments. This is consistent with the canopy trimming experiment of Irlandi and Peterson (1991), which caused a reduction in growth of hard clams commensurate with increased velocities. Predator responses within a vegetated area resulting in decreased predation cannot be ruled out. In Eckman's (1987) 2-yr Long Island study, in one of the years worked, the alternate hypothesis (number 4, Table 8.4) was supported for a location characterized by slow velocities. At a location where velocities were faster, clam growth for both low and high eelgrass shoot density treatments was low. Another possibility was that the seagrass stems provide a refuge from predators which move on the sediment–water surface, e.g. crabs (Pohle et al. 1991; Ambrose and Irlandi 1992).

Bivalve reefs

The potential for suspension feeding bivalves to form dense aggregations, or reefs, occurs in suitable locations throughout the world. The reefs are most common at intertidal or nearshore subtidal locations, particularly in estuaries or other areas of high productivity or tidal energy. Common bivalves which form reefs include the mytilid mussels and oysters (Ostreidae).

Hydrodynamics

We know of no studies that have set out to determine specifically the hydrodynamic variables within a natural bivalve reef, although some have done this incidentally in determining filtration rates of the bivalves. Such studies include those of Fréchette and Bourget (1985) and Fréchette, Butman, and Rockwell Geyer (1989) in the field, Butman et al. (1994) in a laboratory flume, and Monosmith et al. (1990) for model bivalve populations in a flume.

The study by Fréchette et al. (1989) concerned an intertidal mussel reef which was 30 m wide by 100 m long in the St. Lawrence estuary, where the tidal range was ~4.8 m. The average biomass for the *Mytilus edulis* reef was 500 g dry · m^{-2}. Seawater sampling to determine phytoplankton concentration and measure filtration rates by standard physiological methods was at three heights above the reef: 1.0, 0.5, and 0.02–0.05 m. Samples were drawn continuously by a centrifugal pump at 3.5 L · min^{-1} in a 1.3-cm-inner-diameter (ID) plastic hose. The current velocities were determined by continuously recording Aanderaa current meters fitted with salinity and temperature probes. Current meter rotors measured velocities at 1.0 and 0.01–0.1 m above the reef. By using benthic boundary layer theory (Schlichting 1968) and velocity data, the logarithmic velocity profile was constructed; from it the friction velocity, U^*, and bottom roughness parameter, Z_0, were determined. From their detailed hydrodynamic model analysis, Fréchette et al. (1989) were able to show that the commonly observed tendency for the mussels to form hummocks by overgrowing other mussels may result in bottom roughness parameter increases sufficient to increase turbulent transfer of seston to the reef significantly.

Reef metabolism

Studies which have measured bivalve reef metabolism (Table 8.7) are considered here in detail. The aim is to determine whether bivalve reef oxygen uptake data support the conclusions of Chapters 4 and 7 that increased seawater velocity can enhance bivalve growth by increasing the turbulent supply of sestonic food. The results summarized in Table 8.7 are of total metabolism of the reef, inclusive of photosynthetic oxygen production, e.g. from phytoplankton in seawater or diatoms on the reef surface. The only seasonal study of metabolism of which we are aware is presented by Dame et al. (1992) for a South Carolina oyster reef. This study found net oxygen production in winter when macroalgae were present and denoted by a minus sign in Table 8.7; during the summer, the oyster reef was strongly heterotrophic with a net oxygen demand (BOD). Physiological data for oyster oxygen uptake rates discussed in Dame, Spurrier, and Zingmark (1992) suggest that the oysters are only responsible for 10% of the annual reef oxygen uptake. Other uptake sources are aerobic bacteria, meiofauna, other macrofauna, and the chemical oxygen demand (COD) of the reef sediments.

During summer observations of a Narragansett Bay mussel reef, Nixon et al. (1971) attempted to determine the effect of tidal velocity on reef metabolism. Dissolved oxygen determinations in triplicate were measured by Winckler titration. Seawater samples were obtained upstream of the seaward edge of the reef and downstream in the dominant tidal direction, some 20–25 m away. Depth at which the samples were taken was not stated, and a crude measure of velocity was made by observing dye or drift bottle movements, so that velocities may be inaccurate. The data in Fig. 8.10 were obtained by subtracting the downstream from the upstream dissolved oxygen on flooding tides. The difference – assuming a well mixed tidal prism above the mussel reef – represents the oxygen uptake calculated for the area over which the water mass has moved in its passage between the two sampling points. The instantaneous rate of BOD, I, must be corrected for the volumetric flow rate which passes over the reef (Chap. 2, the section, Field flows).

Do the results of Fig. 8.10 support the conclusions of Chapters 4 and 7 and the thesis that water movement energy is a community metabolism multiplier (Table 8.8, number 1)? Assuming that the reef filtration rates are directly proportional to metabolic uptake of oxygen, the results do support the theoretical curve of Fig. 4.1 for a bivalve with inhalant normal to the flow and also the empirical fits of Figs. 7.7 and 7.8 based on

Table 8.7. *Comparison of independent studies to measure bivalve reef metabolism.*

Variable	Nixon et al. (1971)	Dankers et al. (1989)	Dame et al. (1992)
Location	Narragansett Bay, Rhode Island, U.S.A.	Island of Texel, Netherlands	North Inlet, South Carolina, U.S.A.
Dominant species	*Mytilus edulis*	*M. edulis*	*Crassostrea virginica*
Temperature (°C)	20.1–23.5	14–16	8–30
Salinity (o/oo)	27–30	?	15–35
Reef area (m^2)	100	?	600
Path length of reef measured (m)	20–25	10	10
Bivalve density (m^{-2})	6261	?	?
Bivalve dry meat biomass ($g \cdot m^{-2}$)	1523	818	196
Reef respiration rate ($g O_2 \cdot m^{-2} \cdot h^{-1}$)	0.2–2.8	1.2–3.9	−1.5 to +4.0

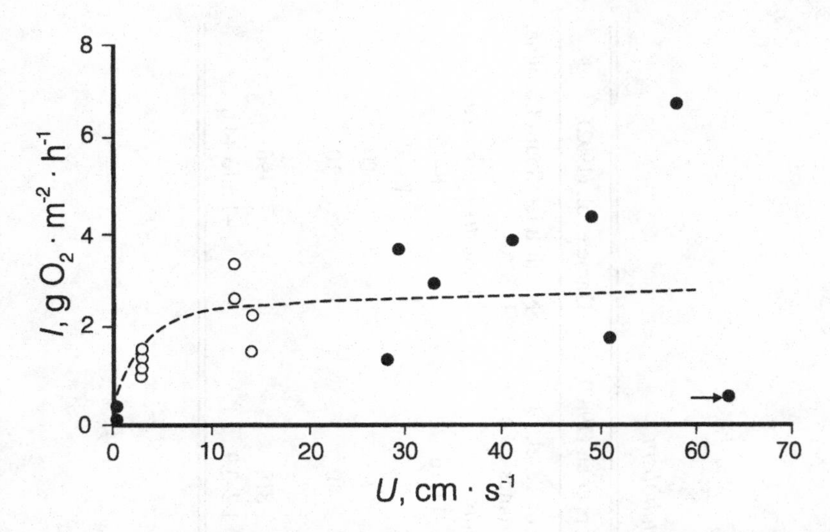

Figure 8.10 Instantaneous rate of oxygen uptake (I) of a Narragansett Bay blue mussel reef as a function of free-stream velocity, U, $cm \cdot s^{-1}$: solid dots, field derived data; open dots, recalculated laboratory data. The fitted hyperbola has $r = 0.83$, with single arrowed point excluded (from Nixon et al. 1971).

population growth of mussels and scallops. There is likely to be an upper limit to the hyperbola of Fig. 8.10 beyond which oxygen uptake of the reef declines. One point – shown in Fig. 8.10 with an arrow and which was omitted from the regression fitting of Nixon et al. (1971) – could indicate a limiting flow consistent with the oxygen uptake's being unimodal. Without knowing the blue mussels' physiological responses (filtration and metabolic rates) to flow, it is impossible to interpret fully the results of Nixon et al. (1971). A further difficulty in interpretation is that the proportion of reef oxygen demand due to mussels was not determined, although because the mussel biomass involved (Table 8.7) is approximately eight times higher than for the oyster reef, it is likely to be higher than the 10% found for the latter.

In the study by Dame et al. (1992), a benthic ecosystem tunnel (Chap. 2, the section, Field flows) was used to obtain metabolic and nutrient flux rates. Typical results of a flood–ebb cycle are shown in Fig. 8.11. During the flood tide ($U \sim 7$–$13\,cm \cdot s^{-1}$), the net uptake of oxygen was 1.1–$3.2\,g\,O_2 \cdot m^{-2} \cdot h^{-1}$, and during the ebb tide ($U \sim 8$–$25\,cm \cdot s^{-1}$), a net release of up to $3.5\,g\,O_2 \cdot m^{-2} \cdot h^{-1}$ occurred. A net release of oxygen did not begin until the ebb velocity exceeded $15\,cm \cdot s^{-1}$. During these observations,

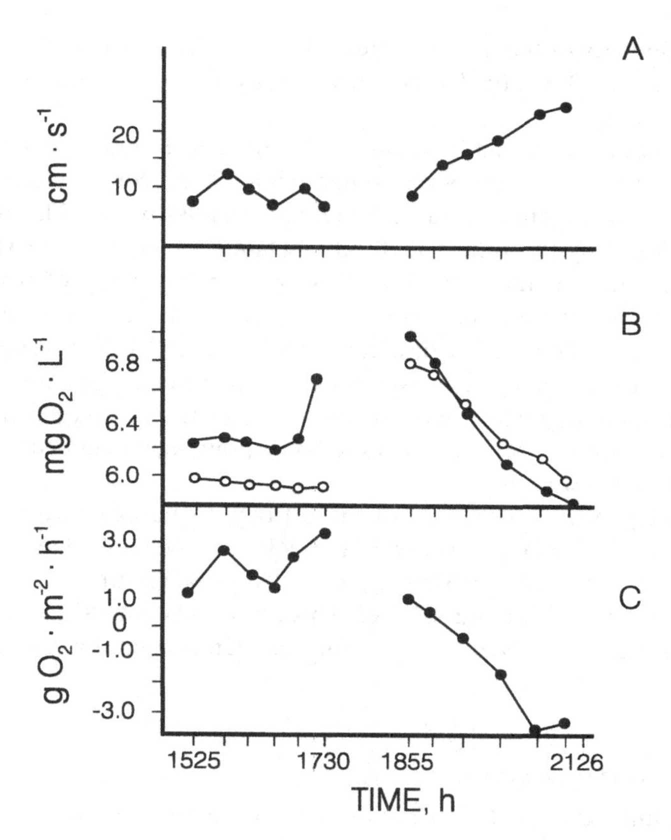

Figure 8.11 Community respiration rate of a South Carolina oyster reef during a flood–ebb cycle in summer 1983: A, tidal velocity; B, measured dissolved oxygen concentration of experimental tunnel, C, I = instantaneous rate of oxygen uptake (Dame et al. 1992): •, inlet; ○, outlet.

oysters fed on the flood tide but not on the ebb, when much of the seston had already been removed by the oysters' filtering. This interpretation by Dame et al. (1992) is consistent with other field studies which linked tidal cycles with feeding as indicated by digestive cycles. Thus, mussels *M. edulis* (Langton 1977) and sublittoral *Pecten maximus* (Mathers 1976) fed on the flood tide. In the latter case, two distinct subpopulations of scallops were present – one facing into and feeding on the flood, the other away from the flooding current and feeding on the ebb current. Histological studies by Mathers (1976) showed that the digestive physiology of each subpopulation was ~6 h out of phase, supporting a tidal periodicity of digestion. Langton and Gabbott (1974) also described a tidal periodicity of feeding in oysters *Ostrea edulis* grown near low tide on the shore at Menai Bridge, North Wales, by monitoring the crystalline style physiology.

In the study described by Dankers, Dame, and Kersting (1989), two methods of measurement were compared: a 200-m^2 concrete tank supplied with mussels and natural seawater and part of a natural blue mussel reef enclosed by the tunnel method of Dame, Zingmark, and Haskin (1984). Results of the former gave average net uptakes of 0.5–1.0 g $O_2 \cdot m^{-2} \cdot h^{-1}$, which were somewhat lower than similar values for a natural mussel reef (Table 8.7). The discrepancy is explained by the sensitivity of the results to the timing of measurements. Hence, net oxygen fluxes were low during the day when photosynthetic production of oxygen was occurring. Nighttime observations were correspondingly high because this production was absent.

An independent estimation of mussel oxygen uptake was made by Dankers et al. (1989) from standard physiological data linked to biomass. This yields a value of 0.37 mg $O_2 \cdot g^{-1} \cdot h^{-1}$ equivalent to an uptake of 0.3 g $O_2 \cdot m^{-2} \cdot h^{-1}$ by the mussel reef. This represents only 8% to 25% of the day–night dissolved oxygen flux, as determined by the tunnel method.

Pelagic–benthic coupling

A direct link between the surface primary productivity of the oceans and benthic productivity, or biomass, of the same area was proposed by Rowe (1971). From published data he showed that surface primary productivity ranked second only to depth in controlling benthic biomass. Thus, carbon fixed photosynthetically in surface waters, and after passage through a pelagic food web, reached benthos in the form of faecal pellets (Hargrave 1973). Rowe et al. (1975) also showed that the exchange was a two-way process with benthically regenerated plant nutrients, e.g. various forms of nitrogen, recycled to primary producers in shallow surface coastal waters. It was thus possible to speak of pelagic–benthic coupling involving positive–negative feedback loops between primary and secondary benthic producers.

Typically, energy or carbon budgets are calculated on the basis of unit area – usually 1 m^2 – close to the area actually sampled by a randomly placed grab or corer used in estimating soft sediment benthic biomass and hence secondary benthic production. Primary production, too, is estimated on the basis of unit area or volume. Thus, the first efforts to link pelagic and benthic fluxes were made on the basis of small vertically linked areas. A simple regression model proposed by Hargrave (1973) was successful in predicting the magnitude of the annual sediment oxygen demand on the basis of the amount of pelagic primary production and the mixed layer depth of the water column at the same site.

Yet, some authors found discrepancies in balancing energy budgets if a tight vertical link between pelagos and benthos was assumed. Thus, Bernard (1974), who described a partial energy budget for an oyster *Crassostrea gigas* reef in British Columbia, showed that the energy available from the local primary production was much less than the oysters' allocation of energy to the annual production of gametes. Emerson, Roff, and Wildish (1986) provided a partial carbon budget based on one site in the Bay of Fundy (Fig. 8.12) – a macrotidal estuarine ecosystem in eastern Canada. Pelagic production in the Bay over depositional LaHave clay sediments provided an input of $60 g C \cdot m^{-2} \cdot yr^{-1}$ from a total phytoplankton primary production of $133 g C \cdot m^{-2} \cdot yr^{-1}$ to the benthos. All of this input could not be utilized by macrofauna or accounted for by the measured sediment burial rates. It was suggested (Emerson et al. 1986) that the pelagic input, consisting of equal amounts of detritus, copepod faeces, and microalgae, was tidally transported by tidal excursions of ~20 km to a harder sediment on which a productive horse mussel *Modiolus modiolus* reef was situated (Table 8.8, number 2). The maximum production of the macrofauna here was $148 g C \cdot m^{-2} \cdot yr^{-1}$ at depths of 60–100 m. Assuming a euphotic depth of 20–40 m, this exceeds the mean primary production of phytoplankton in the column of seawater above this part of the lower Bay. It also provides circumstantial evidence for an alternative source of food by turbulent tidal transport, other than the direct vertical link between pelagos and benthos proposed by Rowe (1971). A further example of a partial carbon budget estimate on the tidal flats of Königshafen in the German Wadden Sea was provided by Asmus and Asmus (1990). It also shows the inadequacy of local autochthonous primary production in the seawater above a blue mussel reef to provide sufficient carbon sources for the mussel production realized. Estimates of primary production by local phytoplankton amount to $73 g C \cdot m^{-2} \cdot yr^{-1}$ compared to secondary production of the mussel reef of $264 g C \cdot m^{-2} \cdot yr^{-1}$. This led Asmus and Asmus (1990) to conclude that the high suspension feeder production is maintained by tidal transport of seston from the North Sea or other parts of the Wadden Sea.

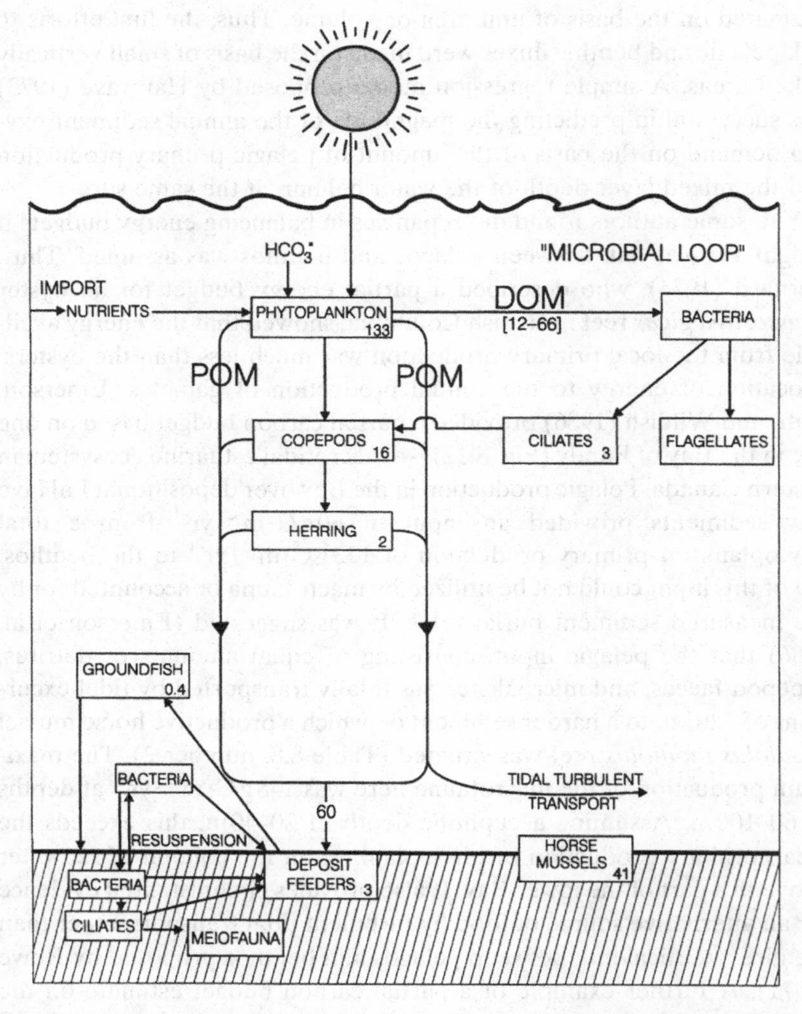

Figure 8.12 Partial carbon budget for the Bay of Fundy ecosystem on LaHave clay sediments at 50–80 m (Emerson et al. 1986): annual production, $g\,C \cdot m^{-2} \cdot yr^{-1}$; bracketed figure estimated by the method of Azam et al. (1983).

The idea that bivalve reefs were an important part of pelagic–benthic coupling (Table 8.8, number 3) was proposed by Dame et al. (1980) and supported by an analysis of input–output energy flow analysis of a South Carolina oyster reef (Dame and Patten 1981). The concept was further developed on a South Carolina oyster reef dominated by *Crassostrea*

Table 8.8. *Ecosystem level hypotheses referable to bivalve reefs and flow.*

No.	H_0/H_1	Description	Reference
1	H_0	Water movement energy has no effect as a community metabolism multiplier	Odum (1971) Nixon et al. (1971)
	H_1	Increasing water movement energy acts as a community metabolism multiplier	
2	H_0	Tidal or wind-driven transport of seston has no effect on bivalve reef production	Odum (1971) Emerson et al. (1986)
	H_1	Tidal or wind-driven transport of seston may support bivalve reef production in excess of locally available resources	
3	H_0	Bivalve reefs are unimportant in materials cycling between pelagic and benthic subsystems	Dame et al. (1980)
	H_1	Bivalve reefs represent a major coupling between pelagos and benthos and hence in materials cycling between these subsystems	
4	H_0	Bivalve reefs have no effect on retaining reactive nutrients in estuaries	Dame and Dankers (1988) Dame and Libes (1993)
	H_1	The net effect of bivalve reefs is to retain reactive nutrients, such as nitrogen, within estuaries	
5	H_0	Bivalve reefs do not influence the development of microalgal blooms in eutrophic estuaries	Cloern (1982) Officer et al. (1982) Nichols (1985)
	H_1	Bivalve reefs influence the development of microalgal blooms in eutrophic estuaries	

virginica as a subsystem of an estuarine ecosystem (Dame et al. 1984). A primary concern in making their measurements was to design samplers which could estimate the flux of materials, particularly the importance of volumetric flow to this calculation. Various benthic tunnels or open flumes for field use have been designed (Chap. 2, the section, Field flows) for this purpose. From tidal velocity measurements made within the tunnel, the volumetric flow could be calculated and hence the flux of materials determined. Dame et al. (1984) showed that the oyster reef reduced particulate organic carbon and chlorophyll a in the flooding tidal prism and added significant amounts of excreted ammonia as it ebbed. Using the benthic ecosystem tunnel, Dame et al. (1992) have determined the annual flux of ammonia to and from the oyster reef with a net annual release of $124.8\,g \cdot m^{-2} \cdot yr^{-1}$. In another study, Dame and Libes (1993) investigated plant nutrient fluxes after experimental manipulations involving removing oysters in some plots. Because the recycling rates of the oysters were faster than the tidal flushing rates, nutrients were recycled in situ. The controlling feedback loop for nitrogen was suggested to be the exchange between phytoplankton and oysters, the latter retaining nitrogen within the body.

Field studies on Dutch and German Wadden Sea blue mussel reefs (Dame and Dankers 1988; Prins and Smaal 1990; Asmus, Asmus, and Reese 1990; Smaal and Prins 1993) have confirmed hypothesis 4 (Table 8.8): that large quantities of seston are filtered from seawater; faecal and pseudofaecal biodeposits are produced in large quantities, the latter are rapidly remineralized, resulting in release of dissolved and inorganic nutrients in seawater, and this results in stimulation of further phytoplankton primary production as well as export of nutrients at certain times of the year. The results are also consistent with H_1 of number 3 (Table 8.8), that is that the reef represents a major coupling between pelagos and benthos.

Nitrogen cycling was investigated by Kaspar et al. (1985) at a green mussel *Perna canaliculus* aquaculture site in the Marlborough Sounds, New Zealand. The main pathways of the nitrogen cycle considered are shown in Fig. 8.13. Losses of nitrogen occurred when mussels were removed at harvest and by denitrification processes in sediments and seawater. Additions occurred from mussel excretion, mainly of ammonia, and the microbial transfer of various organic nitrogen sources which were present within the sediments. Nitrogen fixation may also have been important but was not considered in this study. Although it was clear that there was considerable tidal energy available at the farm, e.g. $U_{max} = 111\,cm \cdot s^{-1}$ and a tidal prism of 3–4 m, no estimate of tidal transport of nitrogen into or out of the farm area was included in this study. By comparison with surface sediments at a nearby reference site, significantly higher denitrification and nitrogen mineralization rates were present at the farm.

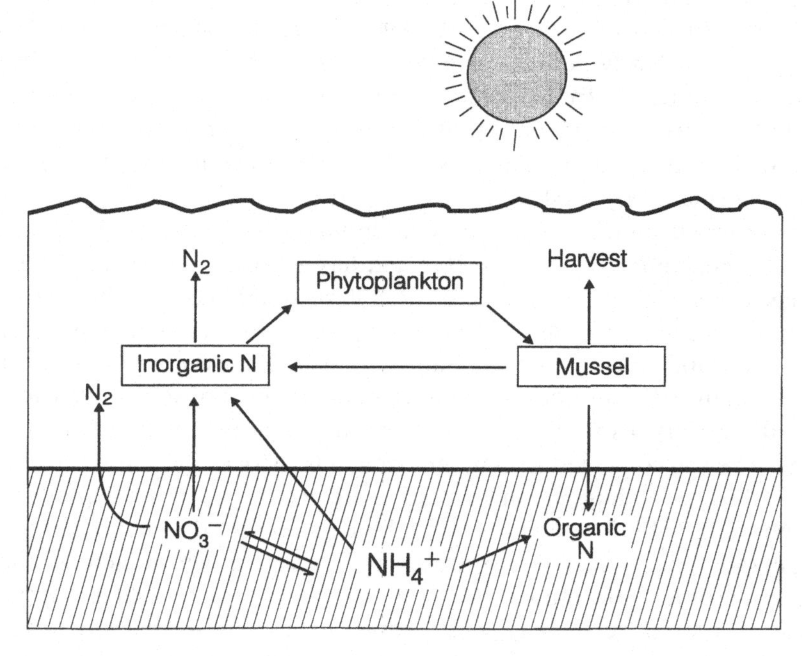

Figure 8.13 Nitrogen cycle of a green mussel *Perna canaliculus* farm, Marlborough Sounds, New Zealand (Kaspar et al. 1985).

Eutrophication and bivalve reefs

In temperate eutrophic estuaries such as San Francisco Bay on the U.S. West Coast, the absence of phytoplankton blooms is contrary to the frequent observation that hypernutrification results in blooms and increases primary production (Neilson and Cronin 1981). Nevertheless, unexpectedly low densities of phytoplankton do occur in some eutrophic estuaries, and it has been proposed (Table 8.8, number 5) that the phytoplankton are limited by the grazing power of extensive reefs of suspension feeding bivalves as in San Francisco Bay. The presence of the introduced soft-shell clam *Mya arenaria* there at densities up to 20,000 clams $\cdot$ m^{-2} was sufficient to filter the whole water column once per day on the basis of laboratory filtering rate estimates (Nichols 1985). It is clear that San Francisco Bay and the Sacramento–San Joaquin river system which drains into it have been highly modified by a range of human activities (Nichols et al. 1986). In 1986, a new invasion of an exotic suspension feeding bivalve, the Asian clam *Potamocorbula amuriensis*, began and rapidly spread throughout the estuary, reaching densities of >10,000 clams·m^{-2} (Carlton et al. 1990). Since the San Francisco Bay ecosystem is clearly not in a steady state, decisions on the original hypothesis number 5 (Table 8.8) are problematic. Multiple causes of chlorophyll a variability in the landward end of San Francisco Bay, and including hydrodynamic causes as well as losses to benthic grazers, have been considered by Jassby and Powell (1994).

Another eutrophic estuary, the 180-km-long Potomac on the U.S. East Coast, opens into Chesapeake Bay and has a freshwater tidal zone, or "stuw zone," of 60 km (Cohen et al. 1984). After the introduction of another exotic, the clam *Corbicula fluminea*, into the freshwater stuw zone, densities reached ~1500 clams·m^{-2} with a maximum biomass of 3 kg·m^{-2}. In 1980 and 1981, a phytoplankton sag curve of 40% to 60% less density than that of adjacent areas spread for 6–8 km within the stuw zone. The sag coincided with the maximum density of the clam distribution, thus providing circumstantial evidence to support hypothesis number 5 in Table 8.8. Other possible explanations for the phytoplankton sag curve, such as decreases in freshwater discharge, which increase the retention time of the stuw zone; the filtering power of zooplankton; the presence of toxic substances, which inhibit phytoplankton growth; or the lack of plant nutrients, which limits phytoplankton growth, were not supported by direct observation. Because all of the water in the affected zone could have been filtered by the clams in 3–4 d, similar to the residence time there, this is circumstantial evidence supporting the hypothesis that the clams' grazing was responsible for the phytoplankton sag curve.

Hypernutrification may result in stimulation of species of phytoplankton that are trophically unacceptable to benthic grazers and hence unable to increase productivity through the pelagic–benthic link. As an example, the "brown tide" blooms along the U.S. East Coast caused widespread mortality rates of blue mussels (Cosper et al. 1987). The brown tide bloom consisted of up to 10^6 cells·ml^{-1} of a 2-μm-long chrysophycean, *Aureoccus anophagefferens*, which was shown by Tracey (1988) to inhibit feeding by *M. edulis* in laboratory tests. Because of the persistence of the brown tide, it was suggested that the mussels died of starvation. Prins, Dankers, and Smaal (1994) have reported reduction in feeding caused by presence of the haptophycean *Phaeocystis* sp., which dominated the phytoplankton for 2 mo in the spring in the Netherlands. Blue mussels placed in a concrete flume 40 × 5 m wide were supplied with natural seawater pumped from the North Sea and showed reduced growth during the *Phaeocystis* bloom.

We should expect more instances where benthic grazers are unable to utilize the enhanced production of eutrophication. Possibly the study reported by Loo and Rosenberg (1989) in the eutrophic Lalholm Bay, Kattegat, is a case in point. The nearshore benthos here was dominated by grazers – *Cardium edule* and *Mya arenaria* – and population filtration experiments with local seawater showed that they filtered at suboptimum rates. Loo and Rosenberg (1989) were unable to pinpoint the cause, although low rates of turbulent transfer from the surface to the bottom were suggested. The study reported by Peterson, Irlandi, and Black (1994) in a coastal lagoon on the south coast of Australia could also be an example. Here, populations of two suspension feeding bivalves, *Katelysia scalarina* and *K. rhytiphora*, suddenly crashed, coincident with the onset of eutrophication. Unfortunately, examination of phytoplankton was not carried out during the study.

Summary

Two common features can be discerned in our review of the three distinct ecosystems we have chosen for inclusion here: kelp forests seagrass meadows, and bivalve reefs. The first is that there is evidence in the perturbation theory for the suggestion by Odum et al. (1979) that ecosystem productivity is a unimodal function of water motion. The second common feature is that each species of kelp, seagrass, and bivalve appears to have a characteristic unimodal response to water movement with distinct tolerance ranges of water movement, as found by Conover (1968) for some species of kelps and seagrasses. In future work, it should be possible to utilize this relationship to predict water movement energies and perhaps also secondary heterotrophic productivity of suspension feeders.

Ecosystem analysis requires consideration and resolution of the temporal and spatial scales to be used in the study. It is simply not sufficient for a biologist to set seasonal and geographic limits arbitrarily. This should be done after a scoping study involving a biologist and a physical oceanographer. This point of view is forced on us by the examples reviewed here, which include the South African kelp ecosystem studies reported by Wulff and Field (1983), the Bay of Fundy horse mussel reef studies reported in Emerson et al. (1986), and the Wadden Sea blue mussel reef studies of Asmus and Asmus (1990). These reefs were shown by Asmus (1994) to depend trophically on turbulent advective transfer of seston from sources external to the reef subsystem. Recent studies, e.g. by Grant et al. (1993), use an interdisciplinary approach and ecosystem simulation modelling, in which the adjacent hydrodynamics is included, to represent fluxes and flows within a blue mussel farm in Nova Scotia, Canada.

Despite the interest in suspension feeding bivalves which occur within seagrass meadows, it has proven difficult to conclude why bivalve growth or productivity is so different within and outside the meadow. One obvious reason for this difficulty is that the basic physiological response of the bivalves present in seagrass meadows to flow is unknown, and this should be rectified. Transplant field experiments of bivalves to seagrass meadows and adjacent reference areas without seagrass may involve complexities which render them difficult to interpret. This is because multiple factors may be involved in each seagrass meadow, so it is likely that determining the cause may prove to be elusive, but possible, if a combination of different hierarchical levels of the scientific method is employed.

We believe that measurement of the instantaneous rate of oxygen uptake of bivalve reefs is a valuable independent metabolic determination for ecosystem analysis. Further development of the methods to be used in its determination is required (see Chap. 2), and its results should be compared with those of the more conventional sediment–oxygen demand enclosed chamber technique. Methodology development is also required in measuring enzyme activity and growth as a means of estimating instantaneous production relatable to instantaneous oxygen uptake rate.

The modelling study of Koseff et al. (1993) shows that a phytoplankton bloom in estuaries like San Francisco Bay depends on a key physical factor – stratification. The latter phenomenon confines phytoplankton cells in the photic zone, where photosynthesis proceeds rapidly, and also limits the vertical transfer of pelagos to the benthos. Some limited wind–wave mixing of stratified surface water is required for optimal bloom formation. During stratification events, the phytoplankton cells are effectively de-coupled from the benthic grazers.

With regard to the hypotheses presented in this chapter (Tables 8.1, 8.2, 8.4), we consider that further experimental work is required to reject H_0 in number 1, Table 8.1, and numbers 5 and 6 in Table 8.4. Some doubt, due to either conflicting results or inconclusive rejection of H_0, means that further experimental work is required with numbers 2, 3, and 4 in Table 8.1; numbers 4 and 5 in Table 8.2; and numbers 2 and 4 in Table 8.4. For the remaining six hypotheses, we conclude that H_1 has been sufficiently supported in experimental tests. Although some evidence in support of H_1 for all hypotheses in Table 8.8 is given, we believe that further experimental conformation is necessary in geographically separate locations before these propositions can be robustly accepted.

References

Ackerman, J. D. 1983. Current flow around *Zostera marina* plants and flowers: implications for submarine pollution. Biol. Bull. 165: 504.

Ackerman, J. D., and A. Okubo. 1993. Reduced mixing in a marine macrophyte canopy. Funct. Ecol. 7: 305–309.

Ambrose, W. G., Jr., and E. A. Irlandi. 1992. Height of attachment on seagrass leads to trade-off between growth and survival in the bay scallop *Argopecten irradians*. Mar. Ecol. Prog. Ser. 90: 45–51.

Anderson, S. M., and A. C. Charters. 1982. A fluid dynamic study of seawater flow through *Gelidium nudifrons*. Limnol. Oceanogr. 27: 399–412.

Asmus, H. 1994. Benthic grazers and suspension feeders: which one assumes the energetic dominance in Köngigshafen? Helg. Meeresunters. 48: 217–231.

Asmus, H., and R. M. Asmus. 1990. Trophic relationships in tidal flat areas: to what extent are tidal flat areas dependent on imported food? Neth. J. Sea Res. 27: 93–99.

Asmus, H., R. Asmus, and K. Reese. 1990. Exchange processes in an intertidal mussel bed: a Sylt-flume study in the Wadden Sea. Ber. Biol. Anst. Helgoland. 6: 1–79.

Azam, F., T. Fenchel, J. G. Field, J. S. Gray, L. A. Meyer-Reil, and F. Thingstad. 1983. The ecological role of water-column microbes in the sea. Mar. Ecol. Prog. Ser. 10: 257–263.

Backus, G. J. 1994. Coral reef ecosystems. A.A. Balkema, Rotterdam.

Bernard, F. R. 1974. Annual biodeposition and gross energy budget of mature Pacific oysters, *Crassostrea gigas*. J. Fish. Res. Board Can. 31: 185–190.

Bertness, M. D. 1984. Ribbed mussels and *Spartina alterniflora* production in a New England salt marsh. Ecology 65: 1794–1807.

Blackburn, T. H., D. B. Nedwell, and W. J. Wiebe. 1994. Active mineral cycling in a Jamaican seagrass sediment. Mar. Ecol. Prog. Ser. 110: 233–239.

Butman, C. A., M. Fréchette, W. Rockwell Geyer, and V. R. Starczak. 1994. Flume experiments on food supply to the blue mussel, *Mytilus edulis* L., as a function of boundary layer flow. Limnol. Oceanogr. 39: 1755–1768.

Carlton, J. T., J. K. Thompson, L. E. Schemel, and F. H. Nichols. 1990. Remarkable invasion of San Francisco Bay (California, USA) by the Asian clam *Potamocorbula amurensis*. I. Introduction and dispersal. Mar. Ecol. Prog. Ser. 66: 81–94.

Cloern, J. E. 1982. Does the benthos control phytoplankton biomass in South San Francisco Bay? Mar. Ecol. Prog. Ser. 9: 191–202.

Coen, L. D., and K. L. Heck. 1991. The interacting effects of siphon nipping and habitat on the bivalve (*Mercenaria mercenaria* (L.)) growth in a subtropical seagrass (*Halodule wrightii* Aschers) meadow. J. Exp. Mar. Biol. Ecol. 145: 1–13.

Cohen, R. R. H., P. V. Dresler, E. J. P. Phillips, and R. L. Cory. 1984. The effect of the Asiatic clam, *Corbicula fluminea*, on phytoplankton of the Potomac River, Maryland. Limnol. Oceanogr. 29: 170–180.

Connell, J. T. 1972. Community interactions on marine rocky intertidal shores. Ann. Rev. Ecol. Syst. 3: 169–192.

Conover, J. T. 1968. The importance of natural diffusion gradients and transport of substances related to benthic marine plant metabolism. Bot. Mar. 11: 1–9.

Cosper, E. M., W. C. Dennison, E. J. Carpenter, V. M. Bricelj, J. G. Mitchell, S. H. Kennster, D. Colflesh, and M. Dewey. 1987. Recurrent and persistent "brown tide" blooms perturb coastal marine ecosystem. Estuaries 10: 284–290.

Dame, R. F., and N. Dankers. 1988. Uptake and release of materials by a Wadden Sea mussel bed. J. Exp. Mar. Biol. Ecol. 118: 207–216.

Dame, R., and S. Libes. 1993. Oyster reefs and nutrient retention in tidal creeks. J. Exp. Mar. Biol. Ecol. 171: 251–258.

Dame, R., F., and B. C. Patten. 1981. Analysis of energy flows in an intertidal oyster reef. Mar. Ecol. Prog. Ser. 5: 115–124.

Dame, R. F., J. D. Spurrier, and R. G. Zingmark. 1992. In situ metabolism of an oyster reef. J. Exp. Mar. Biol. Ecol. 164: 147–160.

Dame, R. F., R. G. Zingmark, and E. Haskin. 1984. Oyster reefs as processors of estuarine materials. J. Exp. Mar. Biol. Ecol. 83: 239–247.

Dame, R., R. Zingmark, H. Stevenson, and D. Nelson. 1980. Filter feeder coupling between the water column and benthic subsystem, p. 520–526. *In* V. S. Kennedy (ed.) Estuarine perspectives. Academic Press, New York.

Dankers, N., R. Dame, and K. Kersting. 1989. The oxygen consumption of mussel beds in the Dutch Wadden Sea. Sci. Mar. 53: 473–476.

Dayton, P. K. 1971. Competition, disturbance and community organization: the provision and subsequent utilization of space in a rocky intertidal community. Ecol. Monogr. 41: 351–389.

1975. Experimental evaluation of ecological dominance in a rocky intertidal community. Ecol. Monogr. 45: 137–159.

1985. The structure and regulation of some South American kelp communities. Ecol. Monogr. 55: 447–468.

Den Hartog, C. 1970. The seagrasses of the world. North Holland, Amsterdam.

Doty, M. S. 1971. Measurements of water movement in reference to benthic algal growth. Bot. Mar. 14: 32–35.

Duggins, D. O., and J. E. Eckman. 1994. The role of kelp detritus in the growth of benthic suspension feeders in an understorey kelp forest. J. Exp. Mar. Biol. Ecol. 176: 53–68.

Duggins, D. O., C. A. Simenstad, and J. A. Estes. 1989. Magnification of secondary production by kelp detritus in coastal ecosystems. Science 245: 170–173.

Eckman, J. E. 1987. The role of hydrodynamics in recruitment, growth and survival of *Argopecten irradians* (L.) and *Anomia simplex* (D'Orbigny) within eelgrass meadows. J. Exp. Mar. Biol. Ecol. 106: 165–191.

Eckman, J. E., and D. O. Duggins. 1991. Life and death beneath macrophyte canopies: effects of understorey kelps on growth rates and survival of marine benthic suspension feeders. Oecologia 87: 473–487.

Eckman, J. E., D. O. Duggins, and A. T. Sewell. 1989. Ecology of understorey kelp environments. I. Effects of kelps on flow and particle transport near the bottom. J. Exp. Mar. Biol. Ecol. 129: 173–188.

Emerson, C. W., J. C. Roff, and D. J. Wildish. 1986. Pelagic–benthic energy coupling at the mouth of the Bay of Fundy. Ophelia 26: 165–180.

Estes, J. A., and J. F. Palmisano. 1974. Sea otters: their role in structuring nearshore communities. Science 185: 1058–1060.

Fonesca, M. S., J. S. Fisher, J. C. Zieman, and G. W. Thayer. 1982. Influence of the seagrass, *Zostera marina* L., on current flow. Est. Coast. Shelf Sci. 15: 351–362.

Fonesca, M. S., and W. J. Kenworthy. 1987. Effects of current on photosynthesis and distribution of seagrasses. Aquat. Bot. 27: 59–78.

Fonesca, M. S., J. Zieman, G. W. Thayer, and J. S. Fisher. 1983. The role of current velocity in structuring eelgrass (*Zostera marina* L.) meadows. Est. Coast. Shelf Sci. 17: 367–380.

Fréchette, M., and E. Bourget. 1985. Energy flow between pelagic and benthic zones: factors controlling particulate organic matter available to an intertidal mussel bed. Can. J. Fish. Aquat. Sci. 42: 1166–1170.

Fréchette, M., C. A. Butman, and W. Rockwell Geyer. 1989. The importance of boundary-layer flows in supplying phytoplankton to the suspension feeder, *Mytilus edulis* L. Limnol. Oceanogr. 34: 19–36.

Gambi, M. C., A. R. M. Nowell, and P. A. Jumars. 1990. Flume observations on flow dynamics in *Zostera marina* (eelgrass) beds. Mar. Ecol. Prog. Ser. 61: 159–169.

Gerard, V. A., and K. H. Mann. 1979. Growth and production of *Laminaria longicornis* (Phaeophyta) populations exposed to different intensities of water movement. J. Phycol. 15: 33–41.

Grant, J., M. Dowd, K. Thompson, C. Emerson, and A. Hatcher. 1993. Perspectives on field studies and related biological models of bivalve growth and carrying capacity. NATO ASI Ser. G33: 371–420.

Hagen, N. T., and K. H. Mann. 1994. Experimental analysis of factors influencing the aggregating behaviour of the green sea urchin *Strongylocentrotus droebachiensis* (Müller). J. Exp. Mar. Biol. Ecol. 176: 107–126.

Hargrave, B. T. 1973. Coupling carbon flow through some pelagic and benthic communities. J. Fish. Res. Board Can. 30: 1317–1326.

Hemminga, M. A., P. G. Harrison, and F. vanLeut. 1991. The balance of nutrient losses and gains in seagrass meadows. Mar. Ecol. Prog. Ser. 71: 85–96.

Irlandi, E. A., and C. H. Peterson. 1991. Modification of animal habitat by large plants: mechanisms by which seagrasses influence clam growth. Oecologia 87: 307–318.

Jackson, G. A. 1984. Internal wave attenuation by coastal kelp strands. J. Phys. Oceanogr. 14: 1300–1306.

Jackson, G. A., and C. D. Winant. 1983. Effect of a kelp forest on coastal currents. Cont. Shelf. Res. 2: 75–80.

Jassby, A. D., and T. M. Powell. 1994. Hydrodynamic influences on interannual chlorophyll variability in an estuary: upper San Francisco Bay–Delta (California, U.S.A.). Est. Coast. Shelf Sci. 39: 595–618.

Jordan, T. E., and I. Valiela. 1982. A nitrogen budget of the ribbed mussel *Geukensia demissa* and its significance in nitrogen flow in a New England salt marsh. Limnol. Oceanogr. 27: 75–90.

Juday, C. 1940. The annual energy budget of an inland lake. Ecology 21: 438–450.

Judge, M. L., L. D. Coen, and K. L. Heck, Jr. 1993. Does *Mercenaria mercenaria* encounter elevated food levels in seagrass beds? Results from a novel technique to collect suspended food resources. Mar. Ecol. Prog. Ser. 92: 141–150.

Kaspar, H. F., P. A. Gillespie, I. C. Boyer, and A. L. MacKenzie. 1985. Effects of mussel aquaculture on the nitrogen cycle and benthic communities in Kenepuru Sound, Marlborough Sounds, New Zealand. Mar. Biol. 85: 127–136.

Kenworthy, W. J., J. C. Zeiman, and G. W. Thayer. 1982. Evidence for the influence of seagrasses on the benthic nitrogen cycle in a coastal plain estuary near Beaufort, North Carolina (USA). Oecologia 54: 152–158.

Kerswill, C. J. 1949. Effects of water circulation on the growth of quahaugs and oysters. J. Fish. Res. Board Can. 7: 545–551.

Koehl, M. A. R. 1986. Seaweeds in moving water: form and mechanical function, p. 603–634. *In* T. J. Givnish (ed.) On the economy of plant form and function. Cambridge University Press, Cambridge.

Koehl, M. A. R., and R. S. Alberte. 1988. Flow, flapping and photosynthesis of *Nereocystis lewkeona*: a functional comparison of undulate and flat blade morphologies. Mar. Biol. 99: 435–444.

Koseff, J. R., J. K. Holen, S. G. Monismith, and J. E. Cloern. 1993. Coupled effects of vertical mixing and benthic grazing on phytoplankton populations in shallow turbid estuaries. J. Mar. Res. 51: 843–868.

Langton, R. W. 1977. Digestive rhythms in the mussel, *Mytilus edulis*. Mar. Biol. 41: 53–58.

Langton, R. W., and P. A. Gabbott. 1974. The tidal rhythm of extracellular digestion and the response to feeding in *Ostrea edulis* L. Mar. Biol. 24: 181–187.

Leigh, E. G., R. T. Paine, J. F. Quinn, and T. H. Suchanek. 1987. Wave energy and intertidal productivity. Proc. Natl. Acad. Aci. U. S. A. 84: 1314–1318.

Lin, J. 1989. Importance in location in the salt marsh and clump size on growth of ribbed mussels. J. Exp. Mar. Biol. Ecol. 128: 75–86.

Lindeman, R. L. 1942. The trophic-dynamic aspect of ecology. Ecology 23: 399–418.

Loo, L., and R. Rosenberg. 1989. Bivalve suspension-feeding dynamics and benthic–pelagic coupling in an eutrophicated marine bay. J. Exp. Mar. Biol. Ecol. 130: 253–276.

Mann, K. H. 1972. Ecological energetics of the seaweed zone in a marine bay on the Atlantic coast of Canada. II. Productivity of the seaweed. Mar. Biol. 14: 199–209.

1973. Seaweeds: their productivity and strategy for growth. Science 182: 975.

Mathers, N. F. 1976. The effects of tidal currents on the rhythms of feeding and digestion in *Pecten maximus* L. J. Exp. Mar. Biol. Ecol. 24: 271–284.

Miller, R. J., and K. H. Mann. 1973. Ecological energetics of the seaweed zone in a marine bay on the Atlantic coast of Canada. III. Energy transformations by sea urchins. Mar. Biol. 18: 99–114.

Menge, B. A., and J. P. Sutherland. 1976. Species diversity gradients: synthesis of the roles of predation, competition and temporal heterogeneity. Am. Nat. 110: 351–369.

Monismith, S. G., J. R. Koseff, J. K. Thompson, C. A. O'Riordan, and H. M.

Nepf. 1990. A study of model bivalve siphonal currents. Limnol. Oceanogr. 35: 680–696.
Munk, W., and G. A. Riley. 1952. Absorption of nutrients by aquatic plants. J. Mar. Res. 11: 215–240.
Myers, A. 1977. Sediment processing in a marine subtidal sandy bottom community. II. Biological consequences. J. Mar. Res. 35: 633–647.
Neilson, B. J., and L. E. Cronin (Editors). 1981. Estuaries and nutrients. Humana Press, Clifton, N. J.
Nichols, F. H. 1985. Increased benthic grazing: an alternative explanation for low phytoplankton biomass in northern San Francisco Bay during the 1976–1977 drought. Est. Coast. Shelf Sci. 21: 379–388.
Nichols, F. H., J. E. Cloern, S. N. Luoma, and D. H. Peterson. 1986. The modification of an estuary. Science 231: 525–648.
Nixon, S. W., C. A. Oviatt, C. Rogers, and K. Taylor. 1971. Mass and metabolism of a mussel bed. Oecologia 8: 21–30.
Odum, E. P. 1961. The role of tidal marshes in estuarine production. Conservationist (New York State Conservation Department, Albany, N. Y.) 15: 1215.
1971. Fundamentals of ecology. 3rd ed. W. B. Saunders, Philadelphia.
1980. The status of three ecosystem level hypotheses regarding salt marsh estuaries: tidal subsidy, outwelling and detritus-based food chains, p. 485–495. *In* V. S. Kennedy (ed.) Estuarine perspectives. Academic Press, New York.
Odum, E. P., J. B. Birch, and J. L. Cooley. 1983. Comparison of giant cutgrass productivity in tidal and impounded marshes with special reference to tidal subsidy and waste assimilation. Estuaries 6: 88–94.
Odum, E. P., J. T. Finn, and E. H. Franz. 1979. Perturbation theory and the subsidy–stress gradient. Bioscience 29: 349–352.
Officer, C. B., T. J. Smayda, and R. Mann. 1982. Benthic filter feeding: a natural eutrophication control. Mar. Ecol. Prog. Ser. 9: 203–210.
Orth, R. J. 1977. The importance of sediment stability in seagrass communities, p. 281–300. *In* B. C. Coull (ed.) Ecology of marine benthos. University of South Carolina Press, Columbia.
Orth, R. J., K. L. Heck, and J. V. van Montfrans. 1984. Faunal communities in seagrass beds: a review of the influence of plant structure and prey characteristics on predator–prey relationships. Estuaries 7: 339–350.
Paine, R. T. 1979. Disaster, catastrophe and local persistence of the sea palm *Postelsia palmaeformis*. Science 205: 685–687.
Patriquin, D. G. 1975. Migration of blowouts in seagrass beds at Barbados and Carracou, West Indies and its ecological and geological implications. Aquat. Bot. 1: 163–189.
Peterson, C. H. 1982. Clam predation by whelks (*Busycon* sp.): experimental tests of the importance of prey size, prey density and seagrass cover. Mar. Biol. 66: 159–170.
Peterson, C. H., and B. F. Beal. 1989. Bivalve growth and higher order interactions: importance of density, site and time. Ecology 70: 1390–1404.
Peterson, C. H., E. A. Irlandi, and R. Black. 1994. The crash in suspension-feeding bivalve populations (*Katelysia* spp.) in Princess Royal Harbour: an unexpected consequence of eutrophication. J. Exp. Mar. Biol. Ecol. 176: 39–52.
Peterson, C. H., and M. L. Quammen. 1982. Siphon nipping: its importance to

small fishes and its impact on growth of the bivalve *Protothaca staminea* (Conrad). J. Exp. Mar. Biol. Ecol. 63: 249–268.
Peterson, C. H., H. C. Summerson, and P. B. Duncan. 1984. The influence of seagrass cover on population structure and individual growth rate of a suspension-feeding bivalve, *Mercenaria mercenaria*. J. Mar. Res. 42: 123–138.
Pohle, D. G., V. M. Bricelj, and Z. Garcia-Esquivel. 1991. The eelgrass canopy: an above-bottom refuge from benthic predators for juvenile bay scallops *Argopecten irradians*. Mar. Ecol. Prog. Ser. 74: 47–59.
Prins, T. C., N. Dankers, and A. C. Smaal. 1994. Seasonal variation in the filtration rates of a semi-natural mussel bed in relation to seston composition. J. Exp. Mar. Biol. Ecol. 176: 69–88.
Prins, T. C., and A. C. Smaal. 1990. Benthic–pelagic coupling: the release of inorganic nutrients by an intertidal bed of *Mytilus edulis*. p. 89–103. In M. Barnes and R. N. Gibson (eds). 24th European Mar. Biol. Symp. Aberdeen University Press.
Raupach, M. R., and A. S. Thom. 1981. Turbulence in and above plant canopies. Ann. Rev. Fluid. Mech. 13: 97–130.
Reusch, T. B. H., A. R. O. Chapman, and J. P. Gröger. 1994. Blue mussels *Mytilus edulis* do not interfere with eelgrass *Zostera marina* but fertilize shoot growth through biodeposition. Mar. Ecol. Prog. Ser. 108: 205–282.
Rowe, G. T. 1971. Benthic biomass and surface productivity, p. 441–454. *In* J. D. Costlow (ed.) Fertility in the sea. Gordon & Breach, New York.
Rowe, G. T., C. H. Clifford, K. L. Smith, and P. C. Hamilton. 1975. Benthic nutrient regeneration and its coupling to primary productivity in coastal waters. Nature 255: 215–217.
Schlichting, H. 1968. Boundary-layer theory. 6th ed. J. Kestin (trans.) McGraw-Hill, New York.
Scoffin, T. P. 1970. The trapping and binding of subtidal carbonate sediments by marine vegetation in Bimini Lagoon, Bahamas. J. Sed. Petrol. 40: 249–273.
Shieh, W. Y., U. Simidu, and Y. Maruyama. 1989. Enumeration of nitrogen-fixing bacteria in an eelgrass (*Zostera marina* L.) bed. Microb. Ecol. 18: 249–259.
Smaal, A. C., and T. C. Prins. 1993. The uptake of organic matter and the release of inorganic nutrients by bivalve suspension feeder beds, p. 271–298. *In* R. F. Dame (ed.) Bivalve feeders in estuarine and coastal ecosystem processes. NATO ASI Ser. Vol. G33. Springer-Verlag, Berlin.
Sorokin, Y. I. 1993. Coral reef ecology. Springer-Verlag, Berlin.
Steever, E. Z., R. S. Warren, and W. A. Niering. 1976. Tidal energy subsidy and standing crop production of *Spartina alterniflora*. Est. Coast. Mar. Sci. 4: 473–478.
Thayer, G. W., S. M. Adams, and M. W. LaCroix. 1975. Structural and functional aspects of a recently established *Zostera marina* community. *In* L. E. Cronin (ed.) Estuarine research. Vol. 1. p. 518–540. Academic Press, New York.
Tracey, G. A. 1988. Feeding reduction, reproductive failure, and mortality in *Mytilus edulis* during the 1985 "brown tide" in Narragansett Bay, Rhode Island. Mar. Ecol. Prog. Ser. 50: 73–81.
Valiela, I., J. M. Teal, S. Volkman, D. Shafer, and E. J. Carpenter. 1978. Nutrient and particulate fluxes in a salt marsh ecosystem: tidal exchanges

and inputs by precipitation and ground water. Limnol. Oceanogr. 23: 798–812.
Wheeler, W. N. 1980. Effect of boundary layer transport on the fixation of carbon by the giant kelp *Macrocystis pyrifera*. Mar. Biol. 56: 103–110.
Wickens, P. A., and J. G. Field. 1986. The effect of water transport on nitrogen flow through a kelp bed community. S. Afr. J. Mar. Sci. 4: 79–92.
Wolaver, T. G., J. C. Zieman, R. Wetzel, and K. L. Webb. 1983. Tidal exchange of nitrogen and phosphorus between a mesohaline vegetated marsh and the surrounding estuary in the lower Chesapeake Bay. Est. Coast. Shelf Sci. 16: 321–332.
Wulff, F. V., and J. G. Field. 1983. Importance of different trophic pathways in a nearshore benthic community under upwelling and downwelling conditions. Mar. Ecol. Prog. Ser. 12: 217–228.

9
Future directions

To get an interesting answer, one needs to pose a pertinent question.

In this final chapter, our aim is two-fold. Firstly, we refine a method for choosing pertinent scientific questions. Secondly, we present some suggestions for interdisciplinary benthic biology/hydrodynamic questions which occur to us. If the latter questions prove to be of no, or less than riveting, interest to those active in benthic biology in the future, then we may console ourselves with the knowledge that our readers should be able to choose pertinent questions for themselves. We firmly believe that the creative natural scientist is rooted in extraordinary powers of observation so that she/he can

Stare in wonder
At this world of beauty

and, by this, be motivated enough to determine how it works.

To recapitulate what is presented in the preceding six chapters (Table 9.1): Each of the subjects reviewed is variable in the attention or success of the studies that are focused on it. This is clearly shown by the rank order of the number of hypotheses associated with each chapter (Table 9.1). Also shown is the associated dominant theory for each subject.

This may also be the best point at which to summarize our scientific judgement on hypotheses dealt with in Chapters 3 through 8 (Table 9.2). Such information is of help in choosing novel questions for further scientific study. Obviously, this will involve hypotheses which do not appear in Table 9.2.

Selecting pertinent scientific questions

The general aspects of asking pertinent scientific questions were considered in Chapter 1. There we learned the difference between proximate

Table 9.1. *Rank order of the number of hypotheses concerning flow and suspension feeders, populations, or ecosystems.*

Number of hypotheses	Subject	Chapter	Associated theory
21	Ecosystems	8	Suspension-feeding benthic communities are important in pelagic–benthic materials cycling
15	Larval biology	3	Passive–active theory of factors influencing the early life history recruitment of benthic suspension feeders
6	Benthic populations	7	Theory of benthic limitation by flow: trophic group mutual exclusion; benthic impoverishment by water movement
5	Initial suspension-feeding responses	4	Unimodal theory of suspension feeding to flow, and bivalve pump theory
4	Behavioral responses to flow	6	Aggregation theory
0	Filtration physiology	5	Particle collection theory

and ultimate, and, at least until now, our questions were limited to the former. In this chapter, perhaps because all we are doing is asking and need not answer them here, we have the luxury of posing both proximate and ultimate questions.

We believe that there are three important additional steps which have to be answered affirmatively in selecting a pertinent scientific question:

- Is the question of personal interest to the scientist?
- Is the question posed testable by scientific methods?
- Will the information gained in the answer be relevant?

The first two points seem self-evident. Because of the considerable amount of work required to complete a research project, an individual with sufficient motivation and commitment is required to complete it. We include here the private, moral responsibility of an individual scientist to ascertain, before it is begun, that the research contemplated is not contrary to the common good. There will always be some questions that

Table 9.2. *Judgements on whether alternate hypotheses are upheld by the experimental evidence presented in Chapters 3–8.*[a]

Table	No.	Hypothesis	Yes	No	Further tests
3.3	1	Estuarine retention of larvae	×		×
	2	Passive shoreward transport by wind	×		×
	3	Passive shoreward transport by internal waves	×		×
	4	Passive shoreward transport by internal tidal bores	×		×
	5	Passive shoreward transport by upwelling	×		×
	6	Plant canopies entrain larvae	×		×
3.6	1	Post-larval dispersal supply	×		×
	2	Post-settlement mortality is flow related			×
3.7	1	Passive larval settlement – soft	×		
	2	Active larval settlement – soft	×		
	3	Passive larval recruitment near small protuberances	×		
3.8	1	Passive larval settlement – hard	×		
	2	Active larval settlement to surface roughness	×		
	3	Active larval settlement to chemical inducers	×		
	4	Benthic density dominated by supply side ecology	×		×
4.5	1	Filtration of sieving suspension feeders a unimodal function of flow			×
	2	Combined passive–active pumping enhancing filtration			×
	3	Energetic costs of filtration low and independent of pump output	×		×
	4	Filtration rate and orientation to flow unrelated in scallops and brachiopods	×		×
6.8	1	Local flow conditions enhancing group living	×		×
	2	Group living allowing protection from flow	×		×
	3	Group living resulting in improved reproductive success	×		×
	4	Group living resulting in better predation protection	×		×

7.2	1	Sestonic food depletion limiting suspension feeding populations	×		
	2	Suspension feeder production positively related to flow	×		×
	3	Energetic water movement impoverishment	×		×
	4	Deposit feeder production inversely related to flow	×		×
	5	Sestonic food depletion controlled by suspension feeder density	×		
	6	Tidal excursions controlling reef path lengths of suspension feeders	×		×
8.1	1	Current velocity inhibited by kelp	×		
	2	Sedimentation greater within a kelp bed	×		
	3	Key predators controlling kelp forests	×		×
	4	Enhanced sedimentation in kelp forests enhancing local production	×		×
	5	Most organic production locally consumed within the kelp forest		×	×
8.2	1	Enhanced wave action increasing gas or nutrient fluxes	×		×
	2	Wave stirring increasing light and therefore photosynthesis of kelp	×		
	3	Wave exposure resulting in shearing stresses which remove kelp plants	×		
	4	Energetic wave action inhibiting grazing by herbivores	×		×
	5	Strong wave action enhancing kelp competitiveness	×		×
8.4	1	Current velocity inhibited by seagrasses	×		
	2	Seagrass primary production a unimodal function of flow	×		
	3	Sediment trapping in seagrass beds enhancing nitrogen availability	×		
	4	Suspension feeder production enhanced in seagrass meadows by flux rates	×	×	×
	5	Suspension feeder production enhanced in seagrass meadows by POM and DOM	×		×
	6	Energetic costs for shore maintenance dependent on velocity	×		×
8.8	1	Benthic plant or bivalve productivity enhanced by water movement	×		×
	2	Seston transport enhancing bivalve reef production	×		×
	3	Bivalve reefs enhancing benthic–pelagic coupling	×		×
	4	Bivalve reefs retaining plant nutrients	×		×
	5	Bivalve reefs influencing microalgal bloom development in eutrophic estuaries	×		×

[a] If further tests are needed as confirmation, this is indicated in the final column.

simply cannot be answered, either by the inductive methods we have eschewed for proximate questions or the deductive ones for ultimate questions (see Chap. 1, the section, Asking the right question). It is therefore important to determine in which category the question lies and abandon it if it is unanswerable.

With the last of the steps, relevancy, involved in examining a potential scientific question, the deliberations become more complex. When we speak of the relevancy of scientific questions, we mean their relevancy either to society as a whole (Peters 1991) or to the particular disciplinary field of science involved – in this case, benthic biology. The former would be classified as an applied, and the latter as a fundamental, question. Fundamental research is equated with curiosity-driven questions, and applied research to questions of obvious practical concern in human society. We believe that Peters (1991) has glossed over the difficulty of recognizing a priori which are the societally relevant questions for *Homo sapiens*. As an example, consider the ecological question addressed in the seston depletion hypothesis of Chapter 7. The origin of this work was clearly curiosity driven and, therefore, was fundamental research. Further development of the concept has generated interest in it as an applied research method to predict the production potential for bivalve aquaculture. This application was not clearly foreseen when this work was initiated in the mid-1970s. For many scientific questions, as for this one, it is often difficult a priori to determine which are applied and which fundamental in nature.

Beyond the difficulty of determining a priori the societal relevance of a particular scientific question, scientists also have a responsibility to their own discipline to maintain high standards in developing a logical and ordered body of knowledge. This would involve all of the types of scientific method outlined earlier in Table 1.4, including those that clearly have no direct societal relevance, e.g. the naming of animals and plants according to Linnaean principles. We consider that such fundamental research in benthic biology is necessary if we aspire to develop practically useful predictive models.

We agree with Peters (1991) that the production of universally applicable deterministic predictive models is the most important sign of a well developed discipline. An example is provided by the discipline of hydrodynamics, which has a respectable 150-yr-long history developed by engineers and scientists, both practically and theoretically. Hydrodynamic theory has accumulated numerous useful models or empirical methods which help manipulate water resources. By contrast, benthic biology is a much younger discipline – 86 years old in 1997, if we accept 1911 as its origin (see Chap. 1). It still has little theory or universally applicable predictive models, which remain to be developed and fully accepted. In order that this can be done, resolution of fundamental questions in benthic biology, as well as repeated efforts to apply this work, must go on hand in hand. We expect that important parts of this effort will be dependent upon hydrodynamic theory and practice.

We suggest that it is a healthy state of affairs that a certain tension between societal and scientific disciplinary relevance questions should be swirling within the mind of each practising scientist. But we believe that scientists themselves should have the ultimate responsibility for framing both fundamental and applied scientific questions.

Proximate fundamental questions

Examples of questions which can be answered by currently available scientific methods are shown in Table 9.3. Here we have selected a representative question, or questions, applicable in each chapter, viz: 3, a and b; 4, c; 5, d; 6, e, f, and g; 7, h and i; 8, j.

Questions a and h in Table 9.3 belong within the theory of mutual exclusion. Wildish (1977) suggested the active selection of certain flow–sediment habitats by larval colonizers, or post-settlement competition among the recruits whose outcome was dependent on flow–sediment interactions. Butman (1987) also hypothesized about this as follows: “Suspension feeders may not co-occur with deposit feeders because the two functional groups have larvae with different fall velocities that are passively deposited in different fluid-dynamic environments.” Thus, questions a and h in Table 9.3 represent a comprehensive coverage regarding the relative importance of flow–sediment interactions as they directly or indirectly affect both passive and active settlement and post-recruitment mortality. As far as we are aware, such questions have not yet been addressed in the literature.

We present in Table 9.4 some generic hypotheses developed from questions a and h of Table 9.3. The field experimental testing of hypotheses 1 and 2 in Table 9.4 might be achieved by providing various hard surfaces and de-faunated soft sediments in core tubes checked daily for new larval settlers. We believe daily observations to be a satisfactory compromise to measure settlement rather than recruitment, although, as pointed out by Olafsson, Peterson, and Ambrose (1994), this measure cannot avoid some very early post-settlement mortality. Although it is

Table 9.3. *Representative fundamental proximate questions in the interdisciplinary field of hydrodynamics/benthic biology.*

No.	Scale		Question
	Spatial (km)	Temporal (y)	
a	0.1–0.01	0.01	Do hydrodynamic–sediment interactions determine the settlement success of deposit versus suspension feeding larvae so that one or the other is variably excluded from the population by the physical conditions?
b	0.1–0.01	0.1	Do hydrodynamic–seston interactions determine the recruitment success of suspension feeders at a particular location?
c	0.1–0.01	0.1	Can we present a model of the internal regulators and external environmental variables which influence filtering, feeding, and growth of a named suspension feeder?
d	—	—	What is the mechanism of filtration in active suspension feeders?
e	0.001	0.001	How do tube-living suspension-feeding epifauna respond behaviorally to flow?
f	—	—	What ethological changes among epifaunal suspension feeders minimize dislodgement by flow?
g	0.001–0.1	>1.0	What causes epifaunal suspension feeders to aggregate at some sites?
h	0.1	>1.0	Do hydrodynamic–sediment interactions determine the post-settlement mortality rate of deposit- versus suspension-feeding populations in such a way that one or the other is variably excluded by the physical conditions?
i	0.1	>1.0	What are the mechanisms of water movement–induced inhibition of suspension feeder production?
j	1–1000	>1.0	How do physical oceanographic currents and wind–wave effects interact to influence pelagic–benthic coupling in a macrotidal estuary?

Table 9.4. *Generic hypotheses referable to the special theory of mutual exclusion.*

No.	H_0/H_1	Hypothesis
1	H_0	The settlement of deposit-feeding larvae is less than expected from physical transport calculations
	H_1	The settlement of suspension-feeding larvae equals that expected from physical transport calculations
2	H_0	The settlement of deposit feeders is greater on soft substrates than on hard ones
	H_1	The settlement of suspension feeders is greater on hard substrates than on soft ones
3	H_0	Post-settlement growth of deposit feeders is inversely related to velocity but unrelated to seston concentration
	H_1	Post-settlement growth of suspension feeders is unimodally related to velocity and seston concentration
4	H_0	The post-settlement mortality rate of deposit feeders is inversely related to velocity
	H_1	The post-settlement mortality rate of suspension feeders is unimodally related to velocity
5	H_0	The post-settlement dispersal of deposit feeders is to lower mean velocity conditions
	H_1	The post-settlement dispersal of suspension feeders is to an optimum velocity range, characteristic for each species

very difficult, it is not impossible, using modern time recording sensors, to obtain a measure of the velocity field in the benthic boundary layer with changing tides so that the passive larval transport rates can be calculated. For hypothesis 2 (Table 9.4), a number of sites with varying flow characteristics could be selected so that a range of velocities and, therefore, passive larval transport rates could be compared. It may be preferable to run both these experiments in laboratory flumes, in which case the hydrodynamic transport calculations with a well defined unidirectional flow would be much easier. Special requirements of such flume experiments would be that they were flowthrough and not recirculating devices and that the incoming seawater was filtered to exclude naturally occurring larvae. It should then be possible to control the absolute densities and proportions of suspension and deposit feeders present during the flume experiments. Settlement responses by the larvae would be indicated by comparing the proportions of the trophic groups which colonize each type of substrate.

Post-settlement growth and mortality of deposit- versus suspension-feeding juveniles (numbers 3 and 4, Table 9.4) probably necessitate a field experimental test. This is because of the great difficulty of simulating the complex water movement patterns present in the sea for the weeks to months required for determining mortality rates. A number of sites with a range of water movement conditions would be required, and various substrates precolonized by representative juveniles of each trophic group placed at each, so that growth and mortality could be assessed. For completeness in Table 9.4, we have included as number 5 a post-settlement dispersal hypothesis since some, although not all, species are potentially able to leave an originally settled, but unsuitable, locality. Almost certainly it would be a field experimental test, perhaps of the type presented by Emerson and Grant (1991).

Question (g) of Table 9.3 may prove to be a special case of (a) in the same table. This is because aggregated larval settlement may be behaviorally or passively controlled by flow. Thus, field or laboratory experiments similar to those offered for hypotheses 1 and 2 (Table 9.4) would serve to sort out the behavioral and/or physical mechanisms involved. Larval responses to constant unidirectional flows in flumes do not, however, bear much resemblance to flow variation in the field, where tidal (regular, hourly) and wind–wave (stochastic, days) effects produce complex and unpredictable patterns. The aggregation may also be induced by behavioral responses to chemical inducers or cues to settlement, as has been demonstrated for some species (see review by Pawlik 1992). This question is of central concern in benthic ecology because of the ubiquitous occurrence of aggregation among suspension feeders at various spatial scales. We suggest that it requires the use of fresh approaches in observing sublittoral spatial distributions of benthos, as is beginning to be done in some field studies discussed in Chapter 6 at SCUBA diving depths. We also suggest that interdisciplinary studies involving biologists, physical oceanographers, and recent sedimentary geologists should be developed to provide accurate observational data over small spatial scales (Table 9.3), but longer, at least seasonal, temporal scales regarding benthic boundary layer flow and sediment movements.

Whether hydrodynamic–seston interactions are important in the success of suspension feeders (Table 9.3, b) provides an alternative hypothesis to that of (a) and (h). The concentration of seston which reaches a particular location, as well as seston quality, will have an effect on active suspension feeder growth directly in a physiological sense (Chap. 4). Indirectly, in an ecological sense (Chap. 7), it will affect population growth by downstream seston depletion effects. Both direct and indirect effects can result in starvation and death, affecting the post-settlement mortality and production of suspension feeders. In an experimental test of this question, the null hypothesis might be that newly settled larval deaths and growth are unaffected by changes in seston concentration or quality. Thus, the alternate hypothesis would be that these seston variables do affect mortality and production. Laboratory experiments are envisaged in which the flow is kept constant while observing test panels with a known density of post-settlement larvae. Different runs of the experiment, preferably in a multiple-channel flume, could be set up with different seston concentrations or qualities, and regular observation of deaths and growth made.

For the physiologically based questions of Table 9.3 (c and d), we suggest that the conceptual model of scallop feeding in Wildish and Saulnier (1993) could be applied more generally to active suspension feeders. With the physiologists' interest in classifying environmental factors (e.g. Fry 1947), this may assist in determining the nature of the external factors which affect filtration/feeding. Although the Fry paradigm (Kerr 1990) is applicable in autecology, we do not believe that it is applicable in synecology for the reasons explained in Chapter 7. Contrary views seem to have been expressed by Kerr (1990) and Jørgensen (1992a). To understand the internal factors which may affect filtration/feeding in bivalve molluscs – inclusive of valve and mantle opening/closing, pumping performance through time, causes of gill bypassing, seston selection at gills or mouth palps, satiation effects, and how growth is controlled – it would be necessary to link them with nervous or neurohormone systems. Detailed knowledge of the integrated control systems involved would, of course, be of potential value in enhancing the culture of commercially important species.

As we have seen in Chapter 5, question (d) of Table 9.3 is unanswered for species of active suspension feeders, such as the bivalve molluscs. All that we can suggest is more intense direct observation of feeding behavior in general, with a video camera, so that it is continuous, or with endoscopes (Chap. 2). Neither of these methods would be able to confirm a mechanism directly, but the clues it could provide, coupled with physical and theoretical modelling of the kind proposed by Jørgensen (1981), might be enough to devise further testable hypotheses.

For the two ethological questions of Table 9.3 (e and f), we believe that both field and laboratory observations of a wide range of species of suspension feeder are required. For locations which are deeper than SCUBA diving depth, underwater, time-lapse photography, from either an anchored vessel or a remotely operated vehicle, may be adequate. In the laboratory, a video camera focused through the acrylic plastic (Plexiglas) walls of the flume may often be suitable. With the latter equipment, it should be possible to record tube-living suspension feeder behavior as a function of unidirectional or oscillatory flows – some results with spionid polychaetes were discussed in Chapter 6. As postulated in Chapter 4, we would expect feeding to be unimodally related to velocity, and at the upper end of this relationship, to see behavioral responses to minimize dislodgement by water movement, e.g. halting feeding and retiring within the tube. Free-living epifauna may also respond in a characteristic way to flows which are nearly strong enough to sweep them away. We would expect that each species could tolerate a particular velocity range and thus would indicate to the experienced benthic biologist which ambient flows were available at the site. The bottom line is that more detailed observations of suspension feeder behavior must be completed before hypotheses and experimentation can be initiated to determine the mechanisms involved.

In some ways, question (i) of Table 9.3 is a continuation of (f), although it is cast at the synecological level. Numerous studies have shown the drastic influence that a violent storm can have on nearshore benthic communities (see Chap. 7). Even if it were possible to simulate wave erosion events in the flume, it would be necessary to provide natural groups of organisms, since it has been established (Chap. 6) that the erosion threshold may be influenced by the presence of other individuals of the same, or different, species. Almost certainly there will be energetic costs associated with maintaining position in flows, whether *behavioral*, e.g. in the ability of sand dollars to burrow prior to storms and scallops to recess into sediments (Chap. 6), or *structural*, e.g. in skeletal stiffening by spicule formation in sponges, deposition of calcium salts in corals, or secretion of adhesives by barnacles (Chap. 6). As far as we are aware, the comparative energy costs associated with this effort have yet to be determined. Such costs would be expected to increase with the water movement experienced by populations at each locality examined.

At subcritical erosion thresholds in either unidirectional or oscillating flows, there is an inhibitory effect on feeding – the c stage of Chapter 4

– due to increasing flows. We believe that behavioral observations, perhaps with the video camera over a sufficiently wide range of velocities, will indicate characteristic changes in behavior which are velocity dependent in both passive and active suspension feeders. One important ecological implication, applicable only to active suspension feeders, is that if they are filtering at less than optimum carrying capacity as a result of flow inhibition, they will effectively share out the available seston along a greater length of the downstream path among a greater number of individuals (Wildish and Kristmanson 1993). This is an example of an "emergent property" perceived at a lower hierarchical level and predicted to occur at the population level. If this prediction is supported by experimental tests, it may help explain why current ecological modelling in this respect is so inaccurate.

The final question, (j), in Table 9.3 is clearly cast at the ecosystem level. We presented in Chapter 8 sufficient evidence to show that the original ecosystem model of pelagic–benthic coupling which assumed a direct spatial link between the pelagos and benthos was often contradicted by the observations. We believe that a new predictive model is required to assist in answering this question. An ecological simulation model of the type required would combine physical oceanographic variables with flows of energy or nutrients. Models, or submodels, of this general type, based on the work described in Chapters 7 and 8, have been designed for the practical purpose of determining carrying capacity and optimal seeding density of commercially important bivalves (e.g. Newell and Shumway 1992; Grant et al. 1992; Heral 1992). A more general submodel is given by Herman (1992) as a means of determining the role of suspension feeders in ecosystems, e.g. in pelagic–benthic coupling, in the context of the framework of holistic estuarine ecosystem models. We expect that an answer to the question of Table 9.3 will require the combined efforts of a physical oceanographer and a marine biologist working at spatial scales determined by the former. The results of such models are site specific since the degree of transport during pelagic–benthic coupling will vary; e.g. in the Bay of Fundy it will be higher than in the nearly tideless Baltic Sea. The variables measured and included in ecosystem simulation modelling are often complex, e.g. in measuring primary production, and suffer some uncertainty regarding their accuracy. The reader should also be aware that ecosystem simulation modelling does not offer the clean kill with Occam's razor between two simple alternatives common at lower hierarchical levels of science

Table 9.5. *Some representative applied benthic biology questions in which ambient hydrodynamics play a part.*

No.	Scale: Spatial (km)	Scale: Temporal (yr)	Question
a	0.001	~1	What are the environmental variables which affect the physiology of feeding/growth in commercially important bivalves?
b	1.0	~1	What is the carrying capacity of blue mussels in suspension culture in a named location?
c	10–100	1–100	What are the benthic and consequent ecosystem effects of landfilling and wharf building at the mouth of an estuary?
d	1–1000	1–100	What are the near and far field ecological effects of building a major tidal power dam?

(reductionism). One reason for this is that the model is a means of deriving hypotheses for testing, but is itself an untested hypothesis (see Jørgensen 1992b).

Proximate, applied benthic biology questions

If the answers to questions clearly have practical value in, for example, aquaculture or environmental management (Table 9.5), it is clear that one is dealing with an applied question. Thus, within Table 9.5, a full answer to (a) will help a culturist to choose the best locality to site a bivalve farm and to (b) to set it up to make the maximum use of the potential available. Ecosystem simulation modelling in relation to (c) and (d) should ideally precede large-scale civil engineering projects. This might include wharf or tidal barrage building and the modelling results used interactively in the debate regarding coastal zone management, but before construction commences.

As we saw in Chapters 4 and 5, the effects of environmental variables on bivalve feeding, particularly flow, have not yet been examined in many species. We suggest that the filtration rate of bivalves needs to be examined in both siphonate and non-siphonate species with respect to velocity (including unidirectional and oscillating flows of different periodicity), seston concentration, seston quality, temperature, and salinity.

Commercially important siphonate bivalves include many species of clams and cockles, while non-siphonate forms include various species of scallops, oysters, and mussels. Some information about the unimodal filtration/feeding response of non-siphonate bivalves is presented in Chapter 4. As far as we are aware, no siphonate bivalves have yet been tested in this way. Thus, it is premature to think of a universal unimodal feeding response to velocity applicable to all bivalves.

For siphonate bivalves, the most important part of the flow response is that which is inhibited, (c) in Fig. 4.1. It needs to be de-limited for a much wider range of species and the mechanism – including the possibility of velocity-induced valve closing – verified. Research necessary for this would include real-time observations of siphon opening/closing in controlled flume flows, as well as experiments to understand the environmental factors acting on the control mechanism. The capacity of bivalves to sense ambient flow pressures, seston concentration, and seston quality also requires more detailed study.

As indicated in Chapter 4, some bivalves may be stimulated or inhibited to filter feed, suggesting that they respond to chemicals in the environment. This work is still in its infancy, yet may provide a mechanism to explain why filtration is inhibited in the presence of some microbial or toxic microalgal species.

As a point of practical importance to those thinking of running filtration/feeding experiments with bivalves, control filtration rates are better obtained with mixed rather than unialgal cultures. Another unsatisfied need in this research is a method of measuring seston as an indication of bivalve food quality.

With regard to (b) in Table 9.5, we mentioned several authors who have already described ecological methods designed to determine the carrying capacity of cultured bivalves. The model of Newell and Shumway (1992) may be described as a physiological approach to a carrying capacity model of blue mussels, and at a smaller spatial scale than, for example, the ecosystem simulation model presented by Grant et al. (1992). The former relies on standard methods of determining filtration rates of individuals, which certainly do not include velocity as the controlling variable we believe it to be. Filtration rates are then linked to scope for growth models in which temperature, sestonic quality, and bivalve density are input variables. The model predicts production for a specified location where turbulent seston flux and uptake by the mussel bed are simulated. The ecosystem simulation model used by Grant et al. (1992) requires extensive site-specific effort in field studies of physical and biological variables in order to determine realistic inputs for the model. This and similar models (e.g. Smaal 1991; Bacher et al. 1991) cannot be directly extrapolated to other localities.

We feel that this still developing field is close to a stage where simple, universally applicable benthic carrying capacity predictions, based on two or three variables, are feasible. Thus, the dependent variable would be bivalve growth or production of a named species for a particular type of culture method. The independent variables might be flow related ($x1$) and food related ($x2$). Possibilities for $x1$ are γU (Chap. 7), U^* (Newell and Shumway 1992), or a diffusion coefficient based on tidal excursion length and residence times. Possibilities for $x2$ are mean seston concentration, perhaps modified by seasonal temperature and seston quality parameters. Further work is required to choose the most efficient predictors, as well as establish the empirical relationship over a wide geographic area. Simple models like this have the potential to be practically useful for all bivalve culturists in, for example, choosing between alternate sites to begin a bivalve culture venture. Because the data are relatively simple to obtain and do not depend on the input from continuously recording electronic sensors, the cost involved would be moderate.

Commonly, a large proportion of the human population of any country lives and works near the sea, often at the mouth of an estuary. The seaward part of the interface from the high tide line on land to the edge of the continental shelf at sea is termed the coastal zone. Humans use the coastal zone for a great variety of purposes. Currently, these include commercial fishing; aquaculture; recreational activities inclusive of whale watching, sport fishing, etc.; source of industrial cooling or process water; receiving environment for industrial and municipal wastes; mineral and hydrocarbon exploration; as well as commercial shipping. It is not surprising that, because of these multiple uses of the coastal zone, there are frequent resource-use conflicts.

In general, coastal zone resource conflicts are well documented (e.g. Wells and Ricketts 1994), although we believe that question (c) in Table 9.5 has received insufficient attention in proportion to the frequency with which landfilling and wharf building are an integral part of modern estuarine developments. Many civil engineering projects, over periods up to 200 yr, accumulate; therefore it is usually impossible to obtain a before/after picture of the estuarine hydrography, sedimentary, and benthic biological environments.

During benthic biological studies to determine the effects of dredging and dumping in the Saint John estuary, New Brunswick, Canada, circumstantial evidence suggested that landfilling and wharf building in the harbour had significant benthic biological effects (Wildish and Thomas 1985). Saint John Harbour serves a small industrialized city of ~120,000 people, and the building of a breakwater there resulted in localized sediment deposition that required maintenance dredging. In sediments well separated from the dredging and dumping areas, Wildish and Thomas (1985) noted benthic impoverishment which seemed to be associated with sediment–tidal current stresses. Whether these stresses were indirectly caused by landfilling and wharf building around the harbour as a result of altered hydrodynamics could not be established during this study, although it seems a likely possibility.

What is required in a definitive study is a multidisciplinary team – physical oceanographer, surficial sedimentary geologist, and biologist – working closely together to document the relevant characteristics of an estuary. The best field site would be one where the initial work could be done before any harbour development, involving wharf building, had been initiated. This would be followed after the harbour had been built to determine what environmental changes resulted from the civil engineering work. Important in this would be to understand any ecosystem level changes caused by the development, and not just those associated with the benthos. These would apply to any possible multiple-use resource conflicts within the estuary, e.g. between commercial fishing and shipping.

The final question, (d), posed in Table 9.5 concerns the ecological effects of building a major tidal power dam. Because of fluctuations in the price of oil, many countries have sought alternate sources of power, inclusive of tidal power generation of electricity. Engineering feasibility and preliminary environmental studies of the potential effects of such a development have been made in the 1920s and 1970s in the Bay of Fundy, Canada. The Bay is actually a large macrotidal estuary with a subtidal area of ~11,149 km^2. Baker (1984) described the preferred tidal dam site in Minas Basin in the upper reaches of the Bay. A finite difference model of the Gulf of Maine/Bay of Fundy was constructed by Greenberg (1979) to investigate tidal phenomena, both with and without the Minas Basin tidal barrage in place. Because of resonance effects in the Bay, it was estimated that tidal differences due to the dam would be felt as far away as Boston, causing possible loss of terrestrial habitat, greater storm surges, and changed sedimentary environments. In regard to the latter, Amos and Greenberg (1980) and Greenberg and Amos (1983) have predicted potential local, near the tidal barrage, and far-field sediment effects. Potential secondary production lost by benthic suspension feeding animals was estimated to be 17% throughout the Bay, which is a non-significant value (Wildish et al. 1986), largely because of the very large variances associated with this form of sampling. Further development of ways to predict changes occurring near the barrage are required, as well as methods for determining ecosystem level effects.

To date, the price of oil has not yet reached levels which mandate the use of alternative energy sources, so the predictions presented have not been tested in a real-world field experiment. Both in Canada and in other parts of the world, only smaller demonstration tidal power projects have so far been initiated.

Ultimate questions

The history of life on planet Earth is considered by Lincoln, Boxshall, and Clark (1982) to have begun some 4500×10^6 yr ago. Throughout this time, tectonic plate movements of all the major land masses and eustatic changes in seawater levels have repeatedly occurred (Myers 1994). Climatic and ocean current changes separate from or associated with these events have also occurred (Berggren and Hollister 1977). Thus, the evidence from geology suggests continuous environmental change through geological time caused by these universal forcing functions. Additional discontinuous events such as meteorite impacts (Swinburne 1993) or violent earthquake activity (Officer 1993) may have sudden and catastrophic effects on animal life on both sea and land. The importance of violent discontinuous events in relation to extinction events is still being debated (e.g. Moses 1989; Quinn and Signor 1989; Serjeant 1990). Because of the adaptational nature of plant and animal life, we should expect continuous changes – evolution or extinction – of these organisms to fit, or exclude, them from such changing environments.

All of the theories or hypotheses which can be formulated for the ultimate questions we have posed in Table 9.6 can only be addressed by deductive methods. In deduction, the formal scientific propositions must be made without direct observation of the natural or experimentally contrived events involved. Because the scientist cannot be present when the evolutionary events associated with the questions in Table 9.6 occur, the only evidence for, or against, an hypothesis is circumstantial. A frequent problem with circumstantial evidence is knowing how much is required to prove that a particular hypothesis is correct or to choose between competing hypotheses when the balance of evidence is nearly

Table 9.6. *Some ultimate questions related to hydrodynamics and referable to suspension feeders or the communities in which they live.*[a]

Discipline	No.	Question
Physiology	1	Why did active suspension feeders evolve from passive forms (or vice versa)?
	2	Why did a ciliary pump evolve?
Ethology	3	Can the fact that some sessile suspension feeders have short, and others long, periods of planktonic life be explained hydrodynamically?
	4	Why do hydrodynamic factors cause epifaunal suspension feeders to aggregate?
	5	What adaptive imperatives resulted in swimming bivalves?
Ecology	6	How did larvae with specific responses to environmental variables, e.g. light, gravity, and flow, evolve?
	7	What causes a mixed benthic assemblage dominated by suspension feeders to evolve?
	8	Why do ecosystems evolve over geological time?

[a] From Table 1.2.

equal between them. Thus, deductive methods lack the power of decision available to those answering proximate questions using inductive methods.

As background to our consideration of ultimate questions (Table 9.6), we discuss three common types of circumstantial evidence which might be available. The first involves the geological record and could include direct preservation of parts of the organism, which gives clues to the species evolutionary history. Alternatively, the geological record could provide inferences about past climates and environment from the types of organisms preserved in ancient sediments, e.g. the palaeoecology of marine sediments (Molfino 1994). Surficial geologists study modern sediments and the benthic flora and fauna they contain to improve their view of the factors which may have operated in the geological past (e.g. Rhoads and Young 1970; Thoms and Berg 1985). Surficial geologists also have independent means of measuring the age of sediments by isotopic decay rates (Molfino 1994). A second general approach is through phylogeny. Modern taxonomy requires determining a wide range of, largely, morphological characters in each species, and this has led to the construction of phylogenetic trees which are thought to indicate the degree of relatedness of each taxon in a common group. Increasingly more important and independent views of genetic relatedness are provided by modern biochemical methods, notably deoxyribonucleic acid (DNA) sequencing of mitochondria or other genetic material (e.g. Field et al. 1988; Powers, Allendorf, and Chen 1990). The third general approach which could provide circumstantial evidence for ultimate questions is biogeographic study of patterns of the distribution of animals and plants. This method allows the palaeoecologist to see how spatial mechanisms in modern environments affect speciation, extinction, and patterns of species richness (Myers 1994), thus suggesting how it could have happened in the past.

To return now to the ultimate questions of Table 9.6: For the first two physiological questions, we propose that active suspension feeders evolved from passive ones to enable the former to colonize a wider range of flow environments than those permissible to passive suspension feeders. Thus, for the passive suspension feeding gorgonian coral *S. suberosa* the range was $7–9\,cm \cdot s^{-1}$; the unimodal peak relating velocity and filtration was sharper, and proportionately more of the (a) part of the response was present (Chap. 4, the section, Background). For scallops as an example of an active suspension feeder, the range is 2 to $>45\,cm \cdot s^{-1}$, the unimodal peak being broader with less (a) and more (c) in the filtration response (Chap. 4, the section, Background). One test of this hypothesis would be to compare the unimodal responses of a wide range of passive versus active suspension feeders over a sufficiently wide range of ambient velocities. We do not mean to imply that the possession of a ciliary pump insulates the active suspension feeder from a need to be in the flow. This is because active suspension feeders commonly form beds of near-monoculture composition which require energetic turbulent fluxes to supply the necessarily large amounts of sestonic food that such populations need to avoid seston depletion effects.

A further aspect of the possession of a ciliary pump is the concomitant energetic costs involved. The following are two aspects of such costs:

- The ontogenetic development of a trophic fluid transport system with an in-line ciliary pump
- The physiological cost of operating a ciliary pump relative to the other physiological activities of a suspension feeder

As far as we are aware, the first of these has not yet been considered. With regard to the latter, Jørgensen, Larsen, and Riisgård (1986) have reviewed the total volumetric flow through active filter feeders in relation to concomitant oxygen consumption. Two methods of determining the relative energy costs of active suspension feeding were considered by Jørgensen et al. (1986). The first involved comparing the product of the operating pressure of the pump and volume of water filtered, from which an estimate of the work done could be obtained. This estimate in relation to the total of all energy expended by various active suspension feeders amounted to ≤1%. A second method involved estimating the energy expenditure of the ciliary pump. In blue mussels, the ciliary beating work done at the gills amounted to 0.7% of the total energy budget. Thus, at a normal operating efficiency of the mussel pump, the energetic cost would be a small percentage of the total metabolic cost (Jørgensen et al. 1986).

The preceding considerations are consistent with the view that the low pressure ciliary pump in bivalves operates at one rate and may be either on (valves open) or off (valves closed). Differences in volumetric flow through the bivalve trophic fluid transport system are controlled by compression or collapse of the system by valve closure, as indicated by the relative size of the exhalant opening. Environmental control of the degree of compression has been shown to include ambient velocity in scallops (Wildish and Saulnier 1993). The large volumetric flow through the trophic fluid system of blue mussels is primarily adapted to the filtration function (the need for large throughputs of ambient seawater from which the gills can obtain sufficient sestonic food to grow). Famme and Kofoed (1980) showed by perfusion experiments through the trophic fluid transport system of mussels that there was no change in oxygen uptake after gill removal and that this process did not occur there but across the general mantle cavity walls. Jørgensen et al. (1986) point out that the diffusive boundary layer in the mantle cavity is ~0.1 mm thick and, because laminar flow is present, only a small fraction of pumped seawater is available for oxygen exchange. This explains the finding (Jørgensen et al. 1986) that oxygen consumption in mussels is an asymptotic function of the volumetric flow rates within the trophic fluid transport system. Maximum oxygen consumption rates are reached at relatively low volumetric flow rates through the mussel, at flows equal to approximately one-fifth of the maximum pump capacity. The limits here are controlled by the diffusion rates across the mantle cavity wall, which vary with the thickness of the boundary layer and, therefore, flow rate.

Another view considered by Jørgensen et al. (1986) was that the curvilinear increase of oxygen uptake as a function of pumping rate was a direct indication of the energetic cost of ciliary pumping (e.g. Verduin

1969; Bayne, Thompson, and Widdows 1976). This interpretation implies that ciliary pumping is a major energetic cost and is at odds with the low energetic costs estimated by Jørgensen et al. (1986), as mentioned. Direct energy measurements of ciliary beating in isolated mussel gills by Clemmesen and Jørgensen (1987) also suggest that costs are low. The explanation for the experimental results obtained by Bayne et al. (1976) could be that the full volumetric capacity of the mussel pump remained untested because the experimental conditions offered resulted in inhibited pumping. Thus, the results were all obtained in the lower range of volumetric pump output, where the thickness of the diffusive boundary layer and mantle cavity flow velocities are the determinants of oxygen uptake.

Bougrier et al. (1995) examined the allometric relationships among temperature, clearance, and oxygen uptake rates for non-reproductive oysters *Crassostrea gigas*. These researchers showed that the latter was a unimodal function of temperature, with a peak of 19°C that clearly was not related to oxygen consumption over the same temperature range.

Clearly, further work is required in this area to determine for a wide range of active suspension feeders what the relative energy costs are, inclusive of ciliary pumping. If the concept of Jørgensen et al. (1986) is upheld, then we can see the low pressure ciliary pump as an energetically efficient organ adapted to suspension feeding in a dilute suspension of seston in seawater. Active suspension feeders should occur in a much wider range of ecological niches than passive forms if the theory of expansion to a wider range of niches is correct.

Turning now to question 3 (Table 9.6), we know that the observations incorporated in it are consistent with the types of larvae associated with benthic suspension feeders (Chap. 3). Previous authors have considered what has caused the larval stages of ontogeny to evolve in such different directions, one to lecithotrophy, the other to planktotrophy (Table 9.7). Vance (1973) proposed models in which natural selection chose among individual parents, specifically those who maximized successful settlement per unit of reproductive effort. Chia (1974) proposed that lecithotrophy developed from planktotrophy as a means of minimizing reproductive energy expenditure. Thus, the few larger eggs of lecithotrophy allowed a lower total reproductive expenditure because far fewer eggs were required to achieve the same settlement rate. Strathman (1974) proposed a concept in which spatially and temporally varying environments were seen as the driving force of selection and, hence, evolution of divergent strategies to lecithotrophy or planktotrophy. By

Table 9.7. *Comparison of two larval strategies in benthic marine suspension feeders.*

Characteristic	Lecithotrophic larva	Planktotrophic larva
Larval type	Larger, non-feeding, food store	Smaller, feeding in plankton, no food store
Larval numbers	Few	Many
Larval period	Short	Long
Larval dispersal distance (maximum)	Short	Long
Inbreeding potential	High	Low
Biogeographic dispersal rate	Slow	Fast
Biogeographic range	Endemic	Pandemic (cosmopolitan)
Geological history of taxa	Short	Long
Speciation rate	High	Low

After Vance (1973); Strathman (1974); Chia (1974); Scheltma (1986).

this proposal, individual suspension feeders that spread sibling larvae evenly over a wide area, inclusive of temporally favorable and unfavorable niches, had higher fitness over many generations. By contrast, lecithotrophic forms produced recruits that varied in settlement only according to the temporal changes that occurred at the parental niche. A review of current ideas relating to the evolutionary ecology of larval types has been presented by Havenhand (1995).

We suggest that the environmental grain postulated in theoretical studies of Strathman (1974) could be an hydrodynamic one. Then, species with lecithotrophic larvae are betting that the best hydrodynamic environment for benthic life is provided close to the parental population. Species with planktotrophic larvae, on the other hand, are betting that there is a better flow environment at some distance from the parental population. Scheltema (1986) has summarized the evidence in the geological record pertaining to the larval strategies of Table 9.7, including biogeographic range, occurrence of extinctions, and speciation rate as indicated by the geological record. Nor is it impossible that rapid larval evolution could occur and be observed directly during the working life of a scientist. As an example, the barnacle *Balanus amphitrite* was introduced to a newly created inland sea in California, where many environmental variables differed from those of the donor coastal populations (Raimondi 1992). The latter author was able to observe the larval phenotypic differences between the inland sea and coastal populations and establish, by field and lab, transplant experiments that the phenotypic divergence was probably due to natural selection.

Fewer lecithotrophic larvae are produced by benthic suspension feeders. They have a shorter larval period and local settlement with dispersal limited to a small ambit near the parental population (Table 9.7). The deleterious consequences frequently associated with inbreeding do not seem to have occurred in some taxa possessing lecithotrophic larvae (e.g. Grosberg 1987). In colonial ascidians, although fertilization barriers which prevent members of the same colony from mating are present, breeding experiments undertaken by Grosberg (1987) suggest that outbreeding is very costly in terms of fitness and that the degree of inbreeding within a local population is high. Grosberg and Quinn (1986) have shown that in at least one colonial ascidian the mechanism of maintaining closely related individuals in a population involves kin recognition at the larval stage (see also Chap. 6, the section, Epifauna and aggregation). In some species of Bryozoa, Keough (1984) showed that kin recognition by larvae enhanced the formation of colonial aggregations consisting of closely related individuals (see also Chap. 3, the section, Hard substrates).

Suspension feeders frequently do form aggregated populations of high density (Chap. 6). The hydrodynamic aggregation theory is that because of localized flow conditions, there are some niches that provide good, and some bad, opportunities for the later benthic stages of life. In some cases, the larval settlement cues are also hydrodynamic ones (see Chap. 3, the section, Hard substrates). Thus, the presence of a living suspension feeder – either larva or adult – on a particular substrate can indicate to a potentially settling larva that it is a suitable niche which could support growth. However, there may be other adaptive advantages associated with aggregation in suspension feeders, besides flow-related ones (see Table 6.8, the section, Epifauna and aggregation). Planktotrophic larval dispersal has also been suggested to be a mechanism carrying the larvae away from the crowding associated with aggregated distributions (Strathman 1974).

The bivalves capable of swimming were discussed in Chapter 6 (the section, Swimming). Among the best swimmers are the scallops, of which ~7000 species of both extant and extinct forms of Pectinidae are known (Waller 1991). The morphological criteria employed in taxonomy have also been used to construct a proposed phylogenetic tree to indicate a geological history beginning in the Oligocene. The available evidence in the fossil record reviewed by Waller (1991) gives little assistance in determining which environmental conditions led to the establishment of swimming in the scallops. Possible factors are predation, active migration, and post-settlement dispersal associated with either improved trophic conditions or more effective reproduction. Some modern scallops do elicit escape swimming responses when tissue extracts of potential predators are introduced (e.g. Stanley 1970). Morton (1980), however, considers that the fast and efficient swimming of *Amusium pleuronectes* is adapted for seasonal migration and reproduction. It is tempting to speculate that swimming in this scallop has evolved from a predator escape mechanism, but any evidence to confirm this suggestion is absent. Some studies of the shell form of extant bivalves have proved to be helpful to geologists in interpreting how fossil bivalves lived, and by inference, the type of environment in which they were found (Stanley 1970).

The evolution of photo-, geo-, and rheotaxes in larvae (number 6, Table 9.6) can best be understood in an ecological context. The use of light or gravity as environmental cues to direct the swimming of planktotrophic larvae could have occurred anywhere on the continental shelf. Planktotrophic larvae originating from sessile benthic suspension feeders all need to swim up to the surface layers of seawater, where microalgal food is available for them to grow. After a period of planktonic development (Chap. 3), the competent larvae need to return again to the bottom, where they find a suitable substrate and complete the settlement process. The benthic stage is the major growth part of the life cycle, and post-settlement success depends critically on wise larval choices made during settlement, particularly for those species lacking a post-settlement dispersal mechanism. The simple behavioral responses, positively phototactic or negatively geotactic, reverse towards the end of larval life and are thus "pre-adapted" to estuarine retention. This mechanism allows competent larvae to be returned to the parental habitat. It is difficult to see how any hypotheses generated from the preceding can be tested by the deductive method. Perhaps number 6 in Table 9.6 qualifies as an unanswerable question – at least by today's standards.

The last two questions of Table 9.6 are both ecological ones and will be dealt with together. For the background information applicable to question 8, a description of the universal forces which result in environmental changes throughout geological time is required (see also the section,

Ultimate questions). As an example of how these forces might operate, we examine the post-glacial formation of the Bay of Fundy. Reviewing the results of other geologists, Fader, King, and MacClean (1977) described the likely history of the Bay of Fundy after the Laurentide Ice Sheet retreated northwards some 13.5×10^3 yr B.P. By 8×10^3 yr B.P., the Bay was a shallow, nearly tideless estuarine bay where emergent Georges and Browns banks protected the Gulf of Maine and Bay of Fundy from the energetic tidal flows of the northwestern Atlantic. Beginning at 6×10^3 yr B.P., and as a result of eustatic sea level rise that exceeded land mass rebound, increases in depth and tidal flows began which have steadily increased to the present. Present-day depths at the mouth of the Bay are >200 m, and maximum tidal current velocities exceed 400 $cm \cdot s^{-1}$ in the upper Bay. The increased tidal flows within the Bay have resulted in changes of the surficial sediments. Initially, a uniform glacial till was present throughout the shallow Bay. In present conditions, three major sedimentary types have formed: tidally reworked sands in the upper Bay, original glacial till in the middle part, and redeposited silt/clay in the deeper parts of the western side of the mouth of the Bay (Fader et al. 1977). Our example suggests the relative speed that the geological imperatives can have in controlling environmental variables such as flow. In this case, the Bay of Fundy went from nearly tideless to some of the most energetic tides of the world in ~6000 years. We can expect that there will be changes involving increases, and decreases, in tidal flows over relatively short geological periods, and that these changes must act as powerful selective forces, for both resident and immigrant suspension feeders.

It should be realized that geologists, like biologists interested in ultimate questions, are forced to use deductive methods. An example of the indecisiveness which such methods can cause is provided in the debate (see the section, Ultimate questions) regarding the relative importance of discontinuous events in causing sudden mass faunal extinctions.

Concerning the question about the causes of deposit versus mixed benthic assemblages (number 7, Table 9.6), the types of benthic community association were described in Chapter 7. If we continue our use of the Bay of Fundy as an example, we can see that there is insufficient time – approximately 6000 generations if the benthic organism is an annual species – for evolutionary processes to produce a suspension feeder from a deposit feeder from scratch. A further point is that the benthic fauna will arrive as immigrants from adjacent seas by one or more of the four methods of dispersal discussed in Chapter 3. The immigrants will include both deposit and suspension feeders, and the environmental conditions that they experience will determine which survive and continue to evolve, as well as those which cannot survive in these conditions. A possible scenario for the Bay of Fundy would be that in the period 13.5 to 6×10^3 yr B.P., the major selection would be for deposit feeders, whereas after 6×10^3 yr B.P., the progressively increasing tidal range and flows would favor the establishment of suspension feeders. The present-day benthic fauna of the Bay of Fundy (Wildish and Peer 1983; Wildish, Peer, and Greenberg 1986) consists mainly of mixed benthic assemblages, where even on net depositional, silt/clay sediments, tube-living suspension feeders (e.g. the amphipod *Haploops fundiensis* and polychaete *Potamilla neglecta*) are present. Only in the most reworked sand sediments of the upper Bay are suspension feeders absent. This sediment is occupied by a specialized – and impoverished – benthic macrofauna, consisting of deposit feeders either able to withstand very high tidal velocities by burrowing deeply below the erosion layer or built massively to avoid being swept away. Because of the brief time available and the known zoogeographic distribution of the named deposit feeders in the section, Water-movement impoverishment, Chapter 7, it is unlikely that they evolved in situ in the Bay of Fundy.

It is clear that approaches to testing some of our hypotheses regarding ultimate questions should be made in collaboration with physical oceanographers, surficial geologists, and palaeoecologists. One obvious question for a palaeoecologist is whether the fossil macrofauna present within the Bay of Fundy during the last 13.5×10^3 yr conforms to the predictions made.

References

Amos, C. L., and D. A. Greenberg. 1980. The simulation of suspended particulate matter in the Minas Basin, Bay of Fundy: a region of potential tidal power development, p. 2–20. *In* Proceedings Canadian Coastal Conference, Nat. Res. Counc., Ottawa.

Bacher, C., M. Heral, J. M. Deslous-Paoli, and D. Razet. 1991. Modele energetique unibiote de la croissance des huitres (*Crassostrea gigas*) dans le bassin de Marennes-Oleron. Can. J. Fish. Aquat. Sci. 48: 391–404.

Baker, G. C. 1984. Engineering description and physical impacts of the most probable tidal power prospect(s) under consideration for the upper reaches of the Bay of Fundy, p. 333–345. *In* D. C. Gordon and M. J. Dadswell (eds.) Update on the marine environmental consequences of tidal power development in the upper reaches of the Bay of Fundy. Can. Tech. Rep. Fish. Aquat. Sci. 1256.

Bayne, B. L., R. J. Thompson, and J. Widdows. 1976. Physiology I, p. 121–206.

In B. L. Bayne (ed.) Marine mussels, their ecology and physiology. Cambridge University Press, London.

Berggren, W. A., and C. D. Hollister. 1977. Plate tectonics and paleocirculation: commotion in the ocean. Tectonophysics 38: 11–48.

Bougrier, S., P. Geariron, J. M. Deslous–Paoli, C. Bacher, and G. Jonquieres. 1995. Allometric relationships and effects of temperature on clearance and oxygen consumption rates of *Crassostrea gigas* (Thunberg). Aquaculture 134: 143–154.

Butman, C. A. 1987. Larval settlement of soft sediment invertebrates: the spatial scales of pattern explained by active habitat selection and the emerging role of hydrodynamical processes. Oceanogr. Mar. Biol. Annu. Rev. 25: 113–165.

Chia, F. S. 1974. Classification and adaptive significance of developmental patterns in marine invertebrates. Thal. Jugosl. 10: 121–130.

Clemmesen, B., and C. B. Jørgensen. 1987. Energetic costs and efficiencies of ciliary filter feeding. Mar. Biol. 94: 445–449.

Emerson, C. W., and J. Grant. 1991. The control of soft-shell clam (*Mya arenaria*) recruitment on intertidal sandflats by bedload sediment transport. Limnol. Oceanogr. 36: 1288–1300.

Fader, G. B., L. H. King, and B. MacLean. 1977. Surficial geology of the eastern Gulf of Maine and Bay of Fundy. Mar. Sci. Pap. 19, Geological Survey of Canada Paper 76–17.

Famme, P., and L. H. Kofoed. 1980. The ventilatory current and ctenidial function related to oxygen uptake in declining oxygen tension by the mussel *Mytilus edulis* L. Comp. Biochem. Physiol. 66A: 161–171.

Field, K. G., G. J. Olsen, D. J. Lane, S. J. Giovannoni, M. T. Ghiselin, E. C. Faff, N. R. Pace, and R. A. Raff. 1988. Molecular phylogeny of the animal kingdom. Science 239: 748–753.

Fry, F. E. J. 1947. Effects of the environment on animal activity. Univ. Toronto Stud. Biol. 55. Publ. Ont. Fish. Res. Lab. 68: 1–62.

Grant, J., M. Dowd, K. Thompson, C. Emerson, and A. Hatcher. 1992. Perspectives on field studies and related biological models of bivalve growth and carrying capacity. NATO ASI Ser. Vol. G33: 371–420.

Greenberg, D. A. 1979. A numerical model investigation of tidal phenomena in the Bay of Fundy and Gulf of Maine. Mar. Geodesy 2: 161–187.

Greenberg, D. A., and C. L. Amos. 1983. Suspended sediment transport and deposition modelling in the Bay of Fundy, Nova Scotia: a region of potential tidal power development. Can. J. Fish. Aquat. Sci. 40: 20–34.

Grosberg, R. K. 1987. Limited dispersal and proximity-dependent mating success in the colonial ascidian *Botryllus schlosseri*. Evolution 41: 372–384.

Grosberg, R. K., and T. F. Quinn. 1986. The genetic control and consequences of kin recognition by the larvae of a colonial marine invertebrate. Nature 322: 456–459.

Havenhand, J. N. 1995. Evolutionary ecology of larval types, p. 79–122. *In* L. McEdward (ed.) Ecology of marine invertebrate larvae. CRC Press, Boca Raton, Fla.

Heral, M. 1992. Why carrying capacity models are useful tools for management of bivalve mollusc culture. NATO ASI Ser. Vol. G33: 455–478.

Herman, P. M. J. 1992. A set of models to investigate the role of benthic suspension feeders in estuarine ecosystems. NATO ASI Ser. Vol. G33: 421–454.

Jørgensen, C. B. 1981. A hydrochemical principle for particle retention in *Mytilus edulis* and other ciliary suspension feeders. Mar. Biol. 61: 277–282.
1990. Bivalve filter feeding: hydrodynamics, bioenergetics, physiology and ecology. Olsen and Olsen, Fredensborg, Denmark.
1992a. Adaptational, environmental and ecological physiology: a case for hierarchical thinking, p. 9–17. *In* S. C. Wood, R. E. Weber, R. R. Hargens, and R. W. Millard (eds.) Physiological adaptions in vertebrates. Marcel Dekker, New York.
Jørgensen, C. B., P. S. Larsen, and H. V. Riisgård. 1986. Nature of relation between ventilation and oxygen consumption in filter feeders. Mar. Ecol. Prog. Ser. 29: 73–88.
Jørgensen, S. E. 1992b. Integration of ecosystem theories: a pattern. Kluwer Academic, Dordrecht.
Keough, M. J. 1984. Kin-recognition and the spatial distribution of larvae of the bryozoan *Bugula neritana* (L.). Evolution 38: 142–147.
Kerr, S. R. 1990. The Fry paradigm: its significance for contemporary ecology. Trans. Am. Fish. Soc. 119: 779–785.
Lincoln, R. J., G. A. Boxshall, and P. F. Clark. 1982. A dictionary of ecology, evolution and systematics. Cambridge University Press, Cambridge.
Molfino, B. 1994. Palaeoecology of marine systems, p. 517–546. *In* P. S. Giller, A. C. Hildrew, and D. G. Raffaelli (eds.) Aquatic ecology. Blackwell, Oxford.
Morton, B. 1980. Swimming in *Amusium pleuronectes* (Bivalvia: Pectinidae). J. Zool. London 190: 375–404.
Moses, C. O. 1989. A geochemical perspective on the causes and periodicity of mass extinctions. Ecology 70: 812–823.
Myers, A. A. 1994. Biogeographic patterns in shallow-water marine systems and the controlling processes at different scales, p. 547–574. *In* P. S. Giller, A. C. Hildrew, and D. G. Raffaelli (eds.) Aquatic ecology. Blackwell, Oxford.
Newell, C. R., and S. E. Shumway. 1992. Grazing of natural particulates by bivalve molluscs: a spatial and temporal perspective. NATO ASI Ser. Vol. G33: 85–148.
Officer, C. 1993. Victims of volcanoes. New Sci. 1861: 34–38.
Olafsson, E. B., C. H. Peterson, and W. G. Ambrose. 1994. Does recruitment limitation structure populations and communities of macro-invertebrates in marine soft sediments: the relative significance of pre- and post-settlement processes. Oceanogr. Mar. Biol. Annu. Rev. 32: 65–109.
Pawlik, J. R. 1992. Chemical ecology of the settlement of benthic marine invertebrates. Oceanogr. Mar. Biol. Annu. Rev. 30: 273–335.
Peters, R. H. 1991. A critique for ecology. Cambridge University Press, Cambridge.
Powers, D. A., F. W. Allendorf, and T. Chen. 1990. Application of molecular techniques to the study of marine recruitment problems, p. 104–121. *In* K. Sherman, L. M. Alexander, and B. D. Gold (eds.) Large marine ecosystems: patterns, processes and yields. Am. Assoc. Adv. Sci., Washington, D.C.
Quinn, J., and P. W. Signor. 1989. Death stars, ecology and mass extinction. Ecology 70: 824–834.
Raimondi, P. T. 1992. Adult plasticity and rapid larval evolution in a recently isolated barnacle population. Biol. Bull. 182: 210–220.
Rhoads, D. C., and D. K. Young. 1970. The influence of deposit-feeding

organisms on sediment stability and community trophic structure. J. Mar. Res. 28: 150–178.
Scheltma, R. S. 1986. On dispersal and planktonic larvae of benthic invertebrates: an eclectic overview and summary of problems. Bull. Mar. Sci. 39: 290–322.
Serjeant, W. A. S. 1990. Astrogeological events and mass extinctions: global crises or scientific chimaerae? Mod. Geol. 15: 101–112.
Smaal, A. C. 1991. The ecology and cultivation of mussels: new advances. Aquaculture 94: 245–261.
Stanley, S. M. 1970. Relation of shell form to life habits of the Bivalvia (Mollusca). Geol. Soc. Am. Inc. Memoir 125.
Strathman, R. 1974. The spread of sibling larvae of sedentary marine invertebrates. Am. Nat. 108: 29–44.
Swinburne, N. 1993. It came from outer space. New Sci. 1861: 28–32.
Thoms, R. E., and T. M. Berg. 1985. Interpretation of bivalve trace fossils in fluvial beds of the basal Catskill formation (late Devonian), eastern U.S.A., p. 13–20. *In* H. H. Curvan (ed.) Biogenic structures: their uses in interpreting depositional environments. Soc. Econ. Paleontol. Mineral Spec. Publ. 35.
Vance, R. R. 1973. On reproductive strategies in marine benthic invertebrates. Am. Nat. 107: 353–361.
Verduin, J. 1969. Hard clam pumping rates: energy requirements. Science 166: 1309–1310.
Waller, T. R. 1991. Evolutionary relationships among commercial scallops (Mollusca: Bivalvia: Pectinidae), p. 1–72. *In* S. E. Shumway (ed.) Scallops: biology, ecology and aquaculture. Elsevier, Amsterdam.
Wells, P. G., and P. J. Ricketts (Editors). 1994. Coastal Zone Canada '94. Cooperation in the coastal zone: conference proceedings, Halifax, N.S.
Wildish, D. J. 1977. Factors controlling marine and estuarine sublittoral macrofauna. Helgolander. Wiss. Meeresunters. 30: 445–454.
Wildish, D. J., and D. D. Kristmanson. 1993. Hydrodynamic control of bivalve filter feeders: a conceptual view. NATO ASI Ser. Vol. G33: 297–324.
Wildish, D. J., and D. Peer. 1983. Tidal current speed and production of benthic macrofauna in the lower Bay of Fundy. Can. J. Fish. Aquat. Sci. 40: 309–321.
Wildish, D. J., D. L. Peer, and D. A. Greenberg. 1986. Benthic macrofaunal production in the Bay of Fundy and possible effects of a tidal power barrage at Economy Point-Cape Tenny. Can. J. Fish. Aquat. Sci. 43: 2410–2417.
Wildish, D. J., and A. M. Saulnier. 1993. Hydrodynamic control of filtration in *Placopecten magellanicus*. J. Exp. Mar. Biol. Ecol. 174: 65–82.
Wildish, D. J., and M. L. H. Thomas. 1985. The effects of dredging and dumping on benthos of Saint John Harbour, Canada. Mar. Environ. Res. 15: 45–57.

Glossary

adhesion number the dimensionless ratio of London attraction and hydrodynamic forces acting on a sestonic particle approaching a cylinder in a flowing fluid.

advection the transport of fluid molecules, solutes, or seston by the bulk motion of the fluid, e.g. by horizontal tidal currents.

aerosol a suspension of solid, or liquid, particles in a gas.

aerosol theory the application of the theory developed for the capture of aerosol particles to sestonic particle capture in water by suspension feeders.

aggregation statistics that part of probability theory concerned with contagious or aggregated distributions in space.

ambient velocity the background flow speed and direction.

anemometer a device for measuring flow speed and sometimes direction in air or water.

asperities surface roughness on a microscopic particle, e.g. seston.

ATP adenosine triphosphate, an energy containing coenzyme present in every living cell.

back pressure that pressure operating in a direction opposite to that being considered, e.g. the difference in external pressure between the outlet and inlet of a hydromechanical pump.

Baconian experimental test a test involving an alternative and a null hypothesis posed in such a way that the experimental result supports or rejects the original proposition.

ball valve a mechanism designed to control fluid flow based on a buoyant ball which acts to open and close an opening by its own weight and fluid pressure.

BBL benthic boundary layer, a discrete layer of flow above a natural benthic substrate. It consists of a viscous sublayer (q.v.) nearest the substrate, followed by a log-linear layer and an outer layer, each with a characteristic velocity profile.

bed load sedimentary particles which are being transported on, or immediately above, the bed.

bed load shear velocity that velocity at which sedimentary particles first begin to be transported as bed load by the flow. Velocity at which critical shear stress is exceeded and resuspension is initiated.

Bernoulli's principle that the sum of the pressure and kinetic energy of a fluid parcel moving along a streamline remains constant. Fluid friction is ignored in this consideration.

bidirectional flow a flow which tends in two directions, as in tidal flows, where the flood and ebb reverse directions.

bilateral the symmetry of an animal in which only one median or sagittal section can yield right and left mirror image halves, e.g. as in a sagittal section of mammals.

bluff body an object which presents a large frontal area to the flow.

BOD biological oxygen demand, a measure of the amount of organic matter in a sample which can be oxidised by aerobic respiration in a defined period.

Boltzmann's constant a thermodynamic expression which is equal to the universal gas constant divided by Avogadro's number.

bottom roughness those projections from the seabed which are sufficient to influence the BBL flow. Usually of sedimentary (e.g. sand ripples) or biological (e.g. epifaunal tubes) origin.

boundary shear the force exerted by the boundary layer at a solid or fluid surface, e.g. by fluid flow at the sediment–water interface.

bulk velocity the mean flow speed in the mainstream direction defined as the volumetric flow rate divided by the cross sectional area occupied by the flow.

buoyant weight a measure of the weight of an object in water, which is less than in air because of the buoyant force exerted by the fluid vertically on the body immersed, or floating, in it.

camber the slightly convex shape, e.g. on the upper surface of a streamlined body which increases flow there and thus enhances lift.

centrifugal pump a device to pump water by accelerating it outward by impeller to a surrounding casing.

COD chemical oxygen demand, a chemical measure of the total amount of oxygen required to oxidise organic matter or reducing compounds present in a sample.

collecting elements the specific surfaces of a suspension feeder on which sestonic food particles are first captured.
collimator a device used in flumes to make steamlines parallel; a flow straightener, e.g. a system of parallel plates placed upstream of the working section.
concentration boundary layer that region of any boundary layer where the concentration of a solute after consumption or discharge at the wall is less, or greater, than that in the bulk flow because of incomplete mixing within the boundary layer.
continental shelf break the underwater topographic point where the continental shelf ends and the continental slope begins as denoted by an increase in gradient (>1 in 20).
convective diffusion the transport of heat or mass by flows resulting from density differences within the fluid due to concentration or temperature differences.
counterion an ion with a charge that is opposite to that of another ion that it is brought into contact with.
creep the slow slipping or bending of solids due to critical hydrodynamic stresses, e.g. skeletal tissue of a coral exposed to a strong flow.
creeping flow flow at very low Reynolds numbers, where inertial forces are negligible with respect to viscous forces.
critical drag that force generated by a flowing fluid which is just sufficient to overcome the static friction acting on an object lying on the bottom, which is required to cause slippage or detachment of it.
current meter a field device for measuring ambient flow speed and sometimes direction.
deposition the process by which organic and inorganic particles suspended in the water column settle onto the sediment.
depth limited boundary layer a benthic boundary layer at a shallow location in which the boundary layer intersects the free water surface, without the presence of a free-stream flow.
detachment force critical drag (q.v.).
dewater a process in which water is removed from a solid by physical process, e.g. filtration of a sediment slurry.
diffusive sublayer a region of the laminar or viscous sublayer (q.v.) where the transport of a solute is dominated by molecular diffusion and turbulent diffusion is unimportant.
dimensionless index a ratio of two mathematical entities used in calculating a single number, without defining units, and which predicts

the importance of specified properties of the system, e.g. SDI (q.v.).

DNA deoxyribonucleic acid.

DOM dissolved organic matter.

downstream vortex an eddy whirlpool water movement pattern downstream of an object in a turbulent flow.

down-welling in physical oceanography, the deflection of seawater downwards towards the sediment.

drag the rate of removal of momentum from a flowing fluid by an immersed object, where this rate depends on the frontal area of the object, the velocity past it (assuming an attached object), and the density and viscosity of the fluid.

drift bottle a buoyant container used by physical oceanographers to track surface or subsurface currents visually.

dynamic pressure a pressure in a moving fluid which is equivalent to the kinetic energy per unit volume at a point defined by Bernoulli's principle, stagnation pressure (q.v.).

dynamic similarity a term used in hydrodynamic modelling where the prototype and model are geometrieally similar. Requires that ratio of forces around prototype and model be the same at equivalent points.

ectocrine a biochemical substance produced externally by a marine organism and which causes a biological response in another of the same or different species, e.g. toxic microalgal metabolites which inhibit bivalve initial feeding, or chemical cues that induce larval settlement.

eddy a turbulent water (or air) current running in a different direction than that of the main current; or flow in the wake of a bluff body different from the mainstream flow lines.

eddy diffusivity the rapid spreading in turbulent flow of heat, mass, or momentum by random movements of the fluid.

electronic particle counter a device for counting and sizing suspended particles down to a micrometer or so in diameter, e.g. seston.

electrostatic force in aerosol theory, the force generated by differences in electrical charges of particle and collecting surfaces.

endoscope a device for microscopic observations in small spaces between animal tissues. Consists of a thin light tube, laser light, and magnifying optics.

engineering efficiency the success of a collecting surface in a flow in capturing particles; the ratio of the rate at which particles are collected to the rate at which they would pass through the space occupied by the collector if it were not there.

entrainment the physical process of transfer of fluid by friction from one water mass to another. Usually occurs between currents moving in different directions.

eustatic the slow, universal changes of sea level which are caused by the melting of continental ice sheets.

fall velocity the rate at which a body passively falls under the influence of gravity through a still fluid; same as settling velocity.

Fick's laws of molecular diffusion the first law states that the flux of molecules is equal to the molecular diffusion coefficient times the concentration gradient. The second law relates the rate of change of concentration at a specified time and position to the diffusion coefficient and the second derivative of concentration with respect to distance.

field flume a flume capable of being used in the field, either in situ, as by surficial geologists, or by benthic biologists where the flume is portable but not necessarily used in situ.

first law of thermodynamics the conservation of energy law: energy may be transformed from one form to another, but cannot be created or destroyed.

flow cytometer a device which uses optical analysis based on light scattering and fluorescence for measuring the size and epifluorescence properties of small particles, e.g. red blood cells or seston.

flow fluorometer a device which measures the fluorescent radiation of a sample after exposure to monochromatic radiation. The sample is continuously brought to the sensor in a pumped flow of seawater, e.g. continuous seawater sampling of chlorophyll a in living phytoplankton cells.

flow straightener a collimator (q.v.).

flow meter a device for measuring the volumetric flow rate of a fluid.

flow vector the direction and speed of a flow at a given point in a fluid.

flowthrough flume a type of flow channel where the water makes only a single pass through the experimental working section.

fluid a liquid or a gas.

flume boundary layer the sheared layer that develops on the side walls and bottom of flumes during passage of water.

force the external agency which changes the state of rest or motion of a body and which is measured in units of dynes in the cm/g/s system.

form drag that portion of drag on an object in a flow caused by the uneven distribution of pressure over it.

free-stream velocity the velocity characteristic of the water body above the BBL; mainstream flow (q.v.).

friction those forces which resist relative motion between surfaces in contact.

frontal area the projected area of a body in a flow where the projection is onto a plane normal to the flow vector.

Froude number a dimensionless expression calculated as the ratio of the square of the relative speed to the product of the acceleration of gravity and a characteristic length of the body.

gypsum flux method a means of measuring water speed by determining weight losses from solid gypsum, after flow exposure and dissolution, which is dependent on exposure time, flow speed, salinity, and temperature.

H/D height/diameter ratio of epifaunal tubes.

Hamaker constant a mathematical expression used to predict the magnitude of the London–van der Waals force between two surfaces in an electrolyte.

head tank a water storage container placed above its delivery point. If it is continuously supplied to maintain a constant head, a constant discharge can be maintained.

hertz (Hz) a measure of electromagnetic wave frequency: one cycle per second.

homeorheostat a type of self-regulating system which is capable of changing its responses with time, e.g. filtration rates in bivalves.

hot film probe a velocimeter which senses flow speed by measuring the temperature change of a thermistor heated at a constant rate and immersed in the flow.

hydrodynamics that branch of physics dealing with fluid motion.

hydrostatic pressure the force per unit area exerted on a surface in a fluid as a consequence of the weight of the fluid above it; also termed gravitational pressure.

inertia the state of rest, or uniform motion, of a body that exists until it is acted upon by an external force.

inference a conclusion reached as a result of the use of either inductive or deductive methods.

internal wave an orbital motion propagating through a water body in which the motion does not reach the water surface and which is generated by a variety of physical means, e.g. tidal currents reflecting off a sloping bottom.

inviscid fluid a theoretical concept of an ideal fluid in which there is no frictional resistance to motion.

invisible college an informal group of people sharing common interests in a subfield of study.

isokinetic equal kinetic motion, as in sampling from an ambient flow with the same velocity in the sampling tube.

kinematic viscosity the absolute viscosity (q.v.) of a fluid divided by its density.

kinetic energy the energy of a body which results from its motion; calculated as the product of mass and velocity squared.

laminar flow an orderly state of flow whereby the fluid moves in layers and any small disturbances are dampened by viscosity.

laminar sublayer that part of the BBL closest to the bottom in which the velocity profile is a creeping flow; i.e. momentum transfer is determined by the viscosity of the fluid. Also viscous sublayer (q.v.).

larval mimic a physical model of a larva designed to have the same physical characteristics in flow, e.g. size and fall velocity, as the prototype.

laser Doppler flow meter a type of velocimeter which utilizes the Doppler shift in the frequency of laser light scattered from particles in the fluid.

lift force the supporting force on a hydrofoil (or airfoil) in a flow of water (or air) which is caused when the pressure on the underside exceeds that on the upper surface, because the velocity on the latter is greater.

lift spoilers structures present on the underside of an air- or hydrofoil which help dissipate the pressure differences between the upper and lower surfaces, thereby decreasing lift or increasing drag.

macrotidal the locations with a large tidal prism.

mainstream flow the free-stream velocity (q.v.).

marine snow an aggregate suspended in seawater consisting of various types of sestonic particles held in a mucilaginous matrix.

matched flows isokinetic flows (q.v.).

mucous net a net made from the protein mucin by some suspension feeders in order to collect sestonic food particles by a sieving mechanism.

multiple-channel flume a flume with two or more flow channels which all share the same water supply.

negative buoyancy a characteristic of a body which is of greater average density than the fluid which surrounds it when the upward pressure exerted by the fluid is insufficient to keep it floating.

normal to a characteristic of a body exposed to flow in which a surface or plane opposes, or is oriented at right angles, to the mainstream flow.

opposed to a characteristic of a body exposed to flow in which a surface or plane is oriented so that it is parallel with the mainstream flow.

oscillating flow the orbital water movement created by wind action on a free water surface.

P*:*B production/biomass ratio, or the annual turnover ratio of a population.

Peclet number a dimensionless group which compares the relative importance of advection and molecular diffusion in the transport of heat or solutes. It is of the form ULk^{-1}, where U is velocity, L a length characteristic, and k^{-1} the molecular diffusivity of heat or solute.

Pitot tube a device which determines the stagnation pressure of a flowing fluid, in a tube pointing into the flow and measured with a pressure gauge, as an indication of its speed.

POM particulate organic matter.

porosity the ratio of the volume of interstitial space to solid, as in a filter or in soil: the void fraction.

positive displacement pump a pump in which a chamber of fixed volume is alternately filled and emptied, often by a rotating or reciprocating piston.

potential energy the energy content of a fluid due to its position in a gravity field; its capacity to do work.

pressure the force per unit area exerted by a fluid as a result of its weight and motion.

pressure head that pressure at a point in a flow which is expressed as its hydrostatic equivalent. The head is equivalent to the height that the fluid would rise in a vertical tube open to the atmosphere and which is connected to a probe in the flow.

pump characteristic the relationship between the pressure difference and the volumetric flow rate, or pump capacity, for a hydromechanical pump.

raceway an open channel, oval-shaped flume which employs a fixed volume of recirculated water.

radial the symmetry of an animal, e.g., a starfish, whose similar parts are arranged in a regular pattern around a central axis and from which any equal section yields mirror image halves.

recirculating flume a type of flume in which the same, fixed volume of water is passed repeatedly through the working section.

reef path length the distance along a tidal current excursion occupied by a suspension feeding reef, e.g. a bivalve bed.

residence time the time spent by a conservative molecule in a tidal body of water before being passed along to an adjacent body by all types of water movement.

resonance a natural phenomenon occurring in the sea where an external periodic driving force enhances the oscillation amplitude because the force approaches the natural free oscillation frequency of a local system, e.g. tides in the Bay of Fundy.

resuspension the process by which previously deposited sedimentary particles are carried into the water column by water movements or biogenic activity.

reworked sediment a sediment that is being or has been transported from a bed either by water movement or by bioturbation activity.

Reynolds number a dimensionless number which expresses the relative magnitude of inertial and viscous forces in a moving fluid. The transition from laminar to turbulent flow occurs at a critical value of this number specific for each flow.

rigid circular cylinder a solid cylinder.

rip current a turbulent, localized counter-current to the ambient flow.

roughness the grain, or flow-induced sculpturing, of a soft sedimentary bed, or sculpturing of a hard substrate as it affects hydrodynamic resistance, characterized by Z_0.

Rouse number a dimensionless number giving the ratio between the sedimentary particle settling rate and the boundary shear velocity. Used to determine whether a particle will resuspend or remain as bed load in a given flow.

sag curve an empirically derived graphic relationship between a spatial variable on the ordinate and an environmental variable on the abscissa, which shows a pronounced dip, e.g. dissolved oxygen in a polluted estuary.

SCUBA self-contained underwater breathing apparatus.

SDI seston depletion index (see page 278).

second law of thermodynamics the entropy (a measure of the randomness) of an isolated system must in time either increase or reach an equilibrium.
sediment–water interface that part of a bed of deposited sediment at the interface between sediment and water.
sediment trap a device for collecting settling sedimentary particles from the water column.
sedimentary geology surficial geology (q.v.).
semi-exposed partly open to wind–wave effects.
seston depletion a process in which the seston concentration within the BBL becomes reduced, as a result of consumption, as it passes over a suspension feeding reef.
settling velocity fall velocity (q.v.).
shear stress the force acting on a unit area between two layers of a fluid which are moving at different velocities and where the force is applied in the flow direction, with the plane of the unit area being tangent to the streamlines.
Shields parameter a dimensionless expression relating the shear stress to the buoyant weight of a particle per unit area of bottom. Used in determining the shear stress needed to move sedimentary particles of a given size and density.
sieving a process in which sestonic particles are removed from a flow of water passed through a mesh, with the minimum particle size trapped being dependent on the mesh spacing.
skimming flow a flow interacting with a regular array of obstacles which at a certain density cause the shear zone of the flow to relocate above the obstacles, thereby protecting the local sediments from erosion.
slip velocity the flow speed at which an epifaunal, free-living organism first begins to be transported downstream by the flow.
stagnation point that point where the central stream line of a steady flow has zero velocity at the surface of a solid body.
stagnation pressure dynamic pressure (q.v.).
stall for a swimming body, e.g. a scallop, the result of a decrease in lift and an increase in drag causing the body to become unstable and fall out of the flow. In scallops the stall is commonly due to a change in the critical angle of attack of the body, which results in a change from a smooth to a separated flow over the body.
static friction coefficient if a solid body is placed on a horizontal plane, and a force applied parallel to the plane, the body will slide at some

critical force. For this force the static friction is the difference between buoyant and gravitational forces on the body, multiplied by the static friction coefficient.

steady flow a flow field which does not change with time.

stochastic a system model in which the relationships are random or probabilistic, so that an input results in many possible outcomes decided by chance. Opposite of **deterministic**, in which an input results in one outcome only.

Stokes law the relation of the total drag, F, on a sphere of diameter, D, moving in a still Newtonian fluid, to its velocity, U, and the viscosity of the fluid, μ, in creeping flow conditions. Thus: $F = 3\pi\mu DU$.

strain gauge a device for measuring the force of attachment of an organism to a hard substrate, by the use of attached pressure sensors which change in electrical resistance as the sensor is strained.

stratification the separation of a water body into discrete horizontal layers of the water column by physical environmental variables, e.g. salinity or temperature.

streamlines the characteristic flow lines of a steady flow as can be visualized by observing paths of neutrally buoyant particles. The streamlines are everywhere tangent to the local velocity vector.

Strouhal number a dimensionless frequency used in analyzing oscillating fluid flows. The ratio of the frequency of the oscillation to UD^{-1}, where U is the velocity and D a dimension used to characterize the flow.

stuw zone the most inland part of some estuaries where freshwater is backed up by the tidal stream during a flooding tide, causing a freshwater tidal rise and fall.

submarine canyon a sharply shelving valley on the seabed.

surf zone that part of the shore between the landward limit of wave uprush and the most seaward breaker.

surface slick that part of the sea surface where up- or down-welling occurs and which is often marked by flotsam and jetsam, e.g. where Langmuir cells meet.

surficial geology the scientific study of unconsolidated sedimentary deposits on the land or seabed.

suspension feeder–biased sampling simulation of feeding by a named suspension feeder. Sampling may be isokinetic or non-isokinetic, but at the same velocity as used naturally by the suspension feeder.

tangential cross-flow filter a type of industrial filter which employs centrifugal force to carry particles to the filter and a cross-flow to remove and carry away the collected particles, so that the filter is in continuous use.

tectonic the regional movements, caused by geological forces, of the large plates of which the Earth's crust is composed.

thermistor a device consisting of a semi-conductor whose resistivity decreases at higher temperatures and which is used in measurement systems, e.g. temperature or velocity. Used as a velocimeter by heating the semi-conductor at a constant rate and then measuring the increase in resistance due to cooling, which is velocity-dependent.

throat section that part of a flume which restricts the flow from a larger to a smaller cross-sectional area.

tidal barrage a civil engineering structure designed to control tidal flows. May be used in flood control or capture of potential energy by passing seawater through turbines to convert it to electricity.

tidal bore the inability of a flooding tide in a shallow estuary to propagate in shallow water as rapidly as required, which causes a discontinuity in level, resolved by a breaking wave advancing rapidly up the estuary.

tidal excursion the net horizontal distance travelled by a conservative particle in a tidal flow during one flood–ebb cycle.

tidal power the electrical power obtained by passing seawater through turbines placed in a tidal barrage.

tidal prism the volume of seawater in a given water body between high and low water levels.

torque the moment of force that causes torsion or twisting in a body.

tow tank a fluid container through which model objects can be mechanically drawn to simulate flow effects, e.g. for testing ship hull design.

trophic fluid transport system an enclosed tubular system for conducting water to, and from, the seston capturing surfaces of a suspension feeder.

turbulence the irregular motion observed in fluids at high Reynolds numbers, characterized by rapid mixing of heat, mass, and momentum.

turbulent boundary layer a discrete layer near a wall where momentum is transferred across the flow by turbulence rather than by viscous stress.

turning vanes flow straighteners designed to assist flume flows turn corners, so that the stream lines remain parallel.

two-dimensional flow flow where it is assumed that its velocity and other properties vary in two of the coordinate directions, but are constant in the third direction.

van der Waals force the weak attractive forces which occur between two atoms or nonpolar molecules.

vectored jet a jet of water whose direction can be adjusted, e.g. by scallops to aid swimming or recessing behaviour.

velocimeter a device for measuring velocity.

velocity calibrator a device for calibrating a velocimeter.

velocity defect layer the region of flow near a solid boundary where the velocity is reduced, e.g. BBL (q.v.).

velocity gradient the rate of change of a velocity component with distance.

velocity profile the distribution of velocity along a transect perpendicular to the flow.

velocity vector the magnitude and direction taken by a flow.

vertical mixing the physical processes by which a water body becomes mixed in the vertical plane with respect to the seabed.

viscosity (absolute) in a Newtonian fluid the shear stress exerted on a fluid is proportional to the gradient of velocity times the absolute viscosity. A measure of the resistance a fluid offers to flow when it is subjected to a shear stress.

viscous dissipation the rate at which turbulent energy is lost as heat as a result of turbulent stresses working against the viscosity of the fluid.

viscous drag that portion of drag on an object in a flow caused by viscous stresses on its surface. Also called skin friction. Total drag in a steady flow is the sum of form drag and viscous drag.

viscous sublayer laminar sublayer (q.v.).

volumetric flow a measure of fluid motion which defines the volume of fluid passing through a reference area in unit time.

wake that flow region in the lee of a submerged body which is altered by the interaction between the flow and the body.

wall effect the property of flowing water constrained by walls to form boundary layers there; also solid boundary effects on the terminal settling velocity of a particle in a fluid.

water tunnel a closed or tubular conduit, oval shaped flume, which employs a fixed volume of recirculated water.

wave an orbital motion induced near the free surface of a fluid as the result of an applied force, e.g. wind stress, and which propagates throughout the water body.

wave exposure a relative measure of the degree to which wind–wave action occurs at a site.

weir a dam in a flow whose purpose is to control or measure the flow speed, e.g. an adjustable weir to regulate stream runoff.

wind–wave the effect of wind action on a water surface in producing orbital wave actions.

Index

Specific plant or animal names and words defined in the glossary are excluded.